Cosmic Rays and Particle Physics–1978
(Bartol Conference)

C. M. G. Lattes chairing the session on new accelerators

Carlo Rubbia summarizing the Conference

AIP Conference Proceedings
Series Editor: Hugh C. Wolfe
Number 49
Particles and Fields Subseries No. 16

Cosmic Rays and Particle Physics–1978
(Bartol Conference)

Editor
T.K. Gaisser
Bartol Research Foundation

American Institute of Physics
New York   1979

Copying fees: The code at the bottom of the first page of each article in this volume gives the fee for each copy of the article made beyond the free copying permitted under the 1978 US Copyright Law. (See also the statement following "Copyright" below). This fee can be paid to the American Institute of Physics through the Copyright Clearance Center, Inc., Box 765, Schenectady, N.Y. 12301.

Copyright © 1979 American Institute of Physics

Individual readers of this volume and non-profit libraries, acting for them, are permitted to make fair use of the material in it, such as copying an article for use in teaching or research. Permission is granted to quote from this volume in scientific work with the customary acknowledgment of the source. To reprint a figure, table or other excerpt requires the consent of one of the original authors and notification to AIP. Republication or systematic or multiple reproduction of any material in this volume is permitted only under license from AIP. Address inquiries to Series Editor, AIP Conference Proceedings, AIP.

L.C. Catalog Card No. 79-50489
ISBN 0-88318-148-7
DOE CONF- 7810116

FOREWORD

About 100 physicists attended the meeting at the Bartol Research Foundation, University of Delaware, Newark, Delaware, 16-21 October 1976. Participants were roughly equally divided between the fields of particle physics and cosmic rays. The theme of the meeting was the interaction between these fields, with emphasis on particle interactions above 10 TeV. The problem of cosmic ray composition above 10^{13} eV was also discussed, and plans for new cosmic ray and accelerator experimental facilities were presented. The schedule of the meeting had three parts: 16-18 October, detailed discussions and talks on emulsion chamber experiments and extensive air showers; 19 October, Workshop on Charm Particle Production and Lifetimes; 20-21 October, Topical Conference on Cosmic Rays and Particle Physics.

Primary support for the meeting was provided by the U.S-Japan Cooperative Science Program of the National Science Foundation and by the Japan Society for the Promotion of Science under a bi-national agreement. A delegation of 13 cosmic ray physicists from Japan and 12 U.S. participants attended the meeting under this program. Publication of the proceedings is supported by the U.S. Department of Energy, which also provided some supplementary funds for the Topical Conference. The meeting was sponsored and also partially supported by the Bartol Research Foundation and by the Physics Department of the University of Delaware through funds from the UNIDEL Foundation.

It is a pleasure to acknowledge the advice and assistance of the advisory committee, the local committee and the session chairmen (all listed on the preceding page) in many aspects of organizing and running the meeting. I am particularly grateful to Professor T. Yuda, the Japanese coordinator, to Professor S. Miyake for a memorable after-dinner toast and to Gaurang Yodh, who, in addition to advising on the program, acted variously as photographer, projectionist and chauffeur. I am also grateful to Mr. David Bartley of the Clayton Hall Conference Management for arrangements during the meeting, and I am especially grateful to Helen Kryka for valuable assistance in editing these Proceedings. Finally, I would like to thank Martin A. Pomerantz for his apt welcoming remarks which opened the meeting and Carlo Rubbia for his valuable summary which ended it.

T. K. Gaisser

January, 1979

Advisory Committee

J. D. Bjorken (SLAC)
D. Cline (Wisconsin)
F. Halzen (Wisconsin)
S. Miyake (ICR, Tokyo)
G. B. Yodh (NSF/Maryland)
T. Yuda (ICR, Tokyo)

Local Committee

M. Barnhill
W. K. Cheng
A. Halprin
D. W. Kent
G. Steigman

Session Chairmen

George Cassiday (Utah)
J. Lord (Washington)
T. Yuda (ICR, Tokyo)
L. W. Jones (Michigan)
W. V. Jones (LSU)
L. Rosenson (MIT)
H. Kasha (Yale)
K. Kamata (ICR, Tokyo)
J. D. Bjorken (SLAC)
H. Bergeson (Utah)
G. B. Yodh (NSF/Maryland)
C. M. G. Lattes (Campinas)
S. Miyake (ICR, Tokyo)

A subset of participants

TABLE OF CONTENTS

16-18 October——Working Sessions

1. EXTENSIVE AIR SHOWERS

   Composition of Primary Cosmic Rays Above 10^{13} eV from the Study of Time Distributions of Energetic Hadrons Near Air Shower Cores (J. A. Goodman et al.)
   G. B. Yodh.................................... 1

   Arrival-Time Distributions of Muons and Electrons in Large Air Showers Observed at 5200 m Above Sea Level (Bolivian Air Shower Joint Experiment)
   K. Suga....................................... 13

   Simultaneous Detection of γ-Ray Bundles and Air Showers (S. Dake et al.)
   M. Sakata..................................... 18

   High Transverse Momentum of γ-Ray Bundles ($>10^{12}$eV) in Air Showers (Y. Yamato et al.)
   M. Sakata..................................... 23

   Preliminary Result for the Abundance of Multicored EAS at Mt. Norikura (Norikura Air Shower Group)
   M. Sakata..................................... 29

   High P_t Cross Section from Multiple Core Rates in Air Showers (W. E. Hazen et al.)
   W. E. Hazen................................... 31

   National Cosmic Ray Facility to Study High Energy Physics and Primary Composition from 10^{14} to 10^{18} eV
   W. V. Jones................................... 41

2. TECHNIQUES

   Technical Aspects——Design, Scanning and Energy Measurement in Small Emulsion Chambers (H. Fuchi et al.)
   K. Niu.. 49

   The Technology of Small Emulsion Chambers: A Review
   R. J. Wilkes.................................. 63

Acoustic Detection: A New Technique
in >10 TeV Experiments
    Theodore Bowen.............................. 72

3. NUCLEUS-NUCLEUS COLLISIONS

Nucleus-Nucleus Collisions at a mean
E = 20 GeV/Nucleon (Phyllis S. Freier
and C. Jake Waddington)
    Phyllis S. Freier........................... 87

4. 10-1000 TeV INTERACTIONS

Multiple Meson Production in the $\Sigma E_\gamma >$
2×10^{13} eV Region (Brasil-Japan Emulsion Chamber Collaboration)
    Y. Fujimoto................................. 94

Computer Simulation of Emulsion Chamber
C-Jets Using Conventional Particle
Physics (R. W. Ellsworth, G. B. Yodh
and T. K. Gaisser)
    R. W. Ellsworth............................ 111

Characteristics of Nuclear Interactions
Around 20 TeV Through the Analysis of
the Monte-Carlo Simulation (H. Sugimoto
and Y. Sato)
    H. Sugimoto................................ 126

Some 10-100 TeV Events (F. Fumuro et al.)
    T. Ogata................................... 133

Baryon Pair Production with Large Decay
Q-Value (Brasil-Japan Emulsion Chamber
Collaboration)
    S. Hasegawa................................ 145

Summary of Expected and Observed Rates
for High Energy Interactions Above
10 TeV
    G. B. Yodh................................. 152

How Fast Does the Multiplicity of Particle Production in p-p Collisions Really
Increase with Primary Energy?
    Erwin M. Friedlander....................... 157

High Energy Interactions and Cosmic Ray
Components in the Atmosphere
    K. Kasahara................................ 162

High Energy Interactions and Cosmic-
Ray Components in the Atmosphere
Y. Takahashi.................................... 166

19 October——"Charm" Workshop

5. EVIDENCE FOR NEW PARTICLES

Cosmic Ray Evidences for Short-Lived
Particles
K. Niu.. 181

Direct Search for Charm in Accelerator
Proton Collisions
Taiji Yamanouchi............................. 199

Observation of Delayed Hadrons in Exten-
sive Air Showers——New Massive Particles?
(J. A. Goodman et al.)
G. B. Yodh................................... 207

Evidence for Prompt Neutrinos and Interpretation of Beam Dump Experiments
G. Conforto.................................. 221

Evidence for Prompt Single Muon Production
in 400 GeV Proton Interactions (B. C.
Barish et al.)
A. Bodek..................................... 240

A Prompt Neutrino Measurement (B. P. Roe
et al.)
L. W. Jones.................................. 246

Hadronic Production of Charmed and Other
Favorite Particles
Francis Halzen............................... 261

Charm Production by Neutrinos and the
Charm Particle Lifetime
David Cline.................................. 281

Charmed Particle Lifetimes
Jonathan L. Rosner........................... 297

20-21 October——Topical Conference

6. INTERACTIONS AROUND 1000 TeV AND ABOVE

   A New Type of Nuclear Interactions in the $\Sigma E_\gamma > 10^{14}$ eV Region (Brasil-Japan Emulsion Chamber Collaboration)
   Y. Fujimoto and S. Hasegawa 317

   Large-Scale Emulsion Chamber Experiment at Mt. Fuji (M. Akashi et al.)
   T. Yuda 334

   High Energy Interactions at Cosmic Ray Energies
   S. Miyake 352

   Speculations Concerning Possible New Physics From Experiments Above 10 TeV
   G. L. Kane 362

   High Energy Cosmic Rays, Particle Physics and Astrophysics
   A. M. Hillas 373

7. NEW COSMIC RAY EXPERIMENTS

   Weak-Interaction Studies with the DUMAND Detector
   Arthur Roberts 391

   The Homestake Neutrino Detector as a Cosmic Ray Composition Analyser (M. Deakyne et al.)
   K. Lande 409

   Report on the Utah Fly's Eye (G. L. Cassiday et al.)
   G. L. Cassiday 417

   Akeno Air Shower Project
   K. Kamata 443

8. NEW ACCELERATOR FACILITIES

   Quark Antiquark Colliding Beam Machines
   D. Cline 451

   ISABELLE
   D. Hywel White 486

9. POST-CONFERENCE COMMENTS ON CENTAURO

Quark Hotspots as a Centauro Scenario
   David Sutherland......................... 503

What is Centauro?
   J. D. Bjorken and L. D. McLerran.......... 509

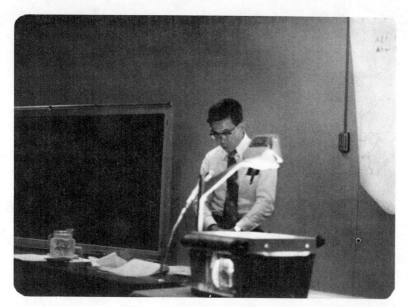

K. Kamata

Erwin M. Friedlander, W. V. Jones, K. Niu,
Y. Takahashi and Jere Lord

Chap. 1 Extensive Air Showers

COMPOSITION OF PRIMARY COSMIC RAYS ABOVE 10^{13} eV FROM THE STUDY OF TIME DISTRIBUTIONS OF ENERGETIC HADRONS NEAR AIR SHOWER CORES

J. A. Goodman[#], R. W. Ellsworth, A. S. Ito[*], J. R. MacFall[**],
F. Siohan[***], R. E. Streitmatter, S. C. Tonwar[+],
P. R. Vishwanath, and G. B. Yodh[++]

Department of Physics and Astronomy, University of Maryland
College Park, Maryland 20742

ABSTRACT

An experimental study of the distribution of arrival time of energetic hadrons relative to associated air shower particles has been made at a mountain altitude of 730 g.cm^{-2}. Monte Carlo simulations of the expermental observations have shown that these observations are sensitive to the composition of primary cosmic rays of energies 10^{13} -10^{15} eV. The energy spectra of primary protons and iron group nuclei required to understand these observations are

$(dN/dE)_{protons} = 1.5 \times 10^4 \, E^{-2.71 \pm 0.06} \, m^{-2} \, sr^{-1} \, sec^{-1} \, Gev^{-1}$

$(dN/dE)_{Fe\;group} = 1.27 \, E^{-2.36 \pm 0.06} \, m^{-2} \, sr^{-2} \, sec^{-1} \, Gev^{-1}$

respectively, where E is energy per nucleon, indicating that iron group nuclei are the dominant component of primary cosmic rays at air shower energies of 10^{14} - 10^{16} eV.

INTRODUCTION

Recent measurements[1-6] of flux and energy spectra of protons and various nuclei in primary cosmic rays for total energies of 100 to 5000 Gev suggest that the iron group nuclei have an energy spectrum considerably flatter than protons and other lighter nuclei. The index of the power law energy spectrum has been measured to be about -2.3 for iron group nuclei, compared to -2.75 for protons. These observations have far reaching significance[7] for theories of origin and propagation of cosmic rays. These results also have interesting implications for interpretations of many air shower observations since they indicate that iron group nuclei dominate

---

\# This work constituted in part the Ph.D. thesis of J. A. Goodman, University of Maryland, August 1978
*Now at SUNY-Stonybrook, Stonybrook, New York
**Now at Pfizer Corp, Columbia, Maryland
***Now at Nuclear Physics Laboratory, Oxford, England
+Now at Tata Institute of Fundamental Research, Bombay, India
++Now on leave at National Science Foundation, Washington, D. C.

the primary cosmic ray flux at air shower energies of $10^{15} - 10^{16}$ eV. Experimental observations such as rapid longitudinal development of air showers[8], high muon to electron ratio[9], and paucity of high energy hadrons[10] in air showers have been interpreted to indicate significant violation of scaling characteristics in high energy interactions[11], implying the onset of some new phenomenon at very high energies. Similar remarks apply to observations of multiple muons[12] and muon charge ratio[13] at high energies. However, these analyses have usually assumed protons to be the dominant component at these energies. Therefore any experimental confirmation of continuation of relative spectral differences between protons and iron group nuclei at air shower energies would be very significant and would lead to better understanding of high energy interactions at these high energies.[11] Direct observations in satellite borne experiments, though desirable, are difficult due to limitations imposed by the size and weight of the instrumentation required.

We present results obtained from a study of arrival time distributions of hadrons in air showers which suggest that the composition of primary cosmic rays is continuing to change with energy above 10^{13} eV. We show that a best fit to the experimental data requires the energy spectrum of iron group nuclei to have a power law exponent of 2.36 ± 0.06 compared to an exponent of 2.71 ± 0.06 for protons in the energy range of $10^{13} - 10^{15}$ eV.

APPARATUS

The experiment was carried out at the Sacramento Ridge Cosmic Ray Laboratory of the University of Maryland located at Sunspot (2900 meters or 730 g.cm^{-2} altitude) in New Mexico during the period April 1975 to May 1976. The experimental arrangement is shown in figure 1 and has been discussed in detail elsewhere[14]. The apparatus basically consists of an ionization calorimeter of area 4 m^2 and depth of 940 g.cm^{-2} of iron absorber. The hadron cascades are sampled by seven layers of liquid scintillation detectors. There are four shower particle detectors. Two of them, T1E and T1W, are placed above the calorimeter and have a total area of 3.2 m^2. The other two shower detectors, TSE and TSW, are located at about 3 meters from the center of the calorimeter on opposite corners. The detectors T1 measure the shower particles density above the calorimeter and time the arrival of shower particles.

A plastic scintillation detector T3 of area 0.55 m^2 and thickness 1 cm viewed by a 56AVP photomultiplier through an air light guide samples the hadron cascades traversing the central area of the calorimeter. The detector T3 is placed under 15.7 radiation length (220 g.cm^{-2} of iron) inside the calorimeter and is thus well shielded from shower particles. This detector is also unaffected by side showers due to its small size relative to the calorimeter area. This detector measures the arrival time of the hadron relative to shower particles as detected by the T1 detector. In cases where two or more hadron cascades traverse the detector T3 in the same air shower, arrival time of the earliest hadron only is measured.

All the calorimeter detectors, T1's, TS's, and T3, are calibrated using near vertical relativistic muons. Their pulse amplitudes are digitized and recorded for selected triggered events.

Four wide gap spark chambers SCB, SC1, SC2, and SC3 placed above and inside the calorimeter provide valuable visual information about hadron cascades in the calorimeter. A transition radiation detector consisting of a 24 layer sandwich of styrofoam radiators and proportional chambers, each of 1 m^2 area, was located above the spark chamber SCB, but was not used in the present experiment and is not shown in figure 1.

DATA

The experimental data was collected basically in two groups with different selection criteria. The group I data selected all showers with particle density Δ_e above the calorimeter greater than 18 m^{-2} which also had 5 or more particles traversing the detector T3. The particle density threshold was lowered to 4 m^{-2} for data group II, but a higher pulse amplitude corresponding to traversal by 25 or more particles was required from T3 for selection. For both data groups a minimum of about 50 Gev energy release in the calorimeter was required. A total of 21,500 events were collected for group I with an exposure area factor of 1.35 x 10^7 m^2 sr sec. The group II data had 9,150 events collected with an exposure area factor of 8.44 x 10^6 m^2 sr sec.

The observed arrival time distribution for hadrons of data group I is shown in figure 2 as a diplot of arrival delay versus the signal amplitude of the detector T3. It is of interest to note that there are no events on the negative delay (tachyon) side, indicating the absence accidental events. This is as expected from calculations using observed detector rates. This time distribution shows a long delay tail with 0.55 ± 0.05 percent of the events having delays larger than 15 ns and pulse amplitude in T3 greater than 5 particles. A similar distribution for events of data group II shows no delayed tail, primarily due to the T3 detector threshold being as high as 25 particles. However, there are 3 events with large delay (~15 ns) and large pulse amplitude (~35 particles) in data group II which are discussed elsewhere[14]. A detailed examination[14] of various electronic and physical effects (like evaporation neutrons, nuclear fragments, stopping antinucleons, etc.) has indicated the background which could generate a delayed event to be negligible (2 x 10^{-5} per event). These delay distributions are very similar to those observed in earlier experiments[15,16], but an exact comparison is not possible due to different selection requirements in different experiments.

MONTE CARLO SIMULATION

For interpretation of experimental results, a detailed simulation of the experiment was carried out using the Monte Carlo method. This 4-dimensional simulation generates atmospheric air showers from primary high energy protons and various nuclei. For generating showers due to primary nuclei (atomic no. A), the superposition model (A nucleons of energy E/A each) is assumed. Hadron-air nuclei interaction cross-sections are considered to be increasing with energy[17]. Hadron-nuclei interactions are assumed to be essentially hadron-nucleon interactions and effects due to intranuclear cascading are ignored. Hadron-nucleon interactions are generated using the independent particle scaling model incorporating various results obtained in accelerator experiments at Fermilab and ISR. The inclusive production cross-section is assumed to have a factorizable form as

$$E \frac{d^3\sigma}{dp^3} \propto e^{-bx} \cdot e^{-ap_t}$$

where x is the Feynman variable (= $2p_\parallel/\sqrt{s}$), $p_\parallel$ and p_t are the cm longitudinal and transverse momenta, and s is the square of the cm energy. The values for the parameter b have been chosen as 11 for baryons, 4.5 for π^+ and K^+, and 5.5 for π^- and K^- for nucleon interactions. The corresponding values for pion and kaon interactions are 11, 2.5, and 3.4 respectively. For the leading nucleons, the x value is picked from a flat distribution between 0 and x_{max}, with negative value assigned to the target nucleon. The x values for the leading pions and kaons are picked from a step distribution with an average x of 0.28 as suggested by observations in Fermilab experiments. Energy and momentum are conserved on the average. Secondaries are produced in the ratio π:K = 0.90 : 0.10, while baryon (p, $\bar{p}$, n, $\bar{n}$) production is assumed to increase with energy as $0.0164 \ln(1 + 0.015 E_{lab})$, where E_{lab} is the energy of the interacting particle in Gev. This form of increase is indicated by accelerator[19] and cosmic ray experiments[10,16]. All hadrons are followed down to 3 Gev unless they decay. For each photon resulting from the decay of a π^o, the contribution to shower size at observational level (730 g.cm^{-2}) is calculated using expressions given by approximation B of cascade calculations. The shower particle density at the location of each hadron at the observational level is calculated using a modified NKG lateral distribution function[20].

To simulate the experiment as closely as possible, each hadron arriving at the observational level is assuumed to be incident on a calorimeter. Its contribution, in terms of number of particles, is computed for a fictitious detector, 'T3', similarly placed as T3. Fluctuations in cascade development in the calorimeter, as observed experimentally[21], are taken into account in this computation. The arrival delay of the hadron relative to super-relativistic particles, as given by the simulation, is also fluctuated according to the observed instrumental time resolution[14].

The energies of all the hadrons incident over a 4 m^2 area around the hadron incident on T3 are summed together. As in the experiment, in case of two or more hadrons incident over the T3 area, the arrival time of only the earliest hadron is considered.

Thus, the simulation gives the number of particles in 'T3', the shower density at the 'T3' location, and the total energy in 4 m^2 area around 'T3' for each hadron arriving at the observational level. This set of computed values allows selection of simulated hadrons with the same criteria as used in the experiment. The number, N(E,A), of hadrons per air shower satisfying the selection requirements of either group I or group II type hadrons is then determined separately for various primary energies, E, and different primary nuclei, A. The variation of N with primary energy for group I type hadrons is shown in figure 3 for various primary nuclei. It should be noted from this figure that at an energy of 20 Tev/n showers initiated by iron nuclei are almost 1600 times more efficient in producing a hadron of group I type than showers initiated by protons. This factor is only 250 for group II type hadrons due to the requirement of higher energy for this group.

Combining these calculated factors, N(E,A), for various energies and various nuclei with extrapolation of measured[2] energy spectral slopes and fluxes normalized at an energy of 1 Tev/n, contributions to the observed number of events by each component have been computed. These calculations show that observed hadrons are generated by primary protons of energies 10 - 100 Tev and by primary iron nuclei of energies 2 - 40 Tev/n for both data groups. The energy range for other nuclei like the CNO group lies between the values for protons and iron nuclei. The energy spectra for α-particles, CNO nuclei, and medium heavy nuclei have been assumed to have the same spectral index as protons, since direct measurements[2-6] have not shown any significant spectral difference between these components at lower energies (10 - 100 Gev/n).

Similar computations have been carried out for various different assumed values of the power law indices assigned to the energy spectra of protons and iron nuclei, again with flux values normalized at an energy of 1 Tev/n to values observed directly or extrapolated from lower energies.

The simulations show that group I contains an appreciable contribution from iron primaries, while events of group II are generated preferentially by proton primaries. However, the relative contributions depend on the values of spectral indices assumed for the energy spectra of proton (and lighter nuclei) and iron nuclei.

In figure 4 is shown the spectral index contour curve for allowed values of spectral indices which are consistant with the observed flux of events of group I. This curve shows, for example, that the observed flux requires the spectral index of iron nuclei to be -2.36 if the proton spectral index is -2.71. A similar curve has also been obtained for events of group II type and predictions from the two data groups agree with each other within a value of 0.05 for the spectral indices.

DELAYED EVENTS

The preceding discussion shows that the observed flux of events of group I or II can only restrict the values of the spectral indices of the two components in primary cosmic rays to the allowed values which parameterize the curve in figure 4. However, an extimate of the spectral indices can be obtained by comparing the observed and expected time delay distributions for hadrons of group I. As mentioned earlier, the observations show 0.55 ± 0.05 % ov the events with >5 particles in T3 to be delayed by 15 ns or more. The simulations show that the proton-initiated showers generate a negligible number of events with these delay and energy characteristics. On the other hand, the iron-nuclei-initiated showers give 1.25 ± 0.4 % of hadrons with these characteristics. Showers initiated by α -particles, CNO group, and medium heavy nuclei contribute relatively a much smaller proportion of delayed hadrons. Using these values for the fraction of delayed events and earlier computations of N and flux for air showers initiated by various nuclei, it is estimated that the iron group nuclei contribute $.40^{+0.2}_{-.1}$ of the observed flux of hadrons in data group I. This value is then used to obtain from the curve in figure 4 the spectral indices for protons (as well as α and CNO) and iron group nuclei to be -2.71 ± 0.06 and -2.36 ± 0.06, respectively. Note that figure 4 implies lower bounds to proton and iron spectral indices of 2.55 and 2.25, respectively, purely from rate considerations.

CONCLUSION

We find that the relative proportion of iron group nuclei in primary cosmic rays at energies above 10 Tev continues to increase with increasing energy. Our observations on the flux of hadrons associated with air showers at mountain altitude and their arrival time distribution when comapred with Monte Carlo simulation strongly suggest that the flux and energy spectra for protons and iron nuclei are well represented by the following expressions:

$(dN/dE)_{protons} = 1.5 \times 10^4 \, E^{-2.71 \pm 0.06} \quad m^{-2} \, sr^{-1} \, sec^{-1} \, Gev^{-1}$

$(dN/dE)_{Fe \, group} = 1.27 \, E^{-2.36 \pm 0.06} \quad m^{-2} \, sr^{-1} \, sec^{-1} \, Gev^{-1}$

for energies of $13^{13} - 10^{15}$ eV, E being the energy per nucleon in Gev.

If these spectra were extended up to 10^{16} eV, the percentage of iron nuclei in primary cosmic rays would be greater than 90 %. An increasing iron component is necessary for scaling to remain valid for high energy interactions in the fragmentation region up to about 10^{16} Ev. These results would also provide reasonable understanding of observations on muon charge ratio and multiple muons.

ACKNOWLEDGMENTS

We wish to thank Dr. T. G. Morrison, Eldon Vann, Ralph Sutton, Harriet Sutton, Geeta Tonwar, Sriram Ramaswamy, Calvin Simpson, and James Schombert for their contribution to various phases of this experiment. Discussions with Drs. J. Ormes, V. K. Balasubrahmanyan, P. H. Steinberg, S. I. Nikolsky, M. Hillas, A. E. Chudakov, V. I. Yakovlev, T. K. Gaisser, and G. A. Snow are gratefully acknowledged.

REFERENCES

1. N. L. Grigorov et al., Proc. 12th Int. Conf. on Cosmic Rays, Hobart, $\underline{5}$, 1946, (1971).
2. M. J. Ryan, J. F. Ormes, and V. K. Balasubrahmanyan, Phys. Rev. Letters, $\underline{28}$, 985, (1972); V. K. Balasubrahmanyan and J. F. Ormes, Ap. J., $\underline{186}$, 109, (1973); J. F. Ormes and V. K. Balasubrahmanyan, Nature Phys. Sci., $\underline{241}$, 95, (1973).
3. L. H. Smith et al., Ap. J., $\underline{180}$, 987, (1973).
4. E. Juliusson, P. Meyer, and D. Muller, Phys. Rev. Letters, $\underline{14}$, 153, (1973); E. Juliusson, Ap. J., $\underline{191}$, 331, (1974)
5. J. H. Caldwell., Ap. J., $\underline{218}$, 269, (1977)
6. C. D. Orth et al., preprint, (1978)
7. See for example, R. Ramaty, V. K. Balasubrahmanyan, and J. F. Ormes, Science, $\underline{180}$, 631, (1973); T. K. Gaisser, Nature, $\underline{248}$, 122, (1974); J. F. Ormes and P. Freier, Ap. J., $\underline{222}$, 471, (1978).
8. C. Aguirre et al., Proc 13th Int. Conf. on Cosmic Rays, Denver, $\underline{4}$, 2598, (1973); M. LaPointe et al., Can. J. Phys., $\underline{46}$, S68.
9. N. N. Kalmykov and G. B. Kristiansen, Proc. 14th Int. Conf. on Cosmic Rays, Munchen, $\underline{8}$, 2861, (1975).
10. R. H. Vatcha and B. V. Sreekantan, J. Phys., 5A, 859, (1971).
11. See e.g., T.K. Gaisser, R.J. Protheroe, K.E. Turver and T.J.L. McComb, Rev. Mod. Phys., $\underline{50}$, 859 (1978) for a review.
12. J. W. Elbert et al., J. Phys. G2, 971, (1976).
13. See for example discussion by R. K. Adair et al., Phys. Rev. letters, $\underline{39}$, 112, (1977).
14. J. A. Goodman, Ph.D. Thesis, Univ. of Maryland, unpublished (1978). See also Goodman et al., this volume.
15. L. W. Jones et al., Phys. Rev., $\underline{164}$, 1584, (1967).
16. S. C. Tonwar et al., Lett. al Nuovo Cimento, $\underline{1}$, 531, (1971); S. C. tonwar and B. V. Sreekantan, J. Phys. A : Gen. Phys., $\underline{4}$, 868, (1971).
17. G. B. Yodh, Yash Pal, and J. S. Trefil, Phys. Rev Lett., $\underline{28}$, 1005, (1972); G. B. Yodh, Proc of the Conf. on Prospects of Strong Interactions at Isabelle, BNL, (1977); U. Amaldi, Phys. Lett., $\underline{66B}$, 390, (1977).

18. F. W. Busser, Phys. Lett., 46B, 471, (1973).
19. M. Antinucci et al., Lett. Nuovo Cimento, 6, 121, (1973).
20. K. Greisen, Progress in Cosmic Ray Physics, Vol. 3, ed.
    J. G. Wilson, (North Holland Publishing Co., Amsterdam, 1956)
21. H. Whiteside et al., Nucl. INstr. Methods, 109, 375, (1973);
    F. Siohan, Ph. D. Thesis, University of Maryland,
    Unpublished, (1976), W. V. Jones, private communication.

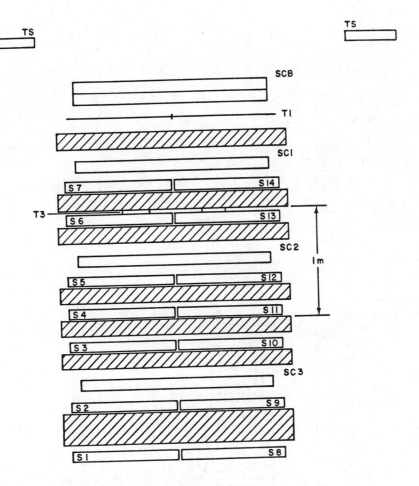

Figure 1. Experimental Apparatus.

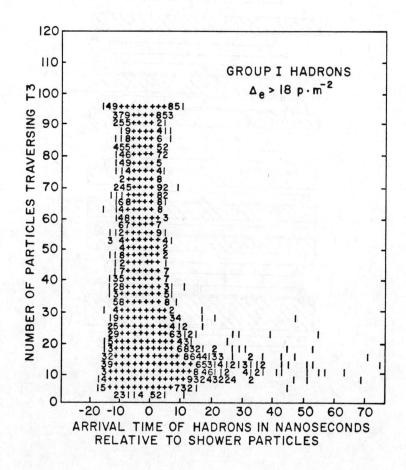

Figure 2. Time delay vs. T3 particle number for Group I.

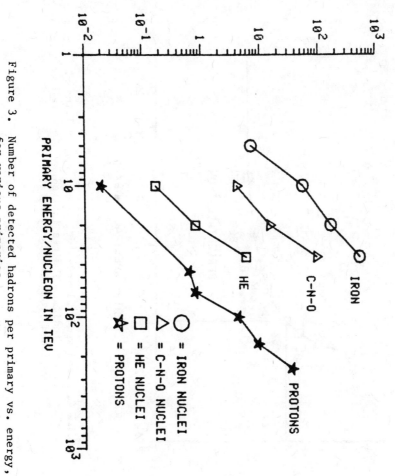

Figure 3. Number of detected hadrons per primary vs. energy, for various primaries.

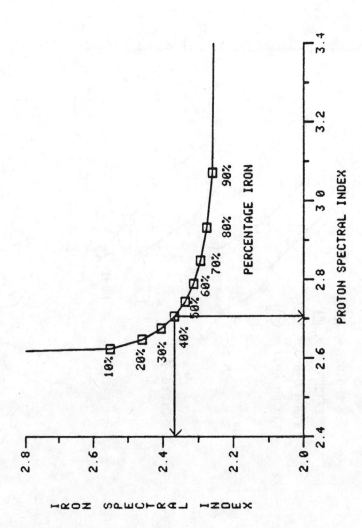

Figure 4. Iron spectral index vs. proton spectral index. Lines indicate indices implied by composition obtained from time delay data.

ARRIVAL-TIME DISTRIBUTIONS OF MUONS AND ELECTRONS

IN LARGE AIR SHOWERS OBSERVED AT 5200m ABOVE SEA LEVEL

Bolivian Air Shower Joint Experiment*

C.Aguirre, R.Anda and A.Trepp: Instituto de Investigaciones Físicas, Universidad Mayor de San Andrés, La Paz, Bolivia
F.Kakimoto, Y.Mizumoto, K.Suga, N.Izu, Y.Kamouchi, N.Inoue and S. Kawai: Department of Physics, Tokyo Institute of Technology,O-kayama, Meguro, Tokyo 152, Japan
T.Kaneko: Department of Physics, Okayama University, Okayama 700, Japan
H.Yoshii: Faculty of General Education, Matsuyama, Ehime 790, Japan
E.Goto, K.Nishi, H.Nakatani, Y.Yamada and N. Tajima: The Institute of Physical and Chemical Research, Wako, Saitama 351, Japan
P.K. MacKeown: Department of Physics, University of Hong Kong, Hong Kong
K.Murakami: Department of Physics, Nagoya University, Chikusa, Nagoya 464, Japan

ABSTRACT

An experiment is going on at Mt. Chacaltaya in Bolivia (550 gcm^{-2} atmospheric depth) to observe arrival-time distributions of muons and electrons in air showers above 10^{17}eV to study the early stage of longitudinal development related directly with the character of nuclear interactions and the composition of primary cosmic rays at these high energies. The preliminary results are presented at the Seminar.

INTRODUCTION

The early stage of longitudinal development of air shower is being studied by several groups, observing the time structure of pulses from deep water Cerenkov detectors (a mixture of muons and electrons)[1], that of pulses from shielded scintillation detectors (muons)[2] and that of the atmospheric Cerenkov light from an air shower[3,4]. All of these experiments are being carried out at sea level. In Fig.1, h_s is the production height of a muon (Cerenkov light by electrons) above sea level. Then, the time delay $(l_s-h_s)/c$ observed at a distance (r) from the shower axis (or the direction of primary cosmic ray) is expressed as $\sim 1/2$ $(r^2/h_s c)$, where c is the speed of light. When the time delay is observed at the same distance (r) at a high altitude such as Mt. Chacaltaya, that is expressed as $(l_c-h_c)/c \sim 1/2$ $(r^2/h_c c)$, where h_c is the production height of the muon (Cerenkov light) above

*All communications should be addressed to K.Suga in Tokyo Institute of Technology.

ISSN 0094-243X/79/49013-12 $1.50 Copyright 1979 American Institute of Physics

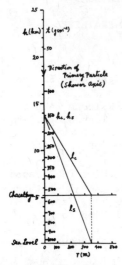
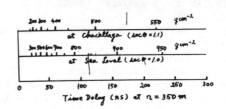

Fig.1 (left). Principle of the experiment.
Fig.2 (above). Relation between the atmospheric depth where the production of muon occurs and the time delay at a distance of 350 m from the shower axis.

the altitude. Since $h_c < h_s$, the time delay observed at the high altitude is larger than that at sea level. Fig. 2 shows a relation between the atmospheric depth where the production of muon (Cerenkov light) occurs and the time delay at a distance of 350 m from the shower axis observed at an altitude of Chacaltaya or sea level. It is clearly seen that the early stage of longitudinal development of air shower is studied with better resolution at a high altitude than at sea level. Therefore, we have been observing the distributions of time delays or arrival-time distributions of muons and electrons in the same air shower, simultaneously, at Mt. Chacaltaya from 1977.

EXPERIMENTAL

The Chacaltaya array[5] to observe air showers consists of 44 unshielded scintillation detectors spread over an area of 700m×700 m. The detector to observe muons above 600 MeV is a matrix of 15 units of $4m^2$ shielded scintillation detector among which 9 units are provided with a fast 5" photomultiplier (Philips XP2040) to observe the arrival-time distributions of muons together with a slow 14" photomultiplier (DuMont K1328). The fast outputs from the 9 detectors are added for recording. An unshielded $0.83m^2$ scintillation detector provided with a fast 5" photomultiplier (XP2040) to observe the arrival-time distribution of electrons is located above the center of the shielded $4m^2$ scintillation detectors with the fast 5" photomultiplier. The arrival-time distributions of muons and electrons are recorded with a Biomation 8100 wave form recorder with a sampling interval of 10 ns. The resultant time resolution is about 20 ns.

PRELIMINARY RESULTS

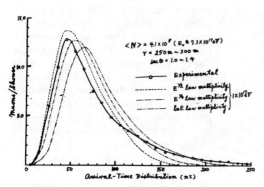

Fig.3. Average arrival-time distribution of muons at distances between 250m and 300m.

Fig. 3 shows an average arrival-time distribution of muons at distances between 250m and 300m. In this figure, three curves obtained from simulations are drawn for an $E^{1/2}$ multiplicity law, an $E^{1/4}$ multiplicity law, both with the CKP energy distribution of produced pions, and the scaling law (lnE multiplicity law). The experimental distribution shows a longer tail than the distributions obtained from the simulations.

Table I shows comparisons of FWHM and t(0.1-0.9) of the observed arrival-time distributions of muons with those obtained from the simulations for showers with $\sec\theta$= 1.0-1.4.

Table I. FWHM and t(0.1-0.9) of the arrival-time distributions of muons

	r = 250m-300m			
	Experimental (<N>=4.1x10^8)	Simulation (1x10^17 eV)		
		$E^{1/2}$	$E^{1/4}$	lnE
FWHM	56ns	63ns	66ns	69ns
t(0.1-0.9)	23ns	23ns	28ns	31ns

	r = 300m-350m			
	Experimental (<N>=1.8x10^8)	Simulation (1x10^17 eV)		
		$E^{1/2}$	$E^{1/4}$	lnE
FWHM	64ns	72ns	77ns	
t(0.1-0.9)	24ns	28ns	32ns	

Fig. 4 shows a comparison of arrival-time distribution of muons with that of electrons in the same shower. The extreme front of electrons is delayed by about 20ns from that of muons. Both of the arrival-time distributions show saddle points at about same time delay from the extreme fronts. This kind of correspondence between the arrival-time distribution of muons and that of electrons is seen in more than 50% of showers with distances from the axis larger than 200m and with $\sec\theta$ of 1.0 to 1.4.

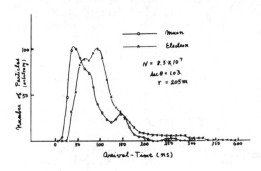

Fig.4. Arrival-time distribution of muons and that of electrons in the same shower

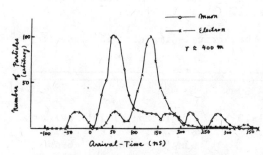

Fig.5. Arrival-time distribution of muons and that of electrons in the same shower.

Fig. 5 shows also a comparison of arrival-time distribution of muons with that of electrons in the same shower. The extreme front of electrons (strictly speaking, soft component absorbed in the PbS shield) precedes the extreme front of muons. Several percent of shower with distances larger than 200m and with $\sec\theta$ of 1.0 to 1.4 indicate this kind of unusual feature not due to an accidental incidence of electrons.

DISCUSSION

The arrival-time distribution of muons obtained in this experiment seems to be inconsistent with those obtained from simulations for primary protons with the scaling law, the $E^{1/4}$ multiplicity law and even with the $E^{1/2}$ multiplicity law, as shown in Fig. 3 and Table I. An electron-size N from a primary proton with 1×10^{17} eV is about $(5-6) \times 10^7$. Then, the arrival-time distributions of muons expected from simulations for $<N> = 4.1 \times 10^8$ in Fig. 1 and Table I and $<N> = 1.8 \times 10^8$ in Table I are broader than those for 1×10^{17} eV. Simulations for iron primaries have been also done with the $E^{1/2}$ multiplicity law. The arrival-time distributions seem to be not inconsistent with those obtained in this experiment. A detailed study has not been completed on the longer tail in the arrival-time distributions as shown in Fig. 3 which is also seen at distances larger than 300m.

The time delay of extreme front of electrons from the extreme front of muons seems to be shorter and the rise-time of electrons seems to be faster than those speculated under our common knowledge. These points will be examined in detail. The correspondence between the arrival-time distribution of muons and that of electrons and the unusual precedence of electrons will be also subject to detailed studies.

REFERENCES

1. M.L. Barret et al, Proc. 15th Int. Cosmic Ray Conf. Plovdiv $\underline{8}$,172 (1977).
2. P.R.Blake et al, Proc. 15th Int. Cosmic Ray Conf. Plovdiv $\underline{8}$,200 (1977).
3. N.N.Kalmykov et al, Proc. 15thInt. Cosmic Ray Conf. Plovdiv $\underline{8}$,244 (1977).
4. R.T. Hammond et al, Proc. 15th Int. Cosmic Ray Conf. Plovdiv $\underline{8}$,287 (1977).
5. C.Aguirre et al, J.Phys. G $\underline{5}$, No.1(1979).

SIMULTANEOUS DETECTION OF γ-RAY BUNDLES AND AIR SHOWERS.

S. Dake*, Y. Hatano[†], K. Jitsuno, M. Hazama,
Y. Nakanishi, K. Nishikawa, M. Sakata, and
Y. Yamamoto

Department of Pysics, Kohnan University, Kobe, Japan.
*Department of Pysics, Kobe University, Kobe, Japan.
[†]Department of Pysics, Kobe University, Now at
   Institute for Cosmic Ray Receach, University of Tokyo.

ABSTRACT

An air shower (EAS) observation system was used with an emulsion chamber, and the correlation between γ-ray bundles and EAS was studied. It was found that nearly 100 % of γ-ray bundles are accompanied by EAS and about 90 % of large EAS with sizes larger than 10^6 have γ-ray bundles in their cores. In this experiment a widely separated double core EAS was observed and the P_t of the subcore was found to be very high, ∼100 GeV/c.

INTRODUCTION

We have continued this experiment at Mt. Norikura (2780 m) since 1968, by using emulsion chambers and air shower (EAS) detectors. The purpose is to study the correlation between the γ-ray bundles and EAS, and to get further data of double core EAS showing high P_T.

APPARATUS

The summary of apparatus, method and results are as follows. Period: 100 days, emulsion chamber: 6.4 m^2 (10.5 r.l.), spark chamber: 18 m^2, scintillation counters for burst: 24 channels (1/4 m^2 each), scintillation cunters for EAS array: 26 channels (1/4 m^2 each), triggering threshold: 2,800 particles in any one of the 24 burst scintillators, total events triggered: 1,450 and number of combined (γ-ray bundles to EAS) events for high bursts (>7,000 particles): 81 out of total 121 events. The details of these data were described in another paper[1].

EVENTS TRIGGERED BY BURST

If the selection threshold is higher than 7,000 particles per burst scintillator, all selected events (121) are accompanied by EAS as seen in Fig. 1. The shower sizes of them are widely distributed from 10^3 to 10^7.

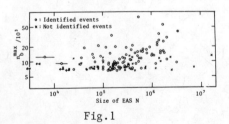

Fig.1

There were 4 showerless (< 100 particles) events out of 200 events with triggered burst size 4,500 ∼ 7,000 (per 1/4 m^2).

COMBINATION OF γ-RAYS AND EAS

The emulsion chamber is very weak in time resolution, so it is not easy to find EAS to which a γ-ray bundle belongs. However the combination was done successfully under the following requirements: i) The high energy γ rays must be located at the peak region of burst density distribution, ii) the arrival direction of the γ rays has to coincide with that of the EAS and iii) the number of electrons under the emulsion chamber estimated from the γ-ray energy (Approximation B) must be consistent with the observed burst density. In Figures 2 and 3 two types of variations for the rate of success in the combinations are shown for EAS with the selected burst size $> 7,000$ (per $1/4$ m^2). The Fig. 2 is concerning to total energies of γ-ray bundle. All γ-ray bundles with energies $\Sigma E_\gamma \geq 8$ TeV are seen to be completely combined but the rate decreases towards the lower energy region because of increasing candidates of EAS.

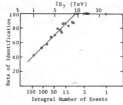
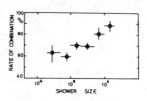

Fig. 2   Fig. 3

Fig. 3 shows the size dependence of the rate and it can be seen that for EAS with sizes $N \geq 10^6$ the rate is about 90 %. Here almost all EAS have the central bursts higher than 7,000 (per $1/4$ m^2).

CORRELATION BETWEEN SHOWER SIZE AND ΣE_γ OR n_γ

The correlation between the shower size and ΣE_γ of γ ray bundle, shown in Fig. 4, is widely distributed but we may find ΣE_γ increasing with shower size. If we extrapolate this tendency up to the energy region $\Sigma E_\gamma \geq 100$ TeV, the EAS accompanying large families observed by Japan-Brazil group[2] and by Mt. Fuji group[3], must have sizes far greater than 10^7 in average.

Fig. 5 shows the correlation between the shower size and number of γ rays, n_γ, in a bundle. It can be seen that the average shower age parameter (NKG) of events located around a line with a certain "c" value

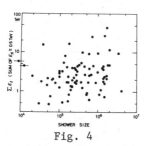
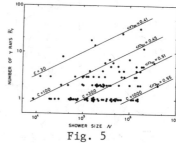

Fig. 4   Fig. 5

increases with c. For small c EAS are young. If 'Andromeda' event of Japan-Brazil group[4] really has the very high n_γ ($\sim 10^4$) primary energy $10^{16} \sim 10^{17}$ eV and does not have so large size EAS, the event is extremely deviated from the average tendency of our data.

A LARGE P_T (~100GeV/c) DOUBLE CORE EVENT

The air shower study groups in Osaka[5] and in Sydney[6] suggested an existence of very high P_T events beyond 100 GeV/c in their observation of multi-cores of EAS. We also observed a very widely separated large double core event. The main core was detected by the EAS array. Fig. 6 shows the density map of shower particles obtained by the spark chambers and Fig. 7 the burst density distribution map under the emulsion chamber with 6 cm lead.

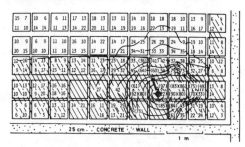

This shower arrived with a zenith angle 26° along the arrow mark in the figure. The three rows of the burst scintillators were shaded by the concrete wall of path length 60 cm (5.5 r.l.) along the arrival direction. The shadow is shown by hatched area. This shower is found to have two largely separated cores as shown in Fig. 8.

Fig. 6

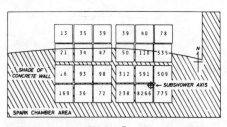

Fig. 7

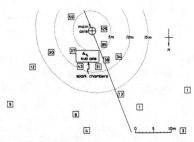

Fig. 8

The lateral density distribution of the main shower is obtained using only the scintillator data except for the area, as shown in Fig. 9. The size and age parameter of the main-shower is $(3.2 \pm 0.4) \times 10^5$ and 0.7 ± 0.1, respectively.

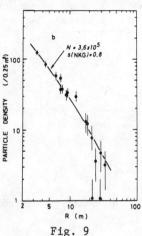

Fig. 9

The lateral distribution of the subshower is given by subtracting the back ground of the main-shower in the shadowless region (Fig. 10). The distribution in the shadow region is shown in Fig. 11 and the result of background subtraction is shown by a solid curve in the figure.

<u>Chance coincidence</u>   The possibility of chance coincidence of two EAS is rejected as follows. The resolving time for the accidental coincidence is 120 μsec at muximum in our scintillation counter system. If within this time interval another EAS with size larger than 5×10^4 hits the sensitive area 15×20 m^2 surrounding the spark chamber area, one recognizes it as another core. So, the intensity of these EAS is 10^{-4}/sec m^2. On the other hand, the number of events which have been examined are 350 showers for

burst sizes larger than
4,000/1/4 m² at the
centre of the shower.
Under this condithion
the expected accidental
chances are calculated
to be smaller than 2.5
× 10⁻³ during our observation period.

Energy of the subshower

To estimate P_t of
the subshower we need to
know the energy flow
of the subshower.

The subshower
particles in the
neighbourhood of the

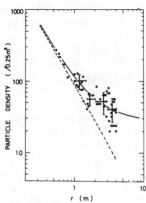

Fig. 10   Fig. 11

axis hits the vertical, thick concrete wall 5.5 r.l., and the total energy
flow decreased somewhat after the wall. The energy flow, E_1, before
the wall must be distinguished from E_2 after the wall.

First, the energy flow distribution after the wall can be deduced
as follows. The rate of multiplication of these subshower particles
in the lead of the emulsion chamber is obtained as a ratio of the burst
density at each burst scintillator to the spark density at the corresponding position. By comparing these ratios with the simulation results
of electron-photon cascade by Messel and Crawford[7], the average energies
of the spark chamber particles can be estimated as shown by squares in
Fig. 12. Then, the lateral distribution of energy flow after the wall
is obtained as shown by open circles in Fig. 12.

Next, we estimate the energy flow, E_1,
before the wall. We assume that the effective
number, n_{eff}, of mono-energetic electrons hit
the concrete wall aming at every burst scintilators. When $n_{eff} = 1$. the minimum energy
flow is expected. But this assumption is unnaturally small, because any peaks of particle
density are not seen at every section of the
spark chambers. Then, we assume n_{eff} to be
5 percent of the spark chamber particles, even
which will give a sufficiently low value of
energy flow compared with true one. The
decrease of energy flow in the wall of these
n_{eff} electrons is given by the calculation by
Dake and Oda[8], and using this result we can
convert E_2 into E_1. The obtained lateral
distribution of energy flow before the wall is

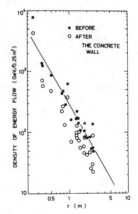

Fig. 12

shown in Fig. 12 by closed squares and total energy flow of the subshower is estimated as 42 TeV by an integration of the average curve
from 0 m to 5 m. The original total energy of this subshower at the
production height must be higher than this value, because the energy
fraction carried by hadronic components is not always fully detected
by the burst scintillators.

P_T of the subshower   The subshower is considered to be originated from a bundle of secondary particles, that is jet or fireball, emitted diretly from the main shower axis at the production height h_{sub}. The P_T of the jet or fireball is given by $P_{TJ} = R \cdot E_{sub}/c \cdot h_{sub}$, where $R = 6 \pm 1$ m and $E_{sub} > 42$ TeV. It is considered that the production height of the subshower is lower than 2 km, since its lateral distribution of particle density, see Fig. 10, is much steeper than that expected from an electromagnetic cascade theory, i.e., the subshower still has the jet structure. Therefore we get $P_{TJ} > 120$ GeV/c.

As another case, it is possible that an energetic hadron produced the subshower after a long travel in the atmosphere without collision. The transverse momentum of the hadron, P_{TH}, is given by $P_{TH} = R \cdot E_{sub}/c \cdot K_\gamma \cdot h_{main}$, where K_γ is the inelasticity of photons at the height h_{sub}. If the main-shower would be originated from a single γ ray, the starting point is estimated to be 6 km (11 r.l.) according to the age parameter and size of the main-shower and its initial energy is about 800 TeV. We take this height as h_{main} which must be higher than the real one, because the single γ approximation gives the highest value. K_γ will be smaller than 0.4, even if the triggering condition was preferable to π^0-rich showers. Then we obtain $P_{TH} > 100$ GeV/c.

In either case, the estimated P_T values are surprisingly great. But, it is opened to future experiments whether such a great P_T event is only a flucuation of P_T spectrum with average value of several GeV/c observed at several hundreds GeV region or a suggestion of a new interaction.

REFERENCE

1. S. Dake et al., Nuovo Cim. 41, N1, 55 (1977).
2. C.M.G. Lattes et al., Suppl. Prog. Theoret. Phys., No47, 5 (1971).
3. M. Akashi et al., 15th International Cos. Ray Conf. 7, 190(1977).
4. The same reference as 2.
5. S. Miyake et al., Cnad. J. Phys. 46(1968) s17.
6. A.M. Bakich et al., Canad. J. Phys., 46 (1968) s30.
7. H. Messel and D.F. Crawford, Electron-Photon Shower ... Pargamon.
8. S. Dake and H. Oda, Prog. Theor. Phys. 56 (1976) 1104.

DISCUSSION

S. Miyake.   In your talk on P_T 120 GeV/c, I guess as (below). Why such event becomes 42 TeV ?

If 42 TeV is correct, why not even small cascade is seen in your ECC ?

M. Sakata.   If you take the subshower size $\sim 6.7 \times 10^3$ which may be obtained from Fig. 10 and also take the single γ-ray approximation, you will get roughly 40 TeV by the help of Approximation B. After passing the concrete wall the size of the subshower is not 2,000 but about 1.1×10^4 and the total energy flow is about 20 TeV as we estimated from the rate of particle multiplication in the ECC. Further, we estimated that the half of the subshower energy have been lost mainly in the concrete wall according to the calculation by Dake and Oda and got 42 TeV as the starting energy. If we assume the starting height to be 5 r.l. above the observation level, the subshower after the concrete wall has passed 10.5 r.l. as a whole. Then, it is reasonable that the detection threshold, 0.5 TeV, in the ECC may not allow any cascades by descendant γ rays or electrons to be found.

HIGH TRANSVERSE MOMENTUM OF γ-RAY BUNDLES (>10^{12} eV) IN AIR SHOWERS

Y. Yamamoto, H. Munakata, Y. Nakanishi,
K. Nishikawa and M. Sakata
Department of Physics, Konan University, Kobe, Japan.

S. Dake
Department of Physics, Kobe University, Kobe, Japan.

ABSTRACT

21 γ-ray bundles ($E_\gamma \geq 1$ TeV, $n_\gamma \geq 2$ and the total is 64 γ rays) with air shower data were analyzed and their lateral and RE distributions were made. It was found that the lateral spread of the γ-ray bundles from the air shower axis is much greater than their own spread. By a comparison of the two RE distributions, in which the distance of a γ ray is measured from the air shower axis and from the bundle's energy flow center, $<P_{TF}> \simeq 2.3^{+1.1}_{-0.7}$ GeV/c was given as a mean transverse momentum of "fireball" generating the bundles with the help of theoretical RE formula or of simulation results. The result $<p_{T\pi^0}> \simeq 0.57^{+0.18}_{-0.09}$ GeV/c (at 50 ~ 200 TeV), with respect to collision axis, of parent neutral pions of bundle's γ rays keeps up the increase of $<p_{T\pi^0}>$ with collision energies above 10 TeV.

INTRODUCTION

In the accelerator experiments,[1] it has been established that inclusive cross sections at high transverse momenta (p_T > 1 GeV/c) increase with the (p-p) collision energy up to 2 TeV in the multiple production of hadrons. An interesting probrem is whether this increase of high p_T cross sections continues up to the extensive air shower (EAS) energy regions. Because some indication of high p_T phenomena had been observed in EAS experiments,[2] we have given, before the accelerator experiments, an attention to the lateral spread of high energy (≥ 1 TeV) γ rays in EAS and we observed them at mountain altitude (738 gcm^{-2}).[3-5] Except for a few experiments[3-7] γ-ray bundles have been observed without observation of EAS by many authers[8] using emulsion chambers. From the analysis of these data of γ-ray bundles the p_T distribution of secondary neutral pions in emission from a "fireball" and the "fireball" mass have been deduced,[8] though some analyses of multi-cored type events of bundles, to get p_T value between "fireballs", are now under way. On the other hand, from the γ-ray bundles with EAS data obtained in our experiment can be deduced the p_T distributions, with respect to the air shower axis, of "fireball" itself and of secondary neutral pions.

APPARATUS AND DATA SELECTION

Experimental apparatus was composed of an emulsion chamber (five sensitive layers, 6.4 m^2 area and 6 cm lead in total thickness) and an EAS array. The EAS array consists of 18 m^2 spark chambers covering the emulsion chamber, 24 burst scintillation counters just under the emulsion chamber and of 26 scintillation counters for detection of shower particle density of EAS. The details of the apparatus, experimental procedure and of a method of combination between the observed γ-ray bundles and the EAS events are described in another paper.[5]

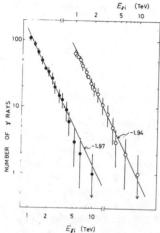

Fig. 1. Integral energy spectra of γ rays. O: selected events (21 bundles, 64 γ rays), ●: all events in the 3/4 area of the emulsion chamber. The selection seems to be nonbias.

In this paper 21 combined bundles whose axes of EAS have been determined are employed for the lateral distribution. The selection of events is performed for γ-ray bundles having two or more members with each energy greater than 1 TeV. The selection bias is tested in energy spectrum of all constituent γ rays of bundles. As shown in Fig. 1 the spectrum of selected combined events (64 γ rays) has a nearly equal slope to that of all events observed in a part of the emulsion chamber. Then, the selection is regarded to be nonbias.

DETERMINATION OF EAS AXIS

We define an air shower axis by a geometrical center of an ellipse, with an eccentricity $\sin\theta$ (θ is a zenith angle), fitted to the equi-density curves at lateral distances $1 \sim 1.5$ m from the density peak on the map of spark chamber data. For the axis position a small ambiguity arises from the error in the fitting and from a systematical shift of the axis with radius of the ellipse. Fig. 2 shows some examples of the ambiguity of the EAS axis thus determined. In Fig. 3 the distribution of this experimental ambiguity is shown, and the result of the simulated air showers by Ueda[9]* is also shown for the differences (ΔR) of EAS axes

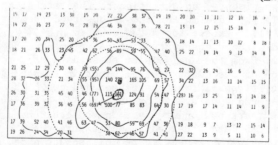

Fig. 2a. Map of spark chamber data of EAS.
The central hatched area is the ambiguity of the EAS axis determination.
◎: position of γ-ray bundle's center. EAS size $N = 1.8 \times 10^5$, zenith angle $\theta = 7.6°$, number of γ rays (≥ 1 TeV) $n_\gamma = 5$, sum of energy $\Sigma E_\gamma = 10.24$ TeV, $\frac{1}{n_\gamma} \Sigma r_i E_{\gamma i} = \langle rE \rangle = 4.13$ cm TeV, $\frac{1}{n_\gamma} \Sigma R_i E_{\gamma i} = \langle RE \rangle = 70.1^{+30.0}_{-22.1}$ cm TeV.

---

* The mean transverse momentum between the forward "fireball" and the survival nucleon is 2 GeV/c in his simulation.

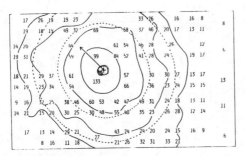

Fig. 2b. Same as Fig. 2a.
$N = 1.0 \times 10^5$, $\theta = 28.3°$, $n_\gamma = 2$, $\Sigma E_\gamma = 2.40$ TeV, $<rE> = 0.340$ cm TeV, $<RE> = 14.3^{+20.3}_{-14.3}$ cm TeV.

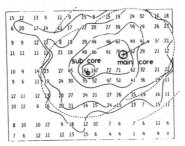

Fig. 2c. Same as Fig. 2a.
$N_{main} = 3.7 \times 10^4$, $N_{sub} \approx 4 \times 10^3$, $\theta = 39.4°$, $n_\gamma = 5$, $\Sigma E_\gamma = 13.3$ TeV, $<rE> = 0.790$ cm TeV, $<RE> = 166.6^{+26.4}_{-31.0}$ cm TeV.

LATERAL AND RE DISTRIBUTION

In our experimental data the two kinds of lateral spread are defined as follows.
r_i: distance from a γ ray to the energy weighted center of the bundle.
R_i: distance from a γ ray to the air shower axis.
As shown in Fig. 4 the R_i disribution shows much wider spread than the r_i distribution, that is, the γ-ray bundle is located far away from the air shower axis as compared with its own spread. This suggests protons to be dominant in the primary composition in $10^{14} \sim 10^{15}$ eV regions, because heavy primaries will symmetrize, due to abundant breakup nucleons, the distribution of γ rays around the air shower axis.

from the incident primary axes when our definition is applied. Because the both have a nearly equal spread, it is considered that a real incidence axis of primary particle lies probably within the experimental ambiguity. It is found in Fig. 3 that the distances ($\bar{R}$) of energy weighted centers of the bundles from the air shower axis are far greater than the above ambiguities. This fact shows that the γ-ray bundles are distributed remarkably asymmetric to the EAS axis.

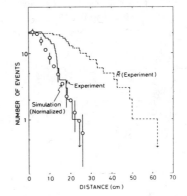

Fig. 3. Ambiguities in EAS axis determination. O: errors of the axis by our method in the map of Ueda's simulation data.[9] Solid line: ambiguities in our experimental data. Dashed line ($\bar{R}$) shows distance between the EAS axis and the bundle's center.

The two kinds of RE distribution, $r_i E_{\gamma i}$ and $R_i E_{\gamma i}$ distributions, are obtained and shown in Fig. 5. In this figure are also shown the results of the simulation by Kasahara[10] for primary proton incidence.

The simulation was calculated in the two collision models, H-quantum (HQ) model and two fireball (2F) model, with a mean transverse momentum $\langle P_T \rangle = 1$ GeV/c, with respect to the collision axis, for both the "fireball" and the survival nucleon. In this simulation it is also assumed for secondary hadrons to be isotropically emitted from the "fireball" with an ordinary low transverse momentum $\langle p_T \rangle \simeq 0.34$ GeV/c.

As shown in Fig. 5 the experimental values of $R_i E_{\gamma i}$ are larger than those of the simulation, in spite of the agreement of $r_i E_{\gamma i}$ with the simulation. This suggests the average value $\langle P_T \rangle$ of "fireball" and/or of the survival nucleon should be greater than 1 GeV/c.

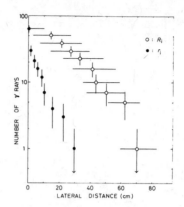

Fig. 4. Integral lateral distribution of γ rays (≥ 1 TeV). R_i (distance from the EAS axis) distribution is much flatter than r_i (distance from the bundle's center) distribution in spite of errors.

MEAN TRANSVERSE MOMENTUM

The RE distribution depends both on the p_T distribution and on the production height distribution through a relation $RE \sim p_T h$. From a simple picture of EAS it is considered that the $r_i E_{\gamma i}$ distribution reflects the p_T distribution of secondaries from a "fireball" and the $R_i E_{\gamma i}$ distribution will mainly reflect that of the most forward "fireball" or survival nucleon, though including the p_T distribution of the secondaries as a minor part.

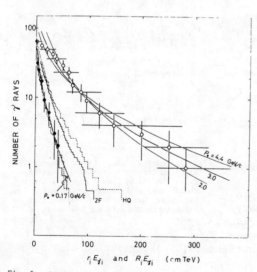

Fig. 5. Integral $r_i E_{\gamma i}$ and $R_i E_{\gamma i}$ distributions. The simulation results by Kasahara[10] are also shown for HQ and 2F models. Solid curves are the theoretical ones.

First, we make a simple estimation of the $\langle P_T \rangle$ of "fireball" or of survival nucleon from a quantity $\langle R_i E_{\gamma i} \rangle / \langle r_i E_{\gamma i} \rangle$. This quantity is roughly proportional to $\langle P_T \rangle / \langle p_T \rangle$ and its value is obtained from Fig. 5 to be about 2 in the simulation and about 6 in the experiment. If we assume $\langle p_T \rangle \simeq 0.34$ GeV/c for secondary hadrons as in the simulation and neglect the difference of the production height distribution between the experiment and the simulation, we can get $\langle P_T \rangle$(experiment) $\simeq 3 \langle P_T \rangle$ (simulation)

= 3 GeV/c.

Next, we deduce more precisely the $\langle P_T \rangle$ by fitting the curves of theoretical RE distributions which have been formulated by us. We assume again that the individual members of the γ-ray bundle are emitted from a "fireball" with $\langle p_T \rangle \simeq 0.34$ GeV/c and the $r_i E_{\gamma i}$ distribution is submitted to this p_T value. Then, the parameter Λ'_γ relating to the production height distribution is chosen to be ~35 gcm^{-2} so the theoretical $r_i E_{\gamma i}$ curve with $2p_0 = 0.34$ GeV/c fits the experimental one in spite of being unknown about the production height of the observed γ rays.*) Using this value $\Lambda'_\gamma \simeq 35$ gcm^{-2} the value of $\langle P_T \rangle = 2P_0$ is deduced by fitting the theoretical $R_i E_{\gamma i}$ curve to the experimental one. The result is given to be $P_0 \sim 3.0^{+1.4}_{-1.0}$ GeV/c.

In the model of EAS for our formula the survival nucleon runs through along the collision axis, that is, $P_{TN} = 0$. If the survival nucleon is recoiled with a same transverse momentum as the "fireball's" one, the above obtained P_0 value must be reduced to about a half ($P_0 \rightarrow \sim \frac{1}{2} P_0$). For, it is considered that both the survival nucleon and the "fireball" can play a nearly equal role to keep the bundles away from the EAS axis. The above reducing factor depends, in more details, on the "fireball" mass, $i.e.$, on the number (n) of secondaries from a "fireball". If we consider $n = 6$ to be reasonable for the most forward cluster in the fragmentation region, the reducing factor is given as 1/2.6 instead of 1/2 in the test of the theoretical formula by the simulation results. Thus we finally obtain $\langle P_{TF} \rangle (\simeq \langle P_{TN} \rangle) \simeq 2.3^{+1.1}_{-0.7}$ GeV/c.

The mean transverse momentum $\langle p_{T\pi^0} \rangle$, with respect to the collision axis, of parent neutral pions of bundle's γ rays is deduced, from the above value of $\langle P_{TF} \rangle$ for $n = 6$, to be $\langle p_{T\pi^0} \rangle \simeq 0.57 \pm^{0.18}_{0.09}$ GeV/c in the collision energy regions of 50 ~ 200 TeV. This is plotted in Fig. 6.[11] Our result seems to keep up the increase of $\langle p_{T\pi^0} \rangle$ with collision energies above 10 TeV.

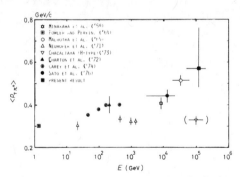

Fig. 6. Mean transverse momentum of neutral pions, with respect to the collision axis, with other experimental data.†)

---

*) In the simulation by Kasahara[10] the $r_i E_{\gamma i}$ distribution is well expressed by our theoretical curve with $2p_0' = \langle p_T \rangle = 0.34$ GeV/c for the value of Λ'_γ given in the simulation.

†) see references in the paper of reference 1 or 2.

REFERENCES

1) For references, see the review by D. Sivers, S. J. Brodsky and R. Blankenbecler, Phys. Reports, 23, 1 (1976).
2) For references, see the report by H. Munakata, M. Sakata, Y. Yamamoto, S. Dake and N. Nii, CRL-Report-55-77-14,(1977);contributing to Prog. T.P.
3) S. Dake et al., Acta Phys. Hungar. 29, Suppl. 3, 671 (1970).
4) Y. Yamamoto et al., *Conference Papers of Int. Conf. on Cosmic Rays, at Munich*, 12, 4369 (1975).
5) S. Dake et al., Nuovo Cimento 41B, 55 (1977).
6) Yu. A. Smorodin et al., *Proceedings of the Int. Conf. on Cosmic Rays, at London*, Vol.2, 827 (1965).
7) T. Matano et al., Acta. Phys. Hungar., 29, Suppl. 3, 451 (1970).
8) For references, see the report in reference 2).
9) A. Ueda, private communication (1968); N. Ogita, M. Rathgeber, S. Takagi and A. Ueda, Canad. J. Phys. 46, s164 (1968).
10) K. Kasahara, private communication (1970); Y. Fujimoto, S. Hasegawa, K. Kasahara, N. Ogita, A. Osawa and T. Shibata, Prog. Theor. Phys. Suppl. 47, 246 (1971).
11) Y. Sato, H. Sugimoto and T. Saito, J. Phys. Soc. Japan, 41, 1821 (1976).

PRELIMINARY RESULTS FOR THE ABUNDANCE OF MULTICORED EAS AT Mt. NORIKURA

NORIKURA AIR SHOWER GROUP*

ABSTRACT

Multicore type EAS was observed by 54 m^2 spark chamber at Mt.Norikura (740 g/cm^2). As a preliminary result, it was found that abundance of multicored EAS is higher and distribution of separation between main and subcores are wider than the results of other groups.

INTRODUCTION

3500 EAS events were observed by the Norikura EAS observation system[1] which is mainly composed from 100 channels of scintillation counters (50x50 cm^2 unit) and 54 m^2 spark chambers. Out of the EAS, 207 events whose axes hit the central one third area of the spark chambers are selected. At present 94 photographs of them are finely analyzed. The size of them is larger than 5×10^4 with average 1.3×10^5.

MULTICORED EXTENSIVE AIR SHOWERS

In order to carry out the multicore analysis, following definitions of five types are introduced for the classification of EAS.

A - type; accompanies subcores which show lateral structures spreading wider than 2 m and the subpeak density Δ_2 larger than one third of the maximum densities Δ_1, i.e. $\Delta_2 \geq (1/3) \cdot \Delta_1$.

B - type; accompanies subcores which show the lateral structures of 1 m or more and also satisfies the criterion $\Delta_2 \geq (1/3) \cdot \Delta_1$.

C - type; accompanies subcores which spread in narrow areas less than 1 m and considered to be hadronic local shower.

D - type; accompanies subcores which show flat top spreading 1 m or more but not satisfing $\Delta_2 \geq (1/3) \cdot \Delta_1$.

E - type; accompanies no subcore.

The number of events and fractions in the total of these types are shown in Table 1.

	Types of EAS	No of events	Percent	Comment
	A	16	20.3%	
	B	26	32.9	
Table 1	C	15	19.0	
	D	1	1.2	
	E	21	26.6	
	$\theta \geq 30°$	15	———	Omitted
	Total	94	100 %	

---

* T.Kitajima and N.Nii: Ashikaga Institute of Technology, Y.Hatano: ICRR (Univ.of Tokyo), T.Yanagita: Osaka Univ., S.Kino: Osaka City Univ., K.Jitsuno, Y.Nakanishi, K.Nishikawa, M.Sakata, Y.Yamamoto, T.Yura and S.Nakamoto: Konan Univ., S.Dake, H. Oda, T.Nakatsuka and T.Sugihara: Kobe Univ., M.Kusunose, N.Ohmori and H.Sasaki: Kohchi Univ.. Reported by M.Sakata: Dept. of Phys. Konan Univ. Okamoto Higashi-nada, Kobe, Japan.

The subcores in types A and B are the same as those appearing in the analysis by the Osaka group[1] at Mt. Norikura. It must be noted that the fraction of these types is a few times larger than the results of the Osaka and the Kobe (also at Mt. Norikura) groups.

The distributions of separation between subcores and maincores are shown in Fig.1, where the correction for finite sensitive area is not performed for our data but done for the Osaka group's by Kino[3]. Inspite of no correction, our data shows flatter distribution than the one of the Osaka group.

These two differences above mentioned are considered to come from the difference in sensitive area, i.e. 54 m^2 for ours and 20 m^2 for both of the other groups. And most of the widely separated subcores are hard to be found out in the observations of the two groups.

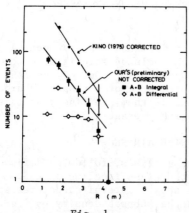

Fig. 1

REFERENCES

1. S. Miyake et al.: Canad. J. Phys. 46 (1968) s25.
2. N. Ohmori et al.: J. Phys. Soc. Japan, 42 (1977) 5, 1439.
3. S. Kino: Uchusen Kenkyu 19 (1975) 3, 370 (in Japanese).

HIGH p_t CROSS SECTION FROM MULTIPLE
CORE RATES IN AIR SHOWERS

W. E. Hazen and A. Z. Hendel
University of Michigan, Ann Arbor, MI 48104

J. F. Foster, B. A. Green, and A. L. Hodson
University of Leeds, Leeds

R. Bull
University of Nottingham, Nottingham

ABSTRACT

The general problem is outlined. The current status of analysis of the Leeds spark chamber observations is given. Comparison is made with the results of other groups.

INTRODUCTION TO THE GENERAL PROBLEM

There have been many observations of multiple cores in air shower detectors. Those made with spark chambers[1-5] have the greatest potential for translation into the lateral structure of the incident shower itself, a problem that has received some attention[3,6,7] but that requires further analysis.

Finally, there is the basic question of testing models of high p_t production by comparison with observed rates of subcores. This has been done preliminarily with available information.[3,4,8,9] In order to make the comparison really meaningful, considerably more analysis is required. The uncertainties in determiniations of the normalizations and of the slopes need further attention.

INTRODUCTION TO THE ANALYSIS PROBLEM

Our group is not yet ready to address the question posed in the program: "Multiple cores and high p_t - can the connection be made?" It is believed that the answer is positive[1-4] and we concur in that belief. But there is still much to be done before models of high energy interactions can be confidently tested.

The expected relative cross section of a few percent[9] is just high enough for statistically significant cosmic-ray observations. But the major problem is to make other uncertainties as small as the statistical uncertainty. (1) First is the question of the effective "beam". The Tokyo groups have made estimates based (a)[2] on simulations by Tanahashi, and (b)[4] on the measured hadron component of air showers at sea level plus the measured attenuation length in air. Both methods have large uncertainties. Updated simulations are needed for improvement here. In principle, fluctuation effects can also be obtained from simulations. (2) Second is the interpretation of the role of the atmosphere as analyzer, which is coupled to the question of the nature of the high p_t "particle". If it is a pion, simulations have shown that a π^o is the most likely source of a subcore.[1,10] If it is a jet, a jet with a π^o as the leading pion is the most likely source. Therefore, analyses made to date[2,4] have assumed π^o sources. The analytical solutions of Nishimura and Kidd[11] for electromagnetic showers have been used. If it seems to be desirable to consider fluctuations, it is probably necessary to do simulations here also. (3) Third is the question of interpreting the detector response. We are using limited-current spark chambers, which give the best representation of the number of shower particles of any feasible detector. However, there is an observable transition effect of the electromagnetic radiation produced by roof beams above the spark chambers, seen at Leeds. (4) The final question on this list is that of efficiency of detection of (i) subcores in the main shower, and apparent subcores due (ii) to fluctuations in main shower particles and (iii) to hadron interaction in overlying materials.

In this paper, I shall outline the current status of our work, which means primarily a discussion of (3) and (4). Then there will be a brief comparison to other groups.

PARTICLE DETECTION; SUBCORE DETECTION

(a) The <u>multiparticle detection efficiency</u> of limited-current spark chambers similar to ours has been measured relative to scintillators.[1] However, there are uncertainties of interpretation of scintillator data due to effects such as scattered particles, local interactions, and statistical breadth. Therefore, we have determined the efficiency from studies of spark chamber photos alone. From the statistics of sparklet clusters in test photos with trigger delay, we deduce an efficiency $\simeq$ 100% in spark formation. The ultimate limit is photographability, because the spark brightness diminishes as the available energy per chamber is shared among an increasing number of sparks.

(b) We have measured our scanning efficiency for subcores from comparison of two independent scans of ~ 60,000 photos. The result is 82% efficiency for double scanned photos. However, this is primarily for rather sharply peaked subcores. We are uneasy about our efficiency for detecting broad subcores, particularly in the larger showers, a problem that the Norikura group has found to be serious.[5] We are working on this question.

(c) Correction for subcores that are missed because the array is finite and because subcores cannot be seen in the dense, inner region of showers is made either from the subcore events themselves, or from a larger sample of main showers with the observed radial frequency distribution of subcores folded in. The result is ~ 0.3 for this efficiency factor.

(d) There is a potential contamination factor due to subcores simulated by statistical fluctuations. In order to estimate this, we have first looked for evidence of residual dependence in the lateral positions of the particles resulting from common parentage. This has been done by counting sparks in bins at a given distance from a shower axis and calculating the standard deviation for each set of bins. We have found that $S^2 = (0.94 \pm 0.07)(<n>)^2$, averaged over the samples, with no significant dependence on shower size or distance from the axis. Therefore, we conclude that the statistics of independent events are a good approximation. Our first cut, in the scanning, was for "subcores" with $n_o \geq 30$ in a circle of 25 cm radius <u>above</u> background ≤ 30. The Hillas analysis[12] for the effect of moving a bin to maximize the count can be used to get an upper limit to the probable number of fluctuations to 60 or more within the 25 cm bin. This has been done, using observed background distributions in the calculation. The result is that no more than 9 (to a 90% confidence level) would have occurred. This is a drastic upper limit since (i) 30 is the minimum n_o in our sample, (ii) the central bin is surrounded by bins that are somewhat above average. Our conclusion is that the number of false subcores generated by fluctuations is negligible.

The corrections from (b) and (c) above will be made after a discussion of the transition effect.

TRANSITION EFFECTS

The spark chambers are mounted directly beneath the roof, which is made of wood and fibre glass (Fig. 1).

(a) The average transition effect of the EM Component is expected to be small since the roof averages only 2.6 gm/cm^2 $\simeq$ 0.06 shower units. The effective Z is not much different from air. Hence the total number of particles simply decreases as if the shower had traversed another 20 m of air (we are past the shower maxima). But the lateral structure is altered, particularly by the beams (12 gm/cm^2 $\simeq$ 0.3 shower unit). Observationally, we can barely detect the beam effect, and only within a meter or two of the axes of large showers, $N \geq 10^6$, where pair production dominates over particle absorption.

We conclude that there might be some distortion of subcores by the beams but not by the prisms (skylights).

(b) The hadron component is a different story; it can generate "subcores", particularly in the beams. Our roof geometry is quite like that at Kiel, where this effect was discovered.[6] We observe a qualitatively similar effect of clustering under the beams (Fig. 1).

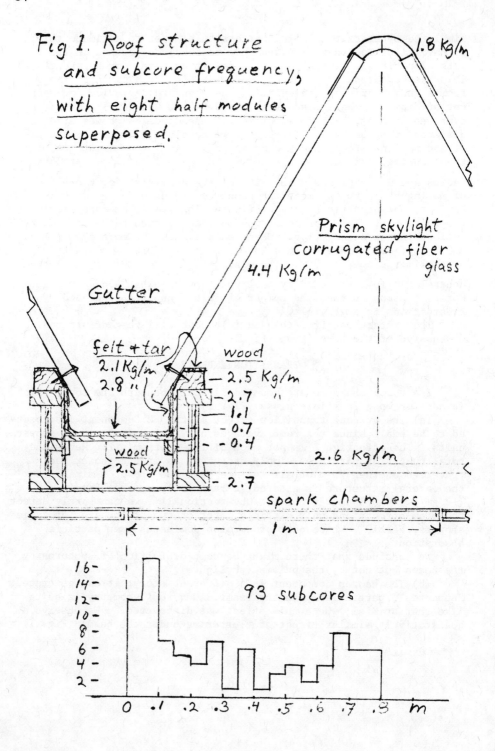

Fig 1. Roof structure and subcore frequency, with eight half modules superposed.

CORRECTION FOR HADRON INTERACTIONS

The expected number of subcores produced by hadron interactions in the roof can be estimated from hadron intensity and scaling of accelerator multiplicity distributions[7] but the input data are uncertain. Therefore, we are trying to rely primarily on our observations by using the observed overlying mass effect on the number and steepness of the subcores.

A subsample of 66 from the 93 observed subcores is ready for this analysis. These are separated into 28 subcores observed under the prisms (1 gm/cm^2) and 38 under the gutters (3.6 gm/cm^2). In each case, the subcores have an air subcore component N_A and a locally produced component N_H. The former is proportional to the area and the latter depends on the mass. Due to secondary interactions, there is a relative factor of about 3 in addition to the mass factor itself.[7] Using the above sort of modeling, we get:

	Prisms	Gutters
N_A	25	20
N_H	3	18

We then turn to the steepness distribution of the subcores (steepness measured by the production height t (s.u.) from Nishimura-Kidd[11] fits) in order to see if there is a correlation with their origins. Figure 2 shows the results. They are not clear cut. However, taken together with the Kiel observation of correlation of steepness and beam effect[6] and unpublished results from FNAL thick-target data[7], the best sorting is based on the assumption of local interactions producing steeper events.

Fig. 2. Frequency vs t, which is the depth in shower units but is also a measure of steepness.

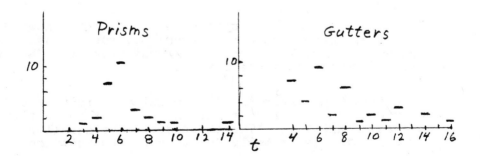

As a sample of high likelihood shower subcores, we select the 25 prism events with t>4 and the 18 gutter events with t>6. The p_t's for these events are obtained with the usual assumption that neutral pions are the most likely source. The Nishimura-Kidd results for gamma rays are used to find E_0, assuming equal energies for the gammas. For the lateral displacement, r, of the subcore from the projected direction of the interacting particle, we follow the Tokyo choice of measuring from the symmetry axis of the main shower. This choice will be tested in the future when we start modeling and simulating. Our p_t distribution is shown in integral form in Fig 3, along with those from Tokyo[2,4].

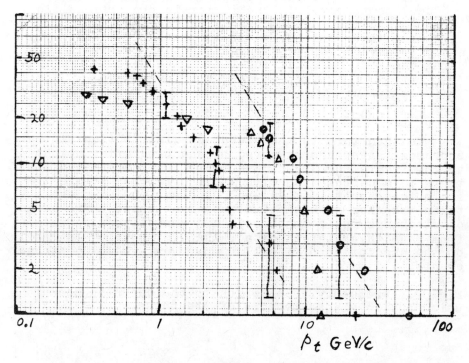

Fig. 3. Integral p_t distributions: +, this experiment; o, Tokyo[2] (shower trigger); △, subcore data, ▽, hadron data, Matano et al[4] (burst trigger). The error bars are only statistical.

The distributions in Fig 3 show only the results of direct reduction of our data with no explicit correction for biases of any kind. There appears to be a fall-off at low p_t, which is probably due to subcores that are lost in main shower background near the main shower axis.

INTERPRETATION

The slope of our distribution is essentially the same as the Tokyo[2] results. Taken at face value, this integral slope of -1.7 favors the gluon exchange model of parton-parton scattering[13] as pointed out by Gaisser[9]. Apparently, it is very unlikely that the (-6) integral slope of the interchange model[14] can describe the data, but one should not be misled into that conclusion by the error bars.

The data themselves suggest biases and uncertainties. Relative bias against observing low p_t events is suggested by the fall-off on observed slope at low p_t, which is very unlikely at production. We deliberately set our trigger requirement at a very low level (central density 5 or $10/m^2$) in order to make any trigger threshold effect negligible. We have measured our scanning efficiency and the results appear to indicate that it is high, right down to the level where small subcores are lost in statistical background from the main shower.

Another conclusion from the data is forthcoming when we turn to rates. The Tokyo array observation[2] was for $20m^2$ x 422da = $8440m^2$da. Our observation reported here is nominally for $35m^2$ x 175da = $6000m^2$da and will probably be reduced to $\sim 5000m^2$da when the average useful area is considered. Fig. 3. displays a factor 7 difference in number which leads to a factor $(5000/8440)7 \simeq 4$ in rate. (Matano et al[4] do not give their running time so we cannot compare to them.) What problems does the rate discrepancy indicate? The absolute rate is required (actually the rate relative to showers) in order to obtain the partial cross section. This has been done preliminarily[2,4,9]; but is it uncertain by a factor 4 as suggested above? Secondly, the disagreement in rate may be indirect evidence of problems that affect the slope significantly, in spite of apparent agreement in slope at the moment.

Trigger: the Tokyo trigger is more selective, which is in the wrong direction for explanation (our dead time is negligible, even at our high rate).

Scanning cut: our size cut is at a lower level, again in the wrong direction for explanation.

Scanning efficiency: our measured efficiency (for what we found by visual scanning) was high. Tokyo does not mention scanning efficiency. Perhaps Tokyo found some subcores by detailed counting studies, as was found necessary by the Norikura group[15]; if so, they do not mention it.

Data cuts: We kept all events above 30 net within a 25 cm circle that we believed were unlikely to be due to statistical fluctuations; Tokyo cut out events that had a production height≤6, because of large uncertainties in p_t. Again, the effect is in the wrong direction.

Event cuts: We cut to ~2/3 of the observed subcores by rempving likely local interaction events originating in the roof structure; Tokyo (private communication) may have cut out 2 or 3 events that could have come from beams. These do not explain the factor 4.

Overlying material effects: We have made the corrections indicated above. The Tokyo roof supports were minor, and the roof was only ~1gm/cm^2. Thus, there would be few roof interactions. There were large glass mirrors above the spark chambers, constituting ~2gm/cm^2. It is unlikely, but perhaps possible, that most of their subcores originated in the mirrors. Our data on roof effects are not very helpful because they are for subcores that are mostly smaller than the Tokyo subcores.

p_t sample cuts: Tokyo made a cut $p_t \geq 5$ GeV/c. We chose to make no p_t cuts, in case lower p_t data helped to determine the slope of the frequency distribution.

Summary: There is no evident potential explanation of the disagreement in rates. We are testing for the presence of "hidden subcores" that we can find only by detailed counts in regions that have only a hint of excess sparks.

RATE IN AIR

The corrections for finite detector size and masking by dense regions near shower axes were made by measuring the observable azimuths at the subcore distances. The result is about a factor 3 for conversion from number of observable subcores to number in the showers whose axes hit the array.

CONCLUSIONS

The results of our first year of operation appear to corroborate the Tokyo results of a p_t integral distribution no steeper than p_t^{-2}. But we have reservations stemming from disagreement in absolute rate. We are looking for the existence of "hidden subcores" that are very difficult to find from visual inspection alone. We are going to examine possible sources of bias that would affect the slope.

We are abstracting data from our second year of operation, taken under a new thin, light roof that minimises local interactions.

We plan to do modelling and simulations in order to sharpen the testing of interaction models by subcore observations.

This work was supported by the Science Research Council of Great Britain, the U. S. Department of Energy, and the Physics Departments of the Universities.

REFERENCES

1. S. Shibata, M. Nagano, T. Matano, K. Suga, and H. Hasegawa, Proc. Int. Conf. Cosmic Rays, London, 2, 672 (1965).
2. T. Matano, M. Nagano, S. Shibata, K. Suga, and G. Tanahashi, Can. J. Phys. 46, 56 (1968).
3. T. Matano, M. Nagano, S. Shibata, K. Suga, G. Tanahashi, and S. Hasegawa, Cosmic Ray Studies 19 No. 3, 370 (1974) (Mimeographed Journal in Japanese).
4. T. Matano, M. Machida, T. Ishizuka, and K. Ohta, Int. Conf. on Cosmic Rays, Munich 12, 4364 (1975).
5. N. Ohmori, K. Jitsuno, M. Sakata, Y. Yamamoto, S. Dake, and Y. Hatano, J. Phys. Soc. Jap. 42, 1439 (1977).
6. E. Boehm, W. Buscher, R. Fritze, U. J. Roose, M. Samorski, R. Staubert, J. Trumper, Can. J. Phys. 46, 41 (1967).
7. W. E. Hazen and D. L. Burke, Jour. of Phys. G 3, 715 (1977).
8. D. Cline, F. Halzen, and J. Luthe, Phys. Rev. Letts. 31, 49 (1973).
9. T. K. Gaisser, Proc. 7th Int. Colloq. on Multiparticle Reactions, Tutzing 521 (1976).
10. B. A. Green (1977), Private communication.
11. J. Nishimura, Hd. d. Physik 46/2 1(1965).
12. A. M. Hillas, Proc. Int. Conf. on Cosmic Rays, Munich 3439 (1975).
13. S. M. Berman, J. D. Bjorken, and J. B. Kogut, Phys. Rev. D4, 3388 (1971).
14. R. Blankenbecler, S. J. Brodsky, and J. F. Gunion, Phys. Letters 42B, 461 (1972).
15. N. Ohmori, K. Jitsuno, M. Sakata, Y. Yamamoto, S. Dake, and Y. Hatano, Jour. Phys. Soc. Jap. 42, 1439 (1977).

NATIONAL COSMIC RAY FACILITY TO STUDY HIGH ENERGY
PHYSICS AND PRIMARY COMPOSITION FROM 10^{14} TO 10^{18} eV*

W. V. Jones
Department of Physics & Astronomy
Louisiana State University, Baton Rouge, LA 70803

ABSTRACT

Within a few years colliding beam accelerators will produce information on particle interactions at energies that overlap lower energy extensive air showers (EAS). Input of the pending accelerator data will permit accurate model calculations, which can be compared directly with EAS measurements. This should allow disentanglement of particle physics from the primary composition. High quality EAS measurements could then be interpreted in terms of the primary mass around 10^{15} eV. At $\sim 10^{16}$-10^{18} eV the primary objective would be to study particle physics in a regime not accessible to the next generation of accelerators. A national facility equipped for measuring individual shower profiles in the atmosphere and, simultaneously, the hadron, electron, and muon components at ground level would provide the EAS data.

INTRODUCTION

Cosmic ray investigators are concerned about the viability of continuing some of their traditional research experiments, in view of approved accelerator programs. The new colliding beam[1,2] machines will have the capability of producing data at energies above 100 TeV with statistical accuracies that cannot be matched with cosmic ray experiments. What, then, does cosmic ray research have to offer? Are there important experiments which are, and will remain, unique to cosmic ray investigations?

These questions were raised at the 15th International Conference on Cosmic Rays in Plovdiv, Bulgaria, and they have been under discussion, both in provate and in a special evening session, at this meeting. Extensive air showers (EAS) originate from particle in the

---

*This suggestion belongs to many members of the cosmic ray community. It was discussed in a community-wide meeting at the 15th International Cosmic Ray Conference, Plovdiv, Bulgaria, 13-26 August, 1977. Further discussions at this workshop have shown there is sufficient interest to proceed with studies for a specific program.

range 10^{14} eV $\leqslant$ E $\leqslant$ 10^{20} eV and perhaps beyond. It is therefore clear that EAS energies extend well beyond those attainable with the proposed machines, and they will be unique to cosmic rays for many years to come. Therefore, the question might be more specifically expressed as: "which are the most basic and important experiments that must be performed in the EAS beam?" This is not to say that other energy regions are closed to cosmic rays. Indeed there are strong arguments for satellite (or space shuttle) exposures of high energy physics experiments which would overlap and extend somewhat beyond the colliding beam energies. The cosmic rays also provide beams for studying the interactions of high energy heavy nuclei which are still limited to comparatively low energies at accelerators.

In this report we will restrict our discussion to the traditional EAS region. It seems to be a general consensus that measurement of the primary composition at energies above 10^{14} eV would be an extremely important contribution to both astrophysics and particle physics. Knowledge of the mass distribution of the primaries, is a long standing objective of high energy astrophysics, because of its implications for sources and acceleration mechanisms. Its importance for particle physics lies primarily in specification of the "beam". The low flux of primaries with energies greater than $10^{14}-10^{15}$ eV precludes their study by observations with conventional balloon or space shuttle detectors that provide direct charge measurements. They can, however, be studied with large area, ground based EAS arrays. A uniquely important goal of EAS experiments is the study of high energy interactions from 10^{16} to the highest energy permitted by the falling cosmic ray flux ($\gtrsim 10^{18}$ eV). In this connection it is essential to note that 10^{17} eV corresponds to a total center of mass energy of about 15 TeV, which is an order of magnitude beyond the highest energy envisioned for the next generation of accelerators.

In this paper we discuss some of the general properties of EAS development. Specific features were described in several reports at this meeting, and they appear in written form in these proceedings.[3] Our discussion is limited to qualitative comments, which are intended to stimulate detailed discussion of a national cosmic ray facility, whose primary objectives will be to measure the primary cosmic ray composition at EAS energies and to do high energy physics around 10^{17} eV in this beam. A basic premise of this suggestion is that the colliding beam facilities will produce particle physics data that can be incorporated into calculations necessary for interpreting the primary composition from simultaneous measurements on the individual EAS components around 10^{15} eV. The increased understanding that should result from solution of the high energy physics/composition problem around 10^{15} eV can be expected to alleviate the problem at higher energies which are of prime interest for doing high energy physics with air showers.

GENERAL EAS CHARACTERISTICS

An extensive air shower is the cascade of particles produced in the atmosphere by a single primary cosmic ray of sufficiently high energy that a coherent flux of cascade particles is observable deep in the atmosphere, essentially down to sea level. Typically, the primary nucleus interacts with a target air nucleus high in the atmosphere. The surviving nucleon (or nuclear fragments and/or nucleons in case of a heavy nucleus primary) and hadrons produced in the interaction undergo additional successive interactions. Neutral pions produced in the interactions rapidly decay into gamma rays, which initiate electromagnetic cascades. Some of the pions, especially those produced near the top of the atmosphere, decay into muons and neutrinos.

The hadronic component of the shower typically comprises the EAS core, which feeds energy into the electromagnetic and muonic components. Multiple cores may result from high transverse momentum of secondaries and/or nuclear fragments. The bulk of the shower particles are electrons resulting from the superposition of electromagnetic cascades beginning at each of the successive interactions. The growth of the electron component, a fundamental characteristic of the shower, reflects the momentum distribution of neutral pions produced in the interactions. The development of the muon component depends on the height of the interactions and on the distribution of energy between charged and neutral pions in the secondaries. Whereas the electron shower rapidly decays after rising to a maximum, the muon component decays very slowly because of the weak interactions of muons. Deep in the atmosphere the ratio of the number of muons to the number of electrons is an indication of the overall shower development.

SOME EAS INTERPRETATIONS

A fundamental problem with EAS interpretation is separation of the astrophysical aspect, i.e. the primary composition, from the particle physics contribution, e.g., scaling and rising cross sections. No model is entirely consistent with all the reported EAS observations.[4] There is considerable evidence that scaling is valid up to about 10^{15} eV and somewhat beyond provided modifications to give a rising plateau in the central region and/or an increasing cross section are considered.[4,5] However, some experimental results, especially the ratio of muons to electrons, may be explained only with heavy primaries. Scaling seems clearly to be violated in the energy range 10^{15}-10^{18} eV if the primaries are essentially protons. Scaling with predominantly heavy (Fe) primaries give much better agreement with measurements.

It is possible that significant violations of scaling occur in the fragmentation region between 10^{12} and 10^{15} eV, which results in discrepancies between measurements and calculations based on proton primaries. For example, Centauro events, which produce many hadrons but no neutral pions, would cause the interactions and the showers to resemble those initiated by heavy primaries. Some of the unexplained features associated with early EAS development could be understood in terms of a significant cross section for Centauro-like interactions.

Some of the discrepancy between model calculations and experimental observations may be attributed to insufficient consideration of fluctuations in EAS development. The more recent EAS experiments emphasize studies of fluctuations as a means of estimating the primary mass. This is accomplished at Haverah Park[6] and the New Akeno array[7] by using a wide range of detectors for measurements on individual showers. The Fly's eye experiment of the Utah group[8] observes the continuous longitudinal development of individual showrs via tracks of scintillation light in the atmosphere. This experiment has the potential for studying showers from about 10^{17} eV up to the highest energies, $\sim 10^{21}$ eV, because of the great distance over which the scintillation light can be observed. The experiment of Orford and Turver[9] studies the longitudinal development by using Cerenkov light produced in the atmosphere by individual showers with energies 10^{17}-10^{18} eV. Crucial to their experiment is measurement of the pulse shapes produced by the Cerenkov light.

INFORMATION GAP IN THE ENERGY RANGE 10^{14}-10^{17} eV

The next generation of accelerators (ISABELLE, $\bar{p}p$ colliding beams, Fermilab collider/doubler) will be able to measure detailed properties of hadronic interactions up to energies approaching the traditional EAS region. One can then infer rather directly the gross features of cosmic ray composition at 10^{14}-10^{15} eV. By comparing observed air showers with model calculations employing known particle physics characteristics, it should be possible to determine whether the primaries are in dominant mass groups (e.g., mainly protons or mainly heavy nuclei) or

mainly heavy nuclei) or whether there is a mixture. It is not clear to what extent it will be possible to separate mass groups in a mixture.

The Fly's Eye experiment is in a unique position to study the composition beyond 10^{17}-10^{18} eV, because the flux at such high energies is extremely low. The Fly's Eye can observe these high energy showers over distances of several tens of kilometers. The distribution of depth in the atmosphere of starting points of the showers will provide information on the composition at the highest energies.

This implies that the least studied region will be between the highest accelerator energy and the energy threshold of the Fly's Eye. It seems important tnat a special effort be made to study this region, since it is the source of several unusual occurrences. It encompasses a kink in the observed primary energy spectrum as well as the source of unusual events like the Centauro and anti-Centauro.[10] It is also clear that there is a great advantage in having measurements which could link the Fly's Eye data at the highest cosmic ray energies to the highest accelerator energies.

BASIC REQUIREMENTS FOR COSMIC RAY FACILITY

Studies of cosmic rays with energies greater than about 10^{14} eV require a large area, ground-based facility. Significant improvements over previous EAS arrays are necessary if quality of the data is to be much improved. Apparatus capable of measuring essentially all aspects of individual showers is needed. It is particularly important to have time resolution among the identifiable EAS components, i.e., hadrons, electrons, and muons. This can be obtained if showers developing in the atmosphere provide triggers for ground based detectors of the individual components. It is crucial to measure both the longitudinal and lateral shower development. Air Cerenkov or Fly's Eye type detectors can provide both signals for timing and the longitudinal profile, which is related to the primary energy. The lateral structure can be obtained from measurements with the component detectors, given the ground-impact position of the core from air Cerenkov or Fly's Eye trajectory information.

As an illustrative example, Fig. 1 shows a schematic diagram of an array that might potentially satisfy these basic requirements. The essential feature of the example is that a mountain altitude lake would be instrumented with photo detectors of Cerenkov light produced in the lake by charged particles comprising the EAS. Cerenkov light produced in the atmosphere would provide a measure of primary trajectory and energy from the longitudinal shower profile, as well as timing information

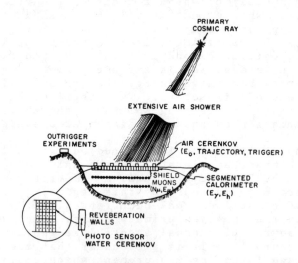

SCHEMATIC
MOUNTAIN LAKE FACILITY

for triggering the water Cerenkov detectors. Hadronic and electromagnetic cascades produced near the surface of the lake would be detected in a matrix of cells separated by thin, light-tight walls, perhaps of plastic sheets. The walls should act as a reverberation chamber so that the directional Cerenkov light will be made nearly uniform throughout a cell. A single photo-sensor at the center of each cell would measure the integrated cascade tracklength in the cell, which is proportional to the energy deposition.

The cross sectional area of the instrumental surface should be about 10^4 m^2 in order to contain the lateral spread of the shower. The depth of the surface portion of the array is governed by the penetration of the hadronic and electromagnetic components. About 20-25m depth should be sufficient. The cell size would determine the energy and position resolution, as well as the total cost.

Hadronic and electromagnetic components should be distinguished, at least partly, by relative signals in adjacent cells in the longitudinal direction. Detection of signals with characteristic 2.2 microsecond delay, which result from the $\pi \to \mu \to e$ decay chain of positively charged stopping pions is a unique hadronic signal.

The muon detector would be made of at least two instrumented planes, separated by, say, 100 m. The top plane must be at sufficient depth, at least 50 m, below the surface so that there is essentially no left-over hadronic or electromagnetic contamination in the muon signal. The basic measurement would be timing between the two planes for coincident signals from a single muon.

APPARENT ATTITUDE OF THE COSMIC RAY COMMUNITY

This mountain lake configuration was suggested in order to initiate discussions that would focus the experience of individual investigators into a planned program, which they as a group could support.

A major objection was that a mountain lake would be expensive to instrument with cell sizes small enough to resolve multiple cores or individual hadrons separated by a few tens of centimeters. Its main advantage is that it offers continuous measurement over the lateral extent of the showers. There are thus differences of opinion regarding the exact approach and configuration. Some investigators stress the need of new detection techniques, e.g., acoustic detection, for significant improvement over previous and existing experiments. Others prefer using conventional methods, perhaps in a novel arrangement.

It has therefore been suggested that the initial step should be a special workshop, lasting about one week, in order to define as well as possible both physics and instrumentation problems as well as to consider possible locations.

CONCLUDING REMARKS

The question posed at the beginning of this report, "whether there are experiments unique to cosmic rays that will provide information of fundamental importance", has been answered in the affirmative. One obvious experiment involves measurement of the primary composition above 10^{14} eV. This is directly relevant to high energy astrophysics. Moreover, it is a prequisite for specification of the beam for fundamental high energy physics experiments at energies beyond the reach of accelerators, up to at least 10^{18} eV lab energy.

We can achieve the goal of making this measurement, if we combine our efforts in order to have the dedicated manpower and expertise required, both for carrying out the experiment and for obtaining adequate funding. A solid program will be supported by the scientific community and should attract qualified young researchers, which are vital to the future of cosmic ray research.

ACKNOWLEDGMENTS

Many of the ideas in this paper were discussed during an evening session at the conference chaired by the author and also in informal conversations during and after the conference. I have tried to reflect these discussions in this paper. I am grateful for discussions with H. E. Bergeson, G. L. Cassiday, T. K. Gaisser, L. W. Jones, G. B. Yodh and many others and to T. K. Gaisser for a critical reading of the manuscript.

REFERENCES

1. D. H. White, these Proceedings.
2. C. Cline, These Proceedings.
3. See, for example A. M. Hillas, these Proceedings and C. Aguirre et al., these Proceedings.
4. For a review see T. K. Gaisser, R. J. Protheroe, K. E. Turver and T. J. L. McComb, Revs. Mod. Phys. 50, 859 (1978).
5. M. Ouldridge and A. M. Hillas, J. Phys. G4, L35 (1978).
6. A. A. Watson and J. G. Wilson, J. Phys. A7, 1199 (1974) and D. M. Edge, et al. Proc. 15th International Cosmic Ray Conference (Plovdiv) 9, 137 (1977).
7. K. Kamata, et al. these Proceedings.
8. G. L. Cassiday, these Proceedings.
9. K. J. Orford and K. E. Turver, Nature 264, 727 (1976) and Nuovo Cimento (to be published).
10. I.e. high multiplicity events with many γ-rays. A recent example is discussed by H. Semba, Institute of Cosmic Rays (University of Tokyo) ICR-Report-65-79-9. Others are "Andromeda" [C. M. G. Lattes et al. Prog. Theo. Phys. 47, 123 (1971)] and "Texas Lone Star" [D. H. Perkins and P. H. Fowler, Proc. Roy. Soc. A278, 401 (1964)].

Chap. 2 Techniques

TECHNICAL ASPECTS --- DESIGN, SCANNING AND ENERGY MEASURMENT IN SMALL EMULSION CHAMBERS.

H.Fuchi, K.Hoshino, S.Kuramata, K.Niu, K.Niwa
H.Shibuya, S.Tasaka* and Y.Yanagisawa
Nagoya University, Nagoya, 464 Japan.

Y.Maeda
Yokohama National University, Yokohama, 240 Japan.

N.Ushida
Aichi University of Education, Kariya, Aichi, 448 Japan.

(Presented by K.Niu)

ABSTRACT

Technical aspects of small emulsion chamber are discussed. Emulsion chamber is a detector which consists of nuclear emulsion plates and other light or heavy material plates. While a pure emulsion stack is likened to a micro bubble chamber, it could be compared to a complex counter detector assembly with super high spatial resolving power. It is possible to measure particle momenta by relative scattering method and trace back γ ray cascades to vertices a few centimeters away. Charged secondary tracks could be examined for sudden changes in direction several centimeters from production vertices. "Invention" of the emulsion chamber enabled one to extend the power of pure emulsions to study very high energy interactions to tens of TeV, and it opend a new widnow for search of short-lived particles.

EMULSION CHAMBER

The emulsion chamber is a complex detector which consists of nuclear emulsion plates and other material plates. It was introduced by Kaplon et al. in 1951[1] to study cosmic ray phenomena. Since then, considerable refinement has been taken place in its design mainly due to the efforts of Japanese physicists[2]. In these years, it has been utilized not only in cosmic ray region but also at sub TeV accelerator region for the study of nuclear interactions and new elementary particles[3].

Unlike the case of a pure emulsion stack (pellicle), nuclear emulsions are used principally as the position detector of tracks of charged particles traversing the materials in the emulsion chamber. By analogical expression, the emulsion chamber corresponds to a complex counter detector assembly while a pure pellicle stack

---

*Present address; Institute for Cosmic Ray Research, University of Tokyo, Tokyo, 188 Japan.

could be compared with a micro bubble chamber.

The emulsion plate coated on both surface of transparent plate is utilized as a two-fold emulsion counter. Spatial resolving power of the emulsion counter is extremely high and of the order of sub-micron. Several types of emulsion counters listed in Table I have been developed for the purpose. Plastic plate or film machined or punched out to have exact rectangular shape with an accuracy of $^{+0}_{-50}\mu$m is adopted as a supporting substrate of emulsion layers. Thickness of each substrate is maesured with an accuracy of $\pm 5\mu$m and registered.

Table I Types of emulsion counters

Supporting substrate				Emulsion	
Material		Thickness μm	Size cm^2	Thickness μm	Coating
Meta-acrylic Lucite	$C_5H_8O_2$	800	20 × 25	50	on both surfaces
"		300	20 × 25	50	"
"		"	12 × 9.5	50	"
Polystyrene	C_8H_8	150	20 × 25	50	"
"		"	12 × 9.5	50	"
"		"	20 × 25	300	"
"		"	12 × 9.5	300	"
"		70	12 × 9.5	300	"

These emulsion counters are combined with thin plates made of light or heavy materials to form a complex detector assembly consisting of a target and a momentum energy analyser.

Fig. 1a

Photograph of an emulsion chamber for observation of cosmic ray jet showers.

Figure 1a shows an unit emulsion chamber for observation of cosmic ray jet showers which is assembled in a container made of black lucite carefully machined and fabricated with an accuracy of 100μm. It measures 20cm×25cm×24cm and weighs about 50kg.

Typical configuration of the emulsion chamber is shown in Figure 1b. The target on the upper part of the chamber is a pile of 50∿80 two-fold emulsion counters, bare lucite plates with thickness of 850μm and some amounts of spacers with 3mm of air gap.

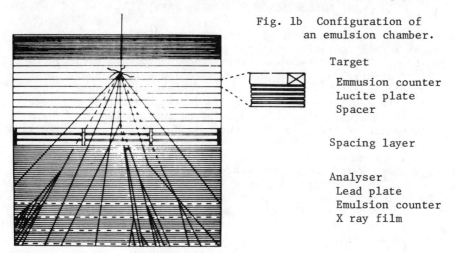

Fig. 1b Configuration of an emulsion chamber.

Target
  Emmusion counter
  Lucite plate
  Spacer

Spacing layer

Analyser
  Lead plate
  Emulsion counter
  X ray film

Total thickness of the target is 14∿16cm. In this part, it is possible to observe cross-sectional view of the secondary tracks each ∿1mm along the shower axis. As the bulk of the material is plastic and air, a nuclear interaction mean free path and radiation length is about 100cm and 50cm respectively. This is very effective to minimize the disturbance due to cascade showers when examinning behaviour of secondary particles. Special care taken in assembling the chamber enabled us, in favorable cases, to measure the relative angle between tracks of charged particles with error less than 10 micro-radians.

The analyser at the lower part of the chamber is a pile of 50 lead plates with thickness of 1mm and the same number of two-fold emulsion counters and some sets of x ray films. Total thickness of this part is 8∿10cm or 10 radiation length. A track of charged particle or a group of cascade electrons could be observed at each 0.2 radiation length. High accuracy in assembling the chamber enabled us to follow very easily even a single track with minimum ionization to the down stream end of the chamber.

An electron is clearly discriminated from other charged particles by inspecting a cascade shower it induced in the lead emulsion pile. The same method is applied to detect a γ ray. Energy and momentum of an electron or a γ ray could be estimated analysing each cascade shower. This is one of the conspicuous advantage of the emulsion chamber superior to the pure emulsion stack.

Another conspicuous advantage of the emulsion chamber is the applicability of relative scattering method to the momentum analysis of charged secondary particles. High precision in measurement of the relative distance between tracks of secondary particles in the same event and high scattering signal due to lead nuclei allow us to extend the applicability of the method up to TeV/c region.

SCANNING OF JET SHOWERS AND FOLLOWING OF SECONDARY PARTICLES.

Emulsion chambers described in the preceding section are exposed to cosmic radiation at airplane or balloon altitude to gather super high energy interactions. The standard exposure duration is 500 hours at airplane altitude ($\sim 260 gr/cm^2$) and 20$\sim$30 hours at balloon altitude ($\sim 10 gr/cm^2$). Length of standard exposure is due to rate of accumulation of background tracks at airplane altitude, but due to limited controlable range at balloon altitude.

Detection of jet showers produced in the target part is carried out as follows. Sakura Type N x ray films interleaved in the lower part of the chamber are scanned by naked eyes for dark spots due to cascade showers induced by secondary γ rays. Figure 2 is a photograph of x ray film after exposure. The detection threshold depends on the film background which is mainly determined by freshness of the film. Fresh x ray films give us the threshold well below 400 GeV, while the films spending 1 year after manufacture give somtimes worse than 600 GeV.

Fig. 2

Dark spots due to cascade showers on a Sakura type N x ray film.

The detected showers are traced back up into the target part using both x ray films and emulsion plates. The trace-back process is facilitated by the careful alignment of plates as well as the fresh high quality Fuji nuclear emulsion Type ET-7B. Figure 3 shows a typical procedure of the trace-back. Finally origin of the jet

Fig. 3. Procedure of tracing back.

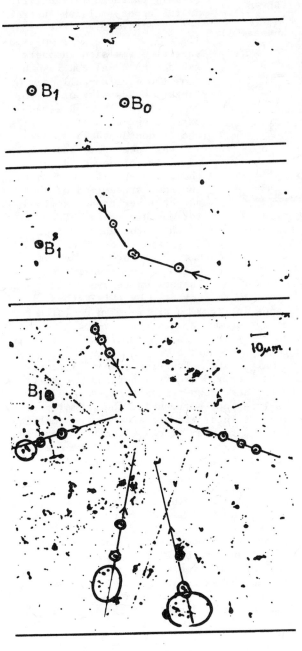

c) An emulsion film just above the interaction vertex.

B_0; primary particle.
B_1; reference particle.

b) Two emulsion films in the up stream part. Two charged particles traced back from the down stream are almost converging to a vertex.

B_1; reference particle.

a) Super imposed photograph of 3 emulsion films in the down stream part (lead emulsion sandwich) of the chamber. 3 cascade showers and 2 single charged particles are seen to be traced back to a common center.

B_1; non interacting reference beam.

Table II Event List of BEC-6

(3 Emulsion chambers exposed at 31km above sea level for 25.5 hours.)

Primary particles	No. of events
Single charged	28
Neutral	3
Heavy primary	12
Unidentified[a]	8

a) Interactions in the side or top plastic plates of the container.

shower is detected and the primary particle is identified in the upper emulsion layers.

Table II gives the event list from exposure of 3 emulsion chambers exposed by a balloon launched from the Sanriku Balloon Center, University of Tokyo, in May 1976. Exposure was carried out at a height of 31km and it continued 25.5 hours.

All secondary charged particles of detected jet showers are followed down to the down stream end of the target layer to be examined for any kink. In the course of this procedure, target diagramms at every emulsion layers are drawn. Any vee which appeared in the inspection volume could be detected by this method.

Several microphotographs in Figure 4 are from an emulsion chamber exposed to high energy beams of accelerator, but give one an idea how the work is carrying out. Each beam track leaves only one dark spot on the photograph because of vertical incidence. The diameter of a spot is only 0.7μm. Therefore, spatial resolving power of emulsion counter is nearly 3 order higher than other track detectors. Proceeding to the next down-stream film, several new spots due to secondary tracks appear surrounding the former position of the primary particle when it interacts between the two emulsion counters, while those spots due to beam tracks without interaction still occupy the same position as before. Relative distances between the secondary spots become larger as proceeding to the following films. Tracing back to the center, position of the interaction vertex is exactly determined. Emitting angle of each secondary track is measured after following them down to a certain depth from the origin. In the fourth film, a new pair of charged particles appeared in the most forward part of the event. They diverge rapidly as proceeding to the down stream, and the vertex point of this V track was estimated to be at 34μm above the film they appeared. Both particles of the vee were identified as hadrons after following them deep into the analysing part of the chamber.

In the analysing part one can also find some electron pairs or cascade showers which are produced by daughter γ rays of neutral mesons.

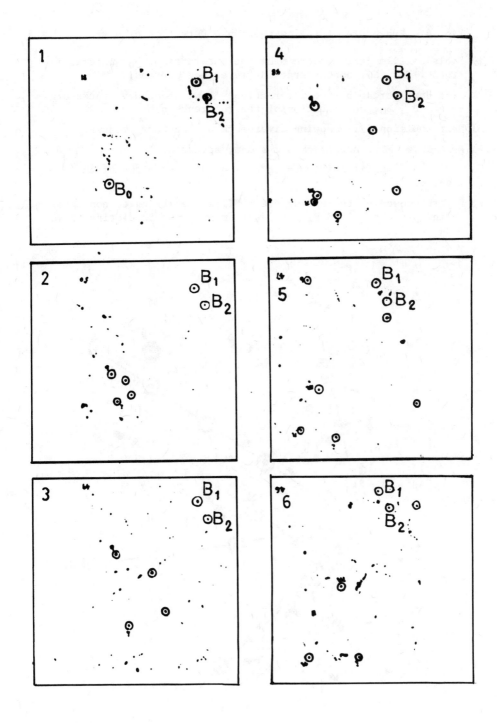

Fig. 4. Procedure of following down. Detection of a vee.

1) Emulsion film just up stream of the interaction. B_0; interacting proton. B_1, B_2; non interacting reference protons.
2) Just down stream of the interaction. Four secondary tracks are seen. B_1, B_2; continuation of the reference protons.
3) Next emulsion film showing divergence of secondary tracks.
4) Next film. Two new tracks, m and n, appeared.
5), 6) These two films show rapid divergence of the two tracks m and n.
7) Super imposed photograph of the 6 films, referring to non interacting protons B_1 and B_2. Production of a vee is distinguished.

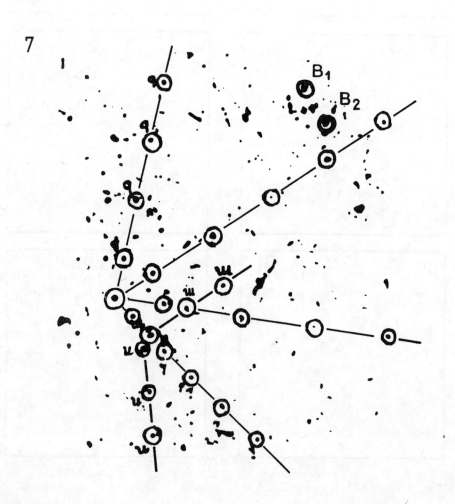

ENERGY MOMENTUM ESTIMATION

Cascade showers induced by tertiary γ rays from neutral pions are followed toward the down stream end of the chamber. Transition of number of shower electrons in a circle with radius of 25μm or 50μm is observed. Energy of a γ ray is estimated by fitting counted data to the curves derived from three dimensional cascade shower theory taking into account the effect of spacing between main scattering substance in the actual chamber[4].

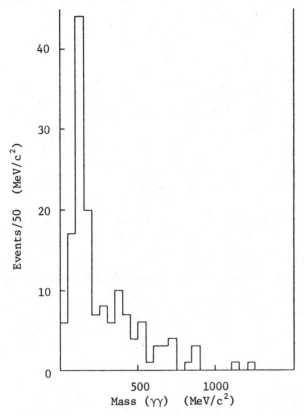

Fig. 5

Invariant mass distribution of any two γ-rays in the same events.

Calibration of this method is made by finding π^0 peak in the invariant mass distribution of any two γ rays in the same events. Figure 5 shows the invariant mass distribution of γ ray pair from 300 GeV/c proton interactions[4]. Clear peak is observed at 134±20 MeV. Another calibration is also carried out exposing same type of emulsion chamber to electron beams of 50 GeV and 200 GeV at Fermilab. Average transition curve of 28 showers of 200 GeV are shown for radii 25μm and 50μm in Figure 6[5]. The agreement of observed data with the theoretical curves is quite satisfactorily.

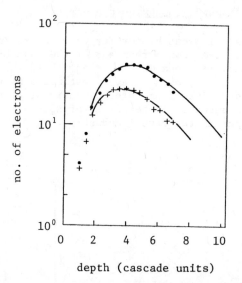

Fig. 6 Transition of a number of electrons (r=25 (+) and 50 (•) μm) for showers initiated by electrons with energy of 200 GeV. ──── Theoretical curve by Nishimura[4]).

In order to get the incident energy of shower as accurately as possible, the total track length of shower electrons within a cylinder with radius of 50μm and length to the down stream end of the chamber is taken into account. The error in energy of γ rays estimated by this method is about 20% at 30 GeV and less at higher energies.

To estimate the momenta of charged secondary particles, relative scattering method on these tracks in the lead emulsion sandwich pile is applied[5)6)]. Our emulsion chamber is horizontally exposed to cosmic rays. Most of the high energy secondary particles of jet showers, therefore, pass through these plates with relatively large dip angles. To apply relative scattering measurement on these tracks, at first, relative distances between two tracks projected on each emulsion surface are measured by microscope with occular micrometer or a precision machine for measurement of x-y coordinates which is specially designed for this purpose. Next, the second difference, D_0, of measured relative distance between tracks are calculated. These processes are shown in Figure 7. Squared mean value of observed second differences $<D_0^2>$ is related to the true scattering signal, D, and to noise due to measurement errors, D_n, by the relation;

$$<D_0^2> = <D^2> + <D_n^2>$$

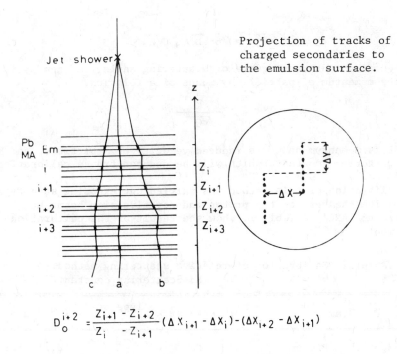

$$D_o^{i+2} = \frac{Z_{i+1} - Z_{i+2}}{Z_i - Z_{i+1}} (\Delta X_{i+1} - \Delta X_i) - (\Delta X_{i+2} - \Delta X_{i+1})$$

Fig. 7   Process of relative scattering measuremtnt.

There are several sources of noise, for example, reading error, error due to distortion, error due to non uniformity of thickness of plates and setting error of plate on the stage of microscope. In our case, average total noise for second difference, $\langle D_n \rangle$, is estimated to be 0.2∼0.4μm or 2μm depending on the method of measurement.

Observations on more than three pairs of tracks allow the individual values of $\langle D^2 \rangle$ to be determined solving following simultaneous equations;

$$\langle D_{ab}^2 \rangle = \langle D_a^2 \rangle + \langle D_b^2 \rangle$$

$$\langle D_{bc}^2 \rangle = \langle D_b^2 \rangle + \langle D_c^2 \rangle$$

$$\langle D_{ca}^2 \rangle = \langle D_c^2 \rangle + \langle D_a^2 \rangle$$

The root mean square of the scattering angle after unit length, $\sqrt{\langle \theta^2 \rangle}$, is related, as follows, to the second difference, D, to the cell length, C, and to the thickness of the main scattering material in each cell, s;

$$\sqrt{\langle\theta^2\rangle} = \sqrt{6\cdot\langle D^2\rangle}\,/\,c\sqrt{s}.$$

After getting root mean square of scattering angle, one can estimate the momentum of particle by means of a relation.

$$P\beta = \frac{K}{\sqrt{\langle\theta^2\rangle}}$$

here, P is momentum and β is reduced velocity of a particle, and K is a scattering constant, which will be deduced from a calibration experiment.

Calibration of this method was carried out exposing same type of emulsion chambers to the proton and negative pion beams at Fermilab and ANL[6]). Table III show the result of the calibration experiments.

Table III Calibration of relative scattering method.
K: Scattering constant.

	Beam	Scatterer	Type of Chamber	K MeV/c/mm
FNAL	303 GeV/c Proton	Pb	II	12.7 ± 1.1
	205 GeV/c Proton	Pb	I	13.5 ± 1.2
ANL	8 GeV/c negative π	Pb	II	10.5 ± 0.5
	8 GeV/c negative π	MA	III	1.06 ± 0.05

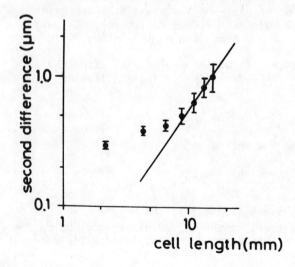

Relation of observed second difference to the cell length is given in Figure 8. From this figure noise value, D_n, is estimated as 0.3∼0.4μm. The contribution of noises become negrigible if one takes cell length longer than 5mm of Pb.

Fig. 8 Second difference versus cell length.

The true second difference due to scattering varies with cell length as $C^{3/2}$, but noises due to errors are almost independent of cell length. Therefore, the practical minimum cell length is given by $D=D_n$, and the maximum detectable momentum, MDM, is given as a function of cell length depending on the noise level as is shown in Figure 9.

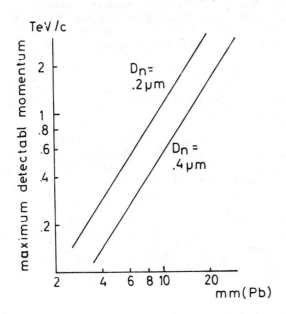

Fig. 9

M.D.M. as a function of cell length.

Statistical error for the obtained value of D is given as $\sim 80/\sqrt{n}\%$, where n is number of independent cells. When the repeat number of the scatterer is 50 and the cell size is 5, number of independent cell is 9 and the statistical error is about $\sim 27\%$. Thus, in our actual case, effective MDM is ~ 2 TeV/c.

Success of application of the relative scattering method added a power to an emulsion chamber designed for the detailed study of high energy jet showers. Including γ rays we have been able to get almost all of the informations about energy and momentum of secondary particle of jet showers, and this makes us possible to analyse fire balls, excited baryons or short-lived new elementary particles much more precisely than before.

We are now constructing semi-automatic machine for measurement on emulsion chambers, and in next year, our analysing power would be greatly increased.

ACKNOWLEDGEMENT

We wish to express our gratitudes to the Fuji Photo Film Co., Ltd. who is providing us high quality nuclear emulsion ET-7B, and to the Konishiroku Photo Industry Co., Ltd. who is supplying us high sensitive N type x ray films. We also thank Japan Air Lines; Sanriku Balloon Center, University of Tokyo; T.I.F.R Balloon Facility, Hyderabad, India; Balloon Launching Station, Mildura, Australia; Fermi National Accelerator Laboratory and Argonne National Laboratory for offering their facilities when we exposed our emulsion chambers to cosmic rays or accelerator beams.

REFERENCES

1. M.F.Kaplon et al., Phys. Rev. $\underline{85}$, 295 (1952).
2. O.Minakawa et al., Suppl. Nuovo Cimento $\underline{8}$, 761 (1958).
   ibid $\underline{11}$, 125 (1959).
   M.Akashi et al., Suppl. Prog. Theor. Phys. $\underline{32}$, 1 (1964).
3. K.Niu et al., Prog. Theor. Phys. $\underline{46}$, 1644 (1971).
   K.Hoshino et al., Prog. Theor. Phys. $\underline{53}$, 1859 (1975).
   H.Fuchi et al., Nuovo Cimento $\underline{45A}$, 471 (1978).
   H.Fuchi et al., submitted to Nuovo Cimento $\underline{A}$.
   N.Ushida et al., submitted to Nuovo Cimento Lett.
4. J.Nishimura, Suppl. Prog. Theor. Phys. $\underline{32}$, 72 (1964).
   J.Kidd and J.Nishimura, Suppl. Nuovo Cimento $\underline{1}$, 1086 (1963).
5. H.Fuchi et al., Nuovo Cimento $\underline{45A}$, 471 (1978).
6. K.Hoshino et al., Proc. Int. Cosmic Ray Symp. on H.E.Phenomena, Cosmic Ray Lab., Univ. of Tokyo, 155 (1974).

THE TECHNOLOGY OF SMALL EMULSION CHAMBERS: A REVIEW

R. J. Wilkes
Visual Techniques Laboratory, University of Washington,
Seattle, Washington USA

ABSTRACT

The materials, construction and analysis techniques used in small emulsion chambers for airplane or balloon-flight cosmic ray experiments are described.

INTRODUCTION

The emulsion chamber is a versatile and relatively inexpensive detector which has played an important role in cosmic ray studies of high-energy particle spectra and interactions. Since its introduction by Kaplon et al. in 1951[1], considerable refinement has taken place in EC design, largely due to the efforts of Japanese workers.[2]

Basically, an emulsion chamber is a shower detector consisting of a block of high density, high-Z material (typically a stack of Pb plates), interleaved with photosensitive layers to sample cascade development. Since only a few thin emulsion plates are used, the total amount of emulsion needed is minimal, and cost is kept low.

As an example of EC design and applications, we will discuss in detail the chambers used in a recent series of US/Japan collaborative experiments to measure the primary electron flux.[3] Later, ECs designed for studying hadronic interactions will be described.

CASCADE DETECTOR DESIGN

Figure 1 shows the chamber configuration for the electron spectrum balloon flights. The Pb component consists of 8.2 radiation lengths of lead in the form of 40 x 50 cm^2 sheets of graduated thicknesses from 0.5 to 5.0 mm (roughly 0.1 to 1.0 r.l.). The 23 photosensitive layers each consist of two sheets of Sakura Type N x-ray film followed by an emulsion plate. The emulsion plates are made by coating layers of Fuji ET7B nuclear emulsion 100 microns thick on both sides of a piece of methacrylate plastic 800 microns thick. The films and emulsion plates are separated from one another and from the Pb plates by thin sheets of paper.

An electron entering the chamber will undergo pair production within a few radiation lengths of the top of the stack. The resulting electromagnetic cascade will develop rapidly in the Pb; as shown in Figure 2, on the average electron showers of energy less than about 1 TeV will be well past shower maximum before the bottom of the stack is reached. Showers of energy greater than about 500 GeV will leave visible spots in the x-ray film. Such high-energy showers can be picked up by scanning films from near the bottom of the stack with a hand magnifier or the unaided eye, and can be rapidly traced back by using the x-ray films as templates for locating the shower in the emulsion plates. Trace-back is facilitated by careful alignment of the emulsion plates. To achieve this, the lead plates are stacked up and their edges are milled to fit inside an alignment box with close tolerances; the methacrylate sheets are similarly machined before emulsion coating. As a result, the emulsion plates are aligned to within ∼100 microns (less than one microscope field of view). The graduated thicknesses of the Pb sheets permits efficient use of the emulsion layers: they are widely spaced near the top of the stack where one must try to follow single tracks. In addition, the double-sided emulsion plates permit accurate determination of particle angles, since the points at which the track enters and exits the plastic base are not affected by processing distortions in the emulsion.

EMULSION PLATES

The emulsion plates are coated under conditions of controlled temperature and relative humidity to assure optimum emulsion quality. The thin plastic substrates are held flat on a vacuum table, and the emulsion gel is spread over the surface. After the gel has set, the plate is dried for ∼24 hours at ∼80% RH. This slow-drying procedure prevents lateral stresses in the drying emulsion and keeps the plate from curling. The second side is then coated and dried; if desired, a reference grid can be printed on the bottom of the first side. Following exposure, the plates are developed using the "Bristol" Amidol-bisulfite formula. Fuji ET7B emulsion typically yields 35-40 grains per 100 microns of relativistic track. Grain size is ∼0.25 micron, comparable to that of Ilford K5 emulsion. With Fuji emulsion, coatings as thin as 50 microns can be used, while 150 microns or more is needed for reasonable track visibility with Ilford emulsion. After development, the plates are cut into 20 x 25 cm^2 segments for easier scanning.

X-RAY FILMS

The x-ray films most commonly used are Sakura Type N or Type RR. Type N consists of 25 micron emulsion layers on both sides of a 175 micron polyester base, and Type RR has 20 micron emulsion layers on the same type of base. Type N is the most sensitive, and fresh film will respond to showers of energy as low as 300 GeV. (The background fog level increases with storage time.) However, the mean grain size for N film is 1.5 microns, making it difficult to resolve closely spaced showers. RR film is somewhat less sensitive, with a threshold on the order of 1-2 TeV, but has a mean grain size of only $\sim$0.2 microns, permitting greater spatial resolution. By using two sheets of film in coincidence in each layer, random dust and pressure marks are readily separated from true shower spots. The films are processed in a commercial developer (Konidol) supplied by the film manufacturer. Development times of 10-20 minutes, much longer than the conventional development times used for medical applications, are used, depending upon the fog level expected.

ENERGY DETERMINATION

The emulsion chamber is primarily a shower detector, and the shower parameter which provides the best energy estimate is the "track length",

$$Z = \int_0^\infty N(t) \, dt$$

where $N(t)$ is the number of charged tracks at depth t in the detector. While the transition curve $N(t)$ is subject to the usual statistical fluctuations, ultimately all of the energy of the shower is dissipated by ionization losses, so that

$$Z = E/\varepsilon$$

where ε is the energy loss per unit length of electron track. Thus Z is in principle (a) directly proportional to E, and (b) not subject to statistical fluctuations. In practice, one cannot, of course, measure Z, but can instead estimate the restricted track length,

$$Z' = \int_0^\infty N(<r,t) \, dt \simeq \sum_i a_i \, N(<r,t_i) + E_{bot}$$

where $N(<r,t_i)$ is the number of tracks within radius r of the shower axis in layer i (located at depth t_i), and the weight factor a_i depends on the effective thickness of the Pb layer traversed just above layer i. E_{bot}

is the punch-through energy. If the chamber is thick enough to contain the shower maximum, E_{bot} can be estimated with sufficient accuracy by extrapolation, assuming an exponential tail on the transition curve. The relationship between the estimated restricted track length Z' and E is obtained using analytical shower theory calculations[4] and/or Monte Carlo calculations.[5]

X-ray film densitometry provides an alternate, although somewhat less exact method for shower energy determination. The film density, $D = -\log_{10}(I/I_0)$ is measured at the shower axis on each film layer, and a density versus t transition curve is plotted, from which D_{max}, the density at shower maximum, can be interpolated. As Figure 3 shows, D_{max} is well correlated with E calculated from the restricted track length.

It should be noted that the gaps introduced by the film layers cause noticeable differences between shower development in an emulsion chamber and the development expected for a homogeneous medium.[6] Multiple scattering effects are greatly reduced in the gaps, so the showers spread, altering the lateral particle density distribution. Figure 4 shows the variation in x-ray spot density (proportional to particle density) caused by this effect; the upper curve is for films at the tops of the 1.8 mm gaps, and the lower curve is for films at the bottoms of the gaps. The shower model used to calculate E as a function of Z' must take this spacing effect into account. Existing calculations and simulation programs have been calibrated up to 300 GeV using emulsion chambers exposed to Fermilab electron beams.

PARTICLE IDENTIFICATION

The small emulsion chambers discussed here are normally exposed by balloon flight. To ensure that only particles entering the apparatus at ceiling altitude are analyzed, the balloon gondola contains a flipper device which keeps the stacks inverted during ascent and descent. Due to the characteristic change in particle density distributions with shower age, it is easy to determine the direction in which a shower is developing, and backward events can be quickly identified and discarded.

Particle identification is possible in emulsion chambers because of the distinctive signatures left by the various particle types. For example, electrons will begin cascade development within a few radiation lengths of the top of the chamber; they are characterized by a single charged incoming particle producing a single-cored shower with starting point near the top of the EC. Gamma rays produce showers with similar characteristics.

For Pb, the nuclear mean free path is on the order of 30 times the radiation length, so protons and pions will have interaction points which are more or less uniformly distributed through the thickness of the chamber. Hadronic interactions typically consist of a single incoming particle producing on the order of 10 charged particles, with no further development for a few radiation lengths, when gamma showers from $\pi°$ decay become visible. It is generally easy to discern a multiple core structure in the developing showers. In order to mimic an electron, an hadronic event would have to have low charged multiplicity, $n_{\pi°} \sim 1$, and K_γ very large. Incoming heavy nuclei can be identified immediately in the emulsion plates.

HADRON DETECTORS

The emulsion chamber design used for the US/Japan e-spectrum project is not an efficient detector for hadron interactions. Figure 5 shows the type of modified EC used for hadron studies. A target segment, or "producer layer" is positioned above the shower detector section. A large gap is provided to permit the narrow jet of produced particles to spread out enough to facilitate multiplicity and angular determinations. In the Seattle design, the target section contains polyethylene, and the gap is simply a layer of styrofoam; plate alignment is achieved through the use of vertical rods which fit through holes in the emulsion plates. To make trace back through the "dead" spacer easier, part of the shower detector is placed above the gap. Nonetheless, many events were very difficult to trace back. The Japanese design represents an improvement in that the spacer segment contains emulsion plates at intervals. The producer layer is simply a stack of emulsion plates coated on thick (1.5 mm) methacrylate sheets. Alignment is via the machined-box technique described earlier.

CONCLUSION

Recent developments will permit improved EC performance in future experiments. More sensitive fine-grained x-ray films are becoming available, which will push the threshold for shower detection down to the 100 GeV region.[7] Fuji ET6 emulsion, similar to Ilford K3, will permit more accurate identification of heavy primaries.

With regard to balloon flight experiments, perhaps the most important feature of the emulsion chamber is that it is in essence a solid block of matter. The sort of "semi-hard" landing that gives a counter physicist ulcers can be accepted with equanimity if the payload is an emulsion chamber.

Discussions with J. Lord, T. Koss, J. Nishimura, K. Niu, T. Taira, I. Ohta, and Y. Takahashi are gratefully acknowledged.

REFERENCES

1. M. F. Kaplon, et al., Phys. Rev. $\underline{85}$, 295 (1952).
2. M. Akashi, et al., Suppl. Prog. Theor. Phys. $\underline{32}$, 1 (1964).
3. M. Matsuo et al., XIV Int. Cosmic Ray Conference (München), $\underline{12}$, 4132 (1975).
4. J. Nishimura, Handbuch der Physik $\underline{46/11}$, 1, Springer Verlag, Berlin (1967).
5. T. Taira, Uchusen Kenkyu, 155 (1967).
6. J. Nishimura, Suppl. Prog. Theor. Phys. $\underline{32}$, 72 (1964).
7. Y. Takahashi, personal communication.
8. I. Ohta, et al., XIV Int. Cosmic Ray Conf. (München), $\underline{9}$, 3154 (1975).
9. T. Shirai and Y. Takahashi, XIV Int. Cosmic Ray Conf. (München), $\underline{9}$, 3149 (1975).
10. K. Hoshino et al., XIV Int. Cosmic Ray Conf. (München), $\underline{9}$, 2330 (1975). [See also the preceding paper-ed.]

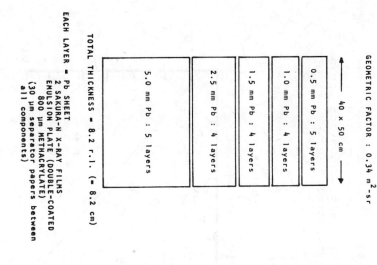

FIG. 1: Emulsion chamber for US/Japan primary electron spectrum project.

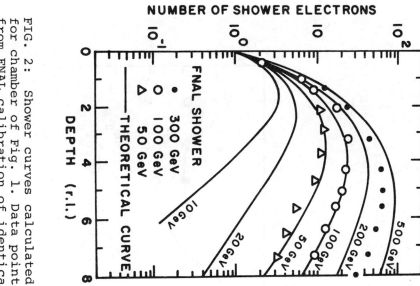

FIG. 2: Shower curves calculated for chamber of Fig. 1. Data points from FNAL calibration of identical chambers.

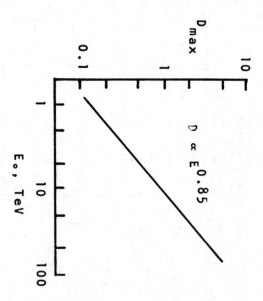

FIG. 3: X-ray densitometric method for energy determination: D_{max} vs E obtained from track counts in emulsion. (ref. 8)

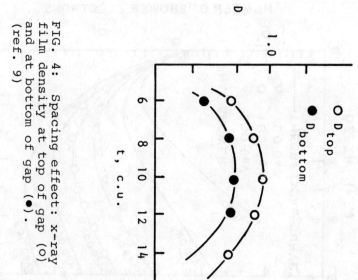

FIG. 4: Spacing effect: x-ray film density at top of gap (o) and at bottom of gap (●). (ref. 9)

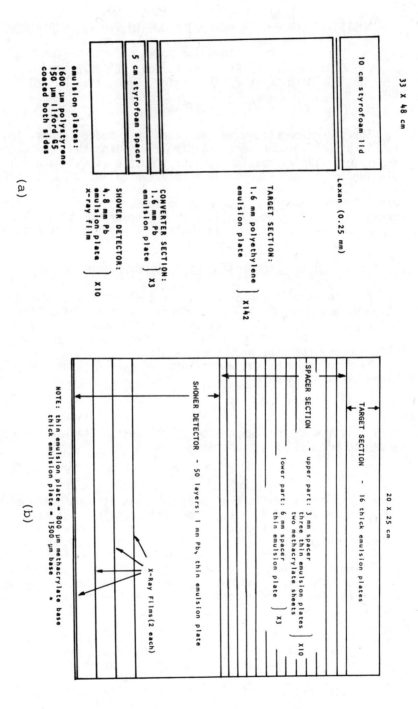

FIG. 5: Emulsion chambers used for hadronic interaction studies:
(a): Seattle chamber; (b): Japanese chamber (ref. 10).

ACOUSTIC DETECTION: A NEW TECHNIQUE IN >10 TeV EXPERIMENTS*

Theodore Bowen
University of Arizona, Tucson, AZ 85721 and
NASA/Goddard Space Flight Center, Greenbelt, MD 20771[†]

ABSTRACT

The thermoacoustic wave generated by a sudden deposition of heat in a fluid is derived in a simple way from the fundamental laws of physics. The expression for the acoustic signal-to-noise ratio relative to thermal vibrations from a typical electrocmagnetic cascade is then developed. The signal-to-noise ratios of several materials are compared; liquid argon appears to be one of the most promising liquids. Observation of air shower cores in a high altitude lake and cascades in an emulsion chamber are discussed as possible applications of acoustic detection.

ELEMENTARY CALCULATION OF THE THERMOACOUSTIC WAVE

The acoustic wave produced in a liquid by the sudden deposition of heat energy will be derived from first principles. This will not only show that few assumptions are needed, but also that functions such as the time integral of the pressure are useful and easier to use than the pressure, itself.[1]

If a volume V has a temperature increase ΔT, the change of volume ΔV is given by

$$\Delta V = \beta \, \Delta T \, V \quad , \qquad (1)$$

where β is the thermal expansion coefficient. The heat energy E_o needed to produce this temperature rise is

$$E_o = C_p \, \rho_o \, \Delta T \, V \quad , \qquad (2)$$

where C_p is the specific heat per unit mass at constant pressure and ρ_o is the liquid density before the deposition of heat. Both Eqs. 1 and 2 are well-known approximate relations which become increasingly exact for small ΔT. These equations assume that the state of the liquid in volume V is specified by a temperature and pressure. By picking the volume V under consideration to be sufficiently small, the assumption of thermodynamic equilibrium can usually be justified because collisions are occuring at such a high rate ($\sim 10^{12}$ per second).

---

*The portion of this work carried out at the University of Arizona was supported by the National Science Foundation.

[†]NASA/National Academy of Sciences Senior Research Associate, on leave from the University of Arizona (9/78-8/79).

ISSN 0094-243X/79/49072-12 $1.50 Copyright 1979 American Institute of Physics

Eliminating ΔT from Eqs. 1 and 2, we obtain

$$\Delta V = \frac{\beta E_o}{C_p \rho_o} . \quad (3)$$

Notice that the volume V has also dropped out. This indicates that the total change of volume of a region due to adding a given amount of heat is <u>independent</u> of the shape or size of the region. Equation 3 continues to be valid even if the heat diffuses, provided it does not diffuse outside the region under consideration. It is the sudden change of volume given by Eq. 3 which generates an acoustic wave.

Now let us consider the changes which must eventually take place at a spherical shell at a radius R surrounding the heated region (R >> size of heated region). Nothing can happen at R until a time R/c has elapsed, where c is the speed of sound in the liquid. Eventually, in order to re-establish pressure equilibrium throughout the liquid, the spherical shell at R must move outward a distance ΔR to accommodate the additional fluid volume ΔV. At radius R we can write

$$\Delta V = 4\pi R^2 (\Delta R)_R . \quad (4)$$

Combining Eqs. 3 and 4, the particle displacement is

$$(\Delta R)_R = 0 \quad , \quad t - R/c < 0 \; ; \quad (5a)$$

$$= \frac{\beta E_o}{4\pi C_p \rho_o} \cdot \frac{1}{R^2} \quad , \quad t - R/c \to +\infty . \quad (5b)$$

Note that ΔR, when considered as a function of time, is a step function, and that the magnitude of the step depends only upon E_o and the distance R for a given liquid. The rise time of the step will depend upon the size of the heated region and upon attenuation, if any, of higher frequencies. We have assumed only that, as D.C. or zero frequency is approached, attenuation vanishes. This must be the case if we assume that the liquid can re-establish a quasi-equilibrium state free of stresses (but not free of temperature variations).

The displacements must be related to the forces on the liquid. A small volume δV is acted upon by a force

$$\delta F = -\nabla p \, \delta V \quad , \quad (6)$$

where p is the pressure. The mass of this small fluid volume is $\rho_o \, \delta V$ and the product of mass times acceleration is given by

$$\delta F = \rho_o \, \delta V \frac{\partial^2 \Delta R}{\partial t^2} . \quad (7)$$

Before combining Eqs. 6 and 7, we note that the outward flow in which we are interested in irrotational, which allows us to introduce a displacement potential Φ such that

$$\Delta R \equiv -\nabla \Phi \quad , \tag{8}$$

Then we obtain from Eqs. 6, 7, and 8, after integrating,

$$\rho_0 \frac{\partial^2 \Phi}{\partial t^2} = p \quad . \tag{9}$$

Equation 9 not only can be viewed as an equation of motion, but also as a way of calculating the pressure p if the displacement potential Φ is known from other considerations.

We return to Eqs. 5a and 5b to find Φ before and after the time $t = R/c$ when the acoustic disturbance arrives at radius R. For $t - R/c < 0$, $\Phi = 0$ correctly gives the result that $\Delta R = 0$. For $t - R/c \to +\infty$, Eq. 5b can be integrated. The results are

$$\Phi = 0 \quad , \quad t - R/c < 0 \quad ; \tag{10a}$$

$$= \frac{\beta E_0}{4\pi C_p \rho_0} \cdot \frac{1}{R} \quad , \quad t - R/c \to +\infty \quad . \tag{10b}$$

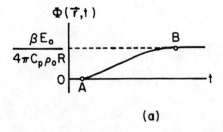

(a)

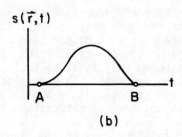

(b)

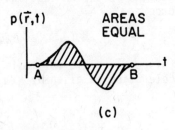

(c)

Fig. 1. Typical pulse shapes.

Viewed as a function of time, Φ is a step function whose amplitude is proportional to the heat deposition energy E_0 and is inversely proportional to the distance R. The rise time of Φ is related to the size of the heated region and to the attenuation of higher frequencies, as discussed above for ΔR.

The fact that Φ is a step function in time (see Fig. 1a) immediately determines that $\partial \Phi/\partial t$ is a unipolar pulse in time (see Fig. 1b) and that the pressure p must be a bipolar pulse as a function of time (see Fig. 1c). It was seen in Eq. 10 that Φ bears a very simple relation to the heat energy deposited, which would be the electromagnetic or hadronic-electromagnetic cascade energy in high energy particle events. However, the observed quantity is usually the pressure, a more complicated appearing function of time. Merely by integrating the pressure twice with respect to time one obtains:

$$\int_0^\infty dt \int_0^t p(t') \, dt' = \frac{\beta}{4\pi C_p} \cdot \frac{E_o}{R} \quad , \tag{11a}$$

$$= -\int_0^\infty t \, p(t) \, dt \tag{11b}$$

Equation 11b indicates that the double integration is equivalent to evaluating the first moment of the pressure signal.

A quantity $s(t)$ which will be useful in the discussion to follow is one integration of the pressure with respect to time, which is also proportional to the time derivative of Φ:

$$s(t) \equiv \int_0^t p(t) \, dt \quad , \tag{12a}$$

$$= \rho_o \frac{\partial \Phi}{\partial t} \quad . \tag{12b}$$

This is a unipolar pulse with a time spread (neglecting high frequency attenuation effects) equal to the length of the heated region along the direction of observation divided by the sound speed, c. If the heated region has a very small volume dV and a heat energy deposition $e_o \, dV$, then $s(t)$ would approach a δ-function with area $(\beta/4\pi C_p)(e_o \, dV/R)$. With this result, a solution for an extended region V, such as a particle cascade, can be constructed:

$$s(t) = \frac{\beta}{4\pi C_p} \frac{\partial}{\partial t} \int_{R<ct} \frac{e_o(\underline{r}_o) \, dV}{R} \tag{13}$$

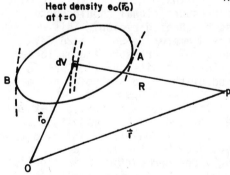

$R = |\vec{r} - \vec{r_o}|$ = distance from point of observation P to volume element dV

Fig. 2. Coordinate system.

where $e_o(\underline{r}_o)$ is the energy deposition per unit volume in the neighborhood of the position $\underline{r}_o$, and where the position vectors $\underline{R}$ and $\underline{r}_o$ are defined by reference to Fig. 2. Equation 13 shows that $s(t)$ gives the profile or projection of the heated region when divided into a series of concentric shells whose center is at the observer (see Fig. 2). When R is large compared to the dimensions of the region, $s(t)$ very closely approximates the perpendicular projection onto the line of observation. If s(t) were available in several different directions, it is clear that not only the total energy deposition E_0 could be obtained, but the size, shape, location and orientation of a cascade could be reconstructed.

EXPERIMENTAL VERIFICATION OF THE THERMOACOUSTIC THEORY

A series of experiments were carried out to observe sound generation in proton beams by Sulak et al.[2] after acoustic detection was suggested at the 1976 DUMAND Summer Workshop as a means of detecting neutrino events in the deep ocean independently by the author[3] and Dolgoshein and Askarian.[4] Equation 10b or 13 in combination with Eq. 9 or 12a, respectively, indicates that, if the pressure is observed at a point well outside the heated region, then the pressure amplitude should be proportional to β/C_p and the total heat energy E_0 and inversely proportional to the distance R from heated region to observer. Experimental results for the pressure amplitude as a function of energy deposition E_0 by proton beams which stopped in water showed linearity over the range of from 10^{15} to 10^{21} eV. The relation of the pressure signal to the expansion coefficient β was checked using 158 MeV protons by varying the temperature of water in the region near the freezing point where β changes sign. The polarity of the pressure signal indeed reversed, but near 6 °C instead of the expected 4 °C. This descrepancy is presumably related to the fact that water is a very complicated substance near the freezing point, so that the effective β for energy deposited uniformily along a straight line differs slightly from β for energy deposited uniformily throughout a three-dimensional region. Proportionality of the pressure signal to β/C_p was confirmed by comparing the signal from protons stopping in various liquids in which β/C_p varied by a factor of 30.

Equations 13 and 12a along with an inspection of Figs. 1 and 2 indicates that the duration of the pressure signal should correspond to the acoustic transit time across the heated region. This was experimentally verified by varying the diameter to which a 28 GeV proton beam was focused. Since the bipolar pressure signal looked approximately like one cycle of a sine wave, its duration was characterized by the apparent period of the bipolar cycle; a linear relationship was observed with the expected slope.

Equation 11b is particularly suitable for a comparison between thermoacoustic theory and experiment of the absolute magnitude of the pressure signal. Equation 11 was confirmed within 15% by data taken with 200 MeV protons. We conclude that the thermoacoustic theory for sound production by singly charged particles in liquids is confirmed by experimental data and should provide a reliable basis for design calculations for various possible applications of acoustic particle detection.

SIGNAL-TO-NOISE RATIO

Since the acoustic pressure amplitude increases with energy deposition, it is clear that at sufficiently high energy depositions an acoustic signal can be detected. (At energy depositions $\sim 10^{20}$ eV by 200 MeV protons in a large water tank, the sound was clearly audible.) In order to obtain a reasonably precise lower limit to the energy deposition which can be acoustically detected, one must apply results of signal theory which are well known in such fields as radar and sonar.

It can be shown[5] that the best possible signal-to-noise ratio for a transient signal of known shape in the presence of white Gaussian stationary noise is obtained from the output of a matched filter. For this case, the signal-to- noise ratio at the output is given by

$$\frac{S}{N} = \frac{2E}{N_0} , \qquad (14)$$

where $\quad E = \int_0^\infty [y(t)]^2 \, dt \quad$ of the signal $\quad y(t)$, $\qquad (15)$

and $\quad N_0 = \frac{d\langle y^2 \rangle}{df} \quad$ of the noise , $\qquad (16a)$

$\qquad\qquad\quad$ = constant if "white" noise. $\qquad (16b)$

Before applying Eqs. 14-16 we must identify the noise background spectrum and, if not already white, prescribe a linear filter which will "whiten" it.

For example, if the noise power rises in proportion to the frequency f squared, an electronic integrator may be inserted ahead of the matched filter. This whitens the noise spectrum because the integrator power gain is porportional to $1/f^2$. Of course, the signal must also pass through the integrator, so the y(t) employed in Eq. 15 must be the form of the signal after passing through the integrator. A block diagram of an electronic analog integrator is shown in Fig. 3. If the integrator input is p(t) and its output is y(t), the noise power spectrum at the output is given by

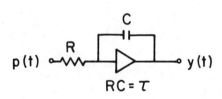

Fig. 3. Block diagram of analog integrator to "whiten" thermal noise spectrum.

$$\frac{d\langle y^2 \rangle}{df} = \frac{1}{4\pi^2 f^2 \tau^2} \frac{d\langle p^2 \rangle}{df} , \qquad (17)$$

and the output signal is

$$y(t) = \frac{1}{\tau} \int_0^t p(t)\, dt \quad , \tag{18a}$$

$$= \frac{1}{\tau} s(t) \quad , \tag{18b}$$

where $s(t)$ is defined in Eq. (12a). The integrator time constant $\tau = RC$ will disappear when the signal-to-noise ratio is formed.

If a more general frequency weighting function is required to whiten the noise, then it is convenient to express Eq. 15 in another form. If $\tilde{y}(\omega)$ is the Fourier transform of the signal $y(t)$, then the integral in Eq. 15 can also be written

$$E = \int_0^\infty y^2\, dt = \int_{-\infty}^\infty \tilde{y}^2\, d\omega \tag{19}$$

Equation 19 shows how different parts of the frequency spectrum contribute to E. Any frequency weighting function required to whiten the noise spectrum can be inserted into the integral over frequency in Eq. 19.

In addition to acoustic noise from sources such as wind, machinery, and aircraft which may be avoidable with appropriate experimental design, noise from thermal fluctuations is always present. In a region whose dimensions are large compared to the wavelengths of interest the noise power rises as f^2 and is given by

$$\frac{d\langle p^2 \rangle}{df} = \frac{4\pi\, kT\, \rho_0}{c} f^2 \quad , \tag{20}$$

where k is the Boltzmann constant and T is the absolute temperature. If the pressure sensor responds uniformily as a function of frequency, then an intergrator is appropriate for whitening thermal noise.

A SIMPLE MODEL OF THE ACOUSTIC SIGNAL FROM PARTICLE CASCADES

The spatial distribution of heat deposition in a particle cascade is, in general, a complicated function which is subject to large fluctuations from one event to another. However, most of the initial energy is dissipated as ionization energy loss which becomes heat energy on a time scale very short compared to the periods of the acoustic frequencies of interest. Therefore, the _total_ heat deposition in the neighborhood of the cascade is expected to fluctuate only a few percent. In order to obtain numerical estimates of S/N, it is convenient to adopt some model giving the approximate spatial distribution of heating in a particle cascade.

The average cascade distribution is expected to be cylindrically symmetric about an axis which is the extension of the trajectory of the initiating particle. The average distribution rises to a broad maximum at the depth where the number of secondary partilces is greatest. Also, the density of the average distribution decreases with distance from the axis. A convenient function having all these properties is a 3-dimensional Gaussian with the scale of one coordinate axis (say the z-axis) elongated so that it has the symmetry of a prolate ellipsoid:

$$e_o(x,y,z)\, dV = \frac{E_o}{(2\pi)^{3/2} (c\tau)^{3/2}\, b}\, e^{-\frac{1}{2(c\tau_o)^2}\left[x^2 + y^2 + \frac{z^2}{b^2}\right]}\, dV \quad (21)$$

where $c\tau_o$ and τ_o characterize the radius of the cascade in units of length and acoustic transit time, respectively, and b is the longitudinal elongation factor.

The function given by Eq. 21 is especially convenient because its projection onto any direction can be calculated analytically, and the result is a 1-dimensional Gaussian. Let the distance R from the cascade to the observer be much greater than the largest cascade dimension:

$$R \gg bc\tau \quad , \quad (22)$$

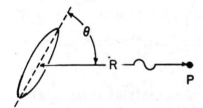

Fig. 4. Sketch showing definition of angle θ to cascade axis.

and let θ specify the orientation of the cascade relative to the line of observation as shown in Fig. 4. Then the integral in Eq. 13 can be approximately evaluated by taking R outside the integral and assuming that the shells of constant R are planes. The result is

$$s(t) \cong \frac{\beta}{4\pi C_p} \cdot \frac{E_o}{R} \cdot \frac{1}{\sqrt{2\pi}\, \tau}\, e^{-\frac{1}{2\tau^2}\left(t - \frac{R}{c}\right)^2} \quad , \quad (23)$$

where

$$\tau = \tau_o [1 + (b^2 - 1) \cos^2\theta]^{1/2} \quad . \quad (24)$$

For purposes of estimation in later sections, parameters appropriate for a pure electromagnetic cascade will be employed. The longitudinal development[6] near the maximum can be fit by a Gaussian with standard deviation given by

$$\bar{bc\tau_o} \approx \left[1.5 \ln\left(\frac{E_o}{\varepsilon_o}\right)\right]^{1/2} \cdot L_{rad} \quad , \quad (25)$$

where E_0 is the total shower energy, ε_0 is the critical energy, and L_{rad} is the radiation length. The lateral spread of an electromagnetic cascade, characterized by the Moliere length r_1, is not so well fit by a Gaussian distribution because the actual lateral distribution rises as $1/r$ for $r \ll r_1$, whereas the Gaussian levels off at small r. For the estimates, the relation

$$c\tau_0 \approx 0.27\ r_1 \tag{26}$$

has been chosen, which gives a rough match of the integral distributions. Since the Moliere length is not listed in readily available tables for various materials, the following relation is convenient:

$$r_1 = [21.2\ \text{MeV}] \left[\left(\frac{dE}{dx} \right)_{plateau} \right]^{-1} \tag{27}$$

where $(dE/dx)_{plateau}$ is the average energy loss per unit length of an electron having energy $\approx \varepsilon_0$ ($\approx (dE/dx)_{min}$). Table I lists τ_0, the maximum signal frequency $f_{max} \equiv 1/2\pi\ \tau_0$, $c\tau_0$, and b for some materials which might be considered for use in high energy cosmic ray experiments.

EFFECT OF ACOUSTIC ATTENUATION UPON THE SIGNAL

For many liquids one can represent the acoustic attenuation as proportional to f^2 while regarding the speed of sound as constant. With these assumptions, the effect of attenuation on pulse shape has been given by Learned.[7] The result is an additional Gaussian spreading so only Eq. 24 needs to be modified to account for the additional width:

$$\tau = \left\{ \tau_0^2 [1 + (b^2-1)\cos^2\theta] + \frac{\alpha R}{2\pi^2} \right\}^{\frac{1}{2}}, \tag{28}$$

where R is the distance from source to observer and α is defined by noting that a plane wave amplitude changes by the factor $\exp(-\alpha f^2\ x)$ in transversing a distance x. The attenuation term becomes significant when the distance from the cascade to the pressure detectors becomes greater than R_{atten}, where

$$R_{atten} = \frac{2\pi^2\ \tau_0^2}{\alpha}, \tag{29a}$$

$$\approx \frac{1.44\ r_1^2}{c^2\ \alpha}. \tag{29b}$$

This distance is given in Table I for some of the materials. Since R_{atten} typically is large compared to the likely size of an experimental apparatus, the attenuation term will be neglected in the estimates in later sections.

SIGNAL-TO-NOISE RATIO FOR ELECTROMAGNETIC CASCADES

Combining Eqs. 15, 18b, 23 and Eqs. 16a, 17, 20 and substituting into Eq. 14, the signal-to-noise ratio for an electromagnetic cascade (or any Gaussian heat source distribution) becomes

$$\frac{S}{N} = \frac{\beta^2 c^2}{16\pi^{3/2} C_p^2 kT} \cdot \frac{1}{\rho_0 \, c\tau} \cdot \left(\frac{E_0}{R}\right)^2 \quad , \qquad (30)$$

where τ is given by Eqs. 28, 25, 26, and 27. The right-hand side of Eq. 30 has been written with three factors: The first factor depends only upon the material. (Even T is severely constrained by the choice of material.) The second factor depends upon (a) cascade size (in g/cm^2), (b) cascade orientation θ, and (c) attenuation (if significant). The third factor depends only upon cascade total energy divided by its distance from the pressure detector.

Returning to the second factor in Eq. 30, it is largest ($\sim 1/\rho_0 \, c\tau_0$) when $\cos^2\theta \leq 1/(b^2-1)$; this maximum value is the same, within a factor of two, for any choice of material (except pure hydrogen) because it depends only upon $(dE/dx)_{min}$ expressed in energy per g/cm^2 (Eqs. 26 and 27). When $\cos^2\theta > 1/(b^2-1)$, the second factor is determined by the longitudinal spread of the cascade in g/cm^2; therefore, materials with high atomic number will give the best S/N under these conditions roughly in direct proportion to atomic number Z.

Going back to the first factor in Eq. 30, notice that it depends upon $(\beta c/C_p)^2(1/T)$. If this is combined with the remarks which indicate that the maximum value of the second factor is proportional to $(dE/dx)_{min}$, a figure of merit, M, for various materials is

$$M = \left(\frac{\beta c}{C_p}\right)^2 \cdot \frac{(dE/dx)_{min}}{T} \quad . \qquad (31)$$

This quantity is listed in Table I for various materials in mixed units determined by the most common units in reference tables, such as calories/g-°C for C_p and $MeV/(g/cm^2)$ for dE/dx. This is done so the reader may more readily add other materials of potential interest to the list. The signal-to-noise ratio may then be calculated from

$$\frac{S}{N} \cong 0.02 \frac{\tau_o}{\tau} \left[\frac{\beta(°C^{-1}) \, c(m/s)}{C_p(cal-g^{-1}-°C^{-1})} \right]^2 \cdot$$

$$\frac{dE/dx(MeV-g^{-1}-cm^2)_{min}}{T(°K)} \cdot \left[\frac{E_o(TeV)}{R(m)} \right]^2 \cdot \quad (32)$$

where
$$1 \geq \frac{\tau}{\tau_o} \approx 1 \quad \text{if } \cos^2\theta \leq 1/(b^2-1) \quad (33a)$$

$$\approx 1/b \cos\theta \quad \text{if } \cos^2\theta > 1/(b^2-1) \quad (33b)$$

and attenuation is assumed to be negligible. Note that the figure of merit M is numerically equal to the maximum possible S/N when $E_o/R \cong 7$ TeV/m.

Since the integral energy spectrum of cosmic rays is roughly proportional to E_o^{-2} and detector sensitive area is usually proportional to R^2, the figure of merit, M, is approximately proportional to the relative event rates which could be realized with various materials. Unfortunately, the fluids available in the natural environment in almost pure form have comparatively low figures of merit, with the exception of air. In air, however, the important signal frequencies are very low (≤ 2 Hz) where wind and aircraft noise is likely to exceed thermal noise by many orders of magnitude. Salt domes, as first suggested by W. V. Jones,[8] may hold considerable promise for detection of cosmic neutrinos, but much must be learned concerning its acoustic properties and noise backgrounds in its high pressure semi-fluid state.

Among the cryogenic liquids, helium is listed for purposes of comparison. Liquid nitrogen and argon are inexpensive and safe, so they may have great potential for use in acoustic detection of high energy cosmic ray events. Storage tanks can be economically constructed using styrofoam insulation. High density liquids such as mercury and low-melting-point lead-tin alloys (not listed in Table I) may be of greatest interest for design of accelerator experiments where the region of interest may be of smaller size compared to cosmic ray experiments. The other liquids listed in Table I seem to have little to recommend them except room temperature operation, since all are more than a factor of 10 poorer than liquid nitrogen or argon and those with the better figures of merit are poisonous.

The longitudinal elongation factor b gives an idea of the length of the cascade relative to its lateral spread. For most applications it is advantageous to have b as small as possible, both to minimize the thickness of the liquid layer which must be provided and to minimize the degradation of S/N as the angle θ at which the cascade is observed is varied from 90°. Hence, liquid argon is a more interesting possibility than liquid nitrogen.

Other liquids whose properties should be considered include inexpensive petroleum products, such as kerosene and heating oil, and fluorocarbons, such as Freon 13 B1. Xenon probably would have almost ideal characteristics, but its cost would be prohibitive for most applications.

COSMIC RAY APPLICATIONS OF ACOUSTIC DETECTION

<u>Detection of High Energy Air Shower Cores in a High Altitude Lake.</u>[9] Suppose we wish to detect the cores of very high energy air showers using a natural body of water. High altitude would be advantageous, as a higher fraction, F, of the initial primary energy would still be concentrated within a few centimeters from the shower axis. Also, for any given F, the solid angle acceptance improves as atmospheric depth is decreased. Since the energy region of potential interest would be $10^{18} - 10^{20}$ eV, a lake might help solve the problem of achieving a sensitive area of many square kilometers. The acoustic detectors might be vertical strings of of hydrophones at spacings $\sim$ 1 km. A reasonable requirement would be S/N $\geqslant$ 100 at each detector. Using Eq. 32 and M from Table I, the condition is imposed that $E_0(TeV)/R(m) \geqslant 2770$, or $E_0(TeV) \geqslant 2.77 \times 10^6$ for $R = 10^3$ m. So the acoustic method would be applicable for $E_0 \geqslant 3 \times 10^{18}$ eV. As can be seen from $R_{atten} = 300$ km in Table I for water, attenuation would be insignificant for distances $\sim$ 1 km. A remaining very serious question concerns the level of background noise to be expected from surface conditions (waves, wind, boats) and from animal life.

<u>Detection of Electromagnetic Cascades in an Emulsion Chamber.</u> If the occurence of high energy cascades, say $E_0 \geqslant 10^{15}$ eV, could be detected and the location given by acoustic techniques, the emulsion plates in that region could be promptly removed and developed before the latent image fades. Unfortunately, more data on the thermal and mechanical properties of nuclear emulsion are needed before a precise evaluation can be made. Also, if high Z plates are sandwiched between layers of emulsion, the plate material should be "phantom" emulsion acoustically; that is, the plate material should match the emulsion in acoustic speed, c, and in acoustic impedance, ρ_0 c. Perhaps this could be done by suspending very finely powdered Pb in gelatin. For a crude evaluation of performance, let us conservatively guess that the emulsion figure of merit is $M \approx 0.01$. Then for S/N $\geqslant$ 100, $E_0(TeV)/R(m) \geqslant 707$. If the acoustic detectors are spaced at 0.3 m intervals, then $E_0 \geqslant 200$ TeV = 2×10^{14} eV. These figures should encourage an effort to at least measure the acoustic properties of nuclear emulsion.

CONCLUSIONS

The applications discussed above are only a small sample of the possible roles for acoustic detection of high energy particle cascades in cosmic ray and high energy physics. Since a large detector area or mass is generally required in very high energy experiments, the relevant physical and acoustic properties of many inexpensive materials should be measured. Among the liquids investigated at present, liquid argon appears to be the most promising for sensitive areas up to ~ 100 m^2.

REFERENCES

1. This derivation was first presented by the author to the 1978 DUMAND Summer Workshop, Session I, July 24-August 4, 1978, Scripps Institution of Oceanography, La Jolla, California.

2. L. Sulak, T. Armstrong, H. Baranger, M. Bregman, M. Levi, D. Mael, J. Strait, T. Bowen, A. E. Pifer, P. A. Polakos, H. Bradner, A. Parvulescu, W. V. Jones, and J. Learned, Nucl. Instr. & Meth. (to be published).

3. T. Bowen, Proc. of 1976 DUMAND Summer Workshop, Sept. 6-19, 1976, Honolulu, Hawaii, p. 523 (referenced below as DUMAND 1976). A corrected version of this work appears in Proc. of 15th Int. Cosmic Ray Conf., 13-26 August 1977, Sophia, Bulgaria, Vol. 6, p. 277.

4. B. A. Dolgoshein, DUMAND 1976, p. 553; G. A. Askarjan and B. A. Dolgoshein, "Acoustic Detection of High Energy Neutrinos at Big Depths," Lebedev Physical Institute, Moscow, preprint no. 160 (1976).

5. C. W. Helstrom, "Statistical Theory of Signal Detection," Pergamon Press, New York, 1960, p. 91.

6. K. Greisen, Ann. Rev. Nucl. Sci. <u>10</u>, 63 (1960).

7. J. G. Learned, "Acoustic Radiation by Charged Atomic Particles in Liquids, An Analysis," Dept. of Physics, University of California, Irvine, Technical Report No. 77-44.

8. W. V. Jones, DUMAND 1976.

9. Recently, acoustic detection of extensive air showers has been discussed by W. L. Barrett, Science <u>202</u>, 749 (1978).

10. American Institute of Physics Handbook, 3rd ed., McGraw-Hill, New York, 1973.

11. Handbook of Chemistry and Physics, The Chemical Rubber Co., Cleveland, Ohio, 1977.

12. Particle Properties Data Booklet from Phys. Lett. <u>75B</u>, No. 1 (1978).

TABLE I

The acoustic-signal and electromagnetic-cascade parameters for selected fluids.[a]

	ρ_o (g/cm³)	Absolute Temp. T(°K)	Acoustic Transit Time τ_o (μs)	Maximum Signal Frequency[b] f_{max} (kHz)	Distance at which Atten. Significant R_{atten} (km)	Lateral Cascade Spread $c\tau_o$ (cm)	Longitudinal Elongation[c] b	Figure of Merit[d] M
Environmental Fluids:								
Air	.00129	273	7.34 × 10⁴	0.0022	2500	52	0.17
Water	0.998	293	19.5	8.2	300	2.9	58	6.5 × 10⁻⁴
Sea Water[e]	1.051	275	17.7	9.0	10	2.8	58	3.0 × 10⁻⁴
Salt Dome[f]	2.17	400	3.6	44	---	1.7	31	0.030
Cryogenic Liquids:								
Helium	0.1251	4.216	1340	0.12	1.4 × 10⁵	24.1	140	400
Nitrogen	0.808	77.3	49.3	3.2	4530	4.0	55	2.16
Argon	1.394	87.3	32.6	4.9	2080	2.8	24	3.65
Dense Liquids:								
Bromine	3.12	293	18.7	8.5	---	1.34	13.5	0.25
Mercury	13.546	293	2.64	60	21	0.38	6.3	0.24
Other Liquids:								
Glycerol	1.261	293	12.1	13	---	2.3	61	0.0165
Methanol	0.7914	293	32.1	5.0	670	3.6	64	0.033
Ethanol	0.7893	293	30.1	5.3	370	3.5	67	0.041
n-Pentane	0.626	293	42.0	3.8	35	4.2	77	0.067
Acetone	0.7899	293	29.9	5.3	15	3.6	67	0.083
Benzene	0.879	293	25.4	6.3	15	3.3	69	0.103
Carbon Tetrachloride	1.594	293	24.2	6.6	21	2.2	28	0.181
Carbon Disulfide	1.263	293	25.6	6.2	---	2.9	29	0.183

FOOTNOTES TO TABLE I

a. Based upon data from AIP Handbook,[10] Handbook of Chemistry and Physics,[11] and Particle Properties Data Booklet.[12]

b. $f_{max} = 1/2\pi \tau_0$

c. Longitudinal elongation $\equiv$ (Longitudinal Spread)/(Lateral Spread) is evaluated at E_0 = 100 TeV.

d. $M \equiv [\beta(°C^{-1}) \ c(m/s)/C_p(cal-g^{-1}-°C^{-1})]^2$
   $[dE/dx(MeV-g^{-1}-cm^2)]_{min}/T(°K)$ = S/N for $E_0/R \cong 7$ TeV/m.

e. Sea water with 3.5‰ salinity at 5 km depth.

f. Acoustic parameters very uncertain for conditions in a salt dome.

Chap. 3. Nucleus-Nucleus Collisions

NUCLEUS-NUCLEUS COLLISIONS AT A MEAN E = 20 GeV/NUCLEON

Phyllis S. Freier and C. Jake Waddington
University of Minnesota, Minneapolis, Mn. 55455

ABSTRACT

Four hundred interactions of cosmic ray nuclei of $Z \geq 8$ in nuclear emulsion have been studied. The nuclei have $E > 7.5$ GeV/n and a mean energy of 21 GeV/nucleon. The mean meson multiplicity is 11.9 for these nucleus-emulsion collisions compared to 6.0 for proton-emulsion interactions at 22.5 GeV. The angular distributions of emitted particles show some interesting effects. There is a statistically significant excess of mesons emitted in close pairs and groups with separation angles $< 5°$. One to two percent of the interactions also show meson emission in "fans" or "jets".

INTRODUCTION

Except at very high energies, accelerators have superseded cosmic rays for providing particles for high energy physics experiments. However, at $E > 10^{15}$ eV, cosmic rays are still the only source of particles. The cosmic ray "beam" is a mixture of protons and heavier nuclei, and it is the consequent uncertainty in the nature of the primary of a specific interaction that may lead to some ambiguity in interpretation of results where the initial interactions are not observed.

At present the only source of artificially accelerated heavy nuclei of relativistic energies is the Lawrence Radiation Laboratories Bevatron/Bevalac accelerator. This accelerator can accelerate nuclei to a maximum energy of about 2 GeV per nucleon with elemental charges up to that of iron. Future plans project the acceleration of nuclei as heavy as uranium by 1981 or 1982. Similar or slightly more energetic machines are in the planning stages in the U.S.S.R. and Europe, but energies higher than about 4.5 GeV per nucleon will not be available until 1985 at the earliest, although discussions have begun for 10 GeV per nucleon machines and even colliding beams at 10 GeV per nucleon in each.

The results of the nuclear physics work obtained on the Bevalac have been summarized in many places,[1,2] and show that interest has been mainly focused into two broad aspects. A large effort has been devoted to studying peripheral interactions of relativistic heavy ions with many different targets by looking at the fragments of the projectile. These studies are related to questions concerning the shell model of nuclear structure and on the role of multi-nucleon clusters as constituents of nuclei. They are also related to the evaluation of fragmentation parameters that are needed by cosmic ray observers in order to correct for the effects of interactions in overlying and intrinsic matter in their detector arrays. However, the major effort, certainly recently, has been directed to the study of central collisions in which total interpenetration occurs between the target and projectile nuclei. Such studies attempt to examine the properties of nuclei under extreme conditions of high temperature and compression. These have the ultimate purpose of obtaining information on the nuclear equation of state[3] during which one may anticipate observing "Lee-Wick" condensations, density isomers or pion condensates.

Evidence for the presence of nonstatistical or collective behavior in the breakup of nuclei at Bevalac energies has not been convincing. It was the prediction of such phenomena for nucleus-nucleus collisions that made us decide to analyze some interactions of cosmic ray nuclei in nuclear emulsion. The mean energy of our particles is an order of magnitude higher than the Bevalac energy. Even though the energy is much lower than air shower energies, we thought additional information on nucleus-nucleus collisions might be valuable in that problem as well.

EXPERIMENT

These nucleus-nucleus interactions were obtained in a 5.2 liter stack exposed at a residual pressure of 3.9 g/cm^2 over India. The geomagnetic cutoff energy there is 7.6 GeV/nucleon and hence, considering the energy spectrum of heavy nuclei, the mean energy is about 21 GeV/nucleon. The nuclei were obtained in several different scans made with varying detection limits, so the charge spectrum is not exactly that of the cosmic rays, but is biased toward charges of 20-26. The median charge of the sample is 12.

With the help of two excellent scanners and a semi-automated computerized microscope, we have measured for each interaction the charges of the incoming nucleus and the charges and emission angles of all fragments, α-particles and protons emitted from the primary. In addition, we have measured the ionization and emission angles for each charged meson as well as for each of the slow particles emitted from the target nucleus.

The fast singly charged particles have been designated protons or mesons by assuming that enough protons must be emitted to conserve the charge of the primary nucleus and that these protons will have less transverse momentum than the mesons produced in the same interaction; hence, the protons are the innermost singly charged particles in every case. This procedure, while not perfect, should correctly identify 90% of the protons.[4]

In this sample of 400 interactions, 6.7% are pure fragmentations with no evidence of breakup of the target nucleus. Fifty-seven percent of the interactions result in at least one fragment of Z > 3, and 2.5% of the interactions have two fragments going forward.

The mean meson multiplicity of these interactions is 11.9 mesons per interaction. This may be compared with a value of 6.0 meson per interaction observed for 22.5 GeV protons in nuclear emulsion.[5] In Figure 1, we compare the normalized multiplicity distribution for these two sets of data. We see that nucleus-nucleus collisions have, as might be expected, a quite different distribution from that produced in nucleon-nucleus collisions. The former have a high probability of having zero multiplicity $n_\pi = 0$, with no apparent mesons

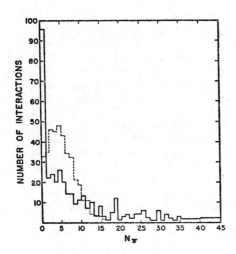

Fig. 1. Multiplicity distributions of mesons from nucleus-emulsion collisions (solid line) compared to proton-emulsion collisions at 22.5 GeV.

being produced. Since events of this kind are also characterized by having very small values for the fraction, f, of the primary charge appearing in singly charged particles, i.e., show little evidence of breakup of the primary nucleus, we consider the peak at $n_\pi = 0$ to be real and not due to misidentification of protons and mesons. In addition, the nucleus-nucleus collisions also show very large values of n_π, much greater than those observed in single nucleon interactions. This presumably reflects the occurrence of multiple nucleon-nucleon collisions in a single event. It may be noted that in those events where n_π is greater than 65, it is almost impossible to make a reliable analysis of the individual emission angles and, hence, we do not have a completely unbiased sample of data in the results we are about to describe.

We have classified the interactions in terms of the parameter f, defined above, because it should be closely related to the impact parameter for the interaction. If f is large, so that the primary nucleus is broken up into singly charged particles, then we are presumably considering central collisions, those with small impact parameters. The same line of argument in the other frame of reference leads us to select events with large numbers of low energy particles being emitted from the target nucleus as those which are central collisions with the target nucleus. However, in this case we have the problem that we do not, a priori, know the charge of the target nucleus, and hence, we can only reliably identify central collisions on heavy, Ag or Br, target nuclei. Individual events are thus best described by both f and N_h, the number of slow prongs, but f alone organizes the data rather well. This is illustrated in Figure 2, which shows n_π as a function of f. There is a well-defined correlation in the data, with large f, and thus presumably small impact parameters, being associated with large n_π. At present, our statistics are too limited to allow us to use the N_h parameter to further subdivide the data, nor can we look carefully at the effects of the differing charges of the primary nuclei, Z_p. However, within our statistical accuracy, we have not seen any dependence of n_π on Z_p, which is consistent with the deduction one can make from Figure 2 that rarely do more than 4 or 5 of the nucleons in a nucleus interact even in central nucleus-nucleus collisions. Hence, if the number of nucleons incident is significantly greater than the number that

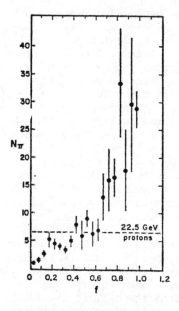

Fig. 2. Number of mesons as a function of the fraction, f, of protons that are emitted from the incident nucleus.

interact, the characteristics of the interaction will be relatively insensitive to the size of incident nucleus.

The characteristics of the individual interactions were studied by constructing target diagrams for each interaction. These show the point of intersection of each fast particle emerging from the interaction with a plane perpendicular to the primary direction and 1.0 mm from the interaction. We have not shown evaporation particles on these diagrams, nor the very rare backward mesons. Visual examination of these diagrams shows interesting evidence for several nonrandom spatial distributions of the mesons.

Firstly, there are many examples of close association of two or more particles, forming pairs or multiple clusters, as illustrated in Figure 3. In our initial attempt to see whether the occurrences of these associations were statistically significant, we have used a simple analysis whose main advantage was that we could rapidly assess the result. In every event having a target diagram, each meson was assigned a radial coordinate r and an azimuthal coordinate ϕ. For each meson appearing at a given r, i.e., with a given transverse momentum, we have looked for any other mesons having $|\Delta r|/r = 0.1$ and for all mesons satisfying this condition determined $\Delta\phi$. Our experimental distribution of $\Delta\phi$ is shown in Figure 4, for the 2900 mesons measured thus far. There is a very significant peak at $\Delta\phi < 5°$, with 200 close pairs being observed, compared with the mean of 61.3 ± 10.1 measured for $\Delta\phi > 30°$. Ninety percent of these 200 pairs are due to clusters of just two particles, 7% are due to clusters of three, and 3% to clusters of four or more.

Possible explanations of this pairing effect include a) Dalitz electron pairs from neutral pions, b) decay of ρ or other short-lived heavy mesons, c) pion correlations due to boson interference.[6] From the 2900 charged mesons we have observed, we would expect fewer than 20 Dalitz pairs. However, in order to study these pairs more closely, we have begun to trace a certain fraction of the individual particles through the emulsions in an attempt to determine whether they could be electrons.

A second apparently nonrandom effect that we have seen in these target diagrams is illustrated in Figures 5 and 6. Here we see examples of interactions of quite small f and n_π in which the mesons appear to all lie along a single line. Some 1 to 2% of the interactions show such 'fan-like' emission of the mesons. These events all appear to be peripheral interactions, with an appreciable frag-

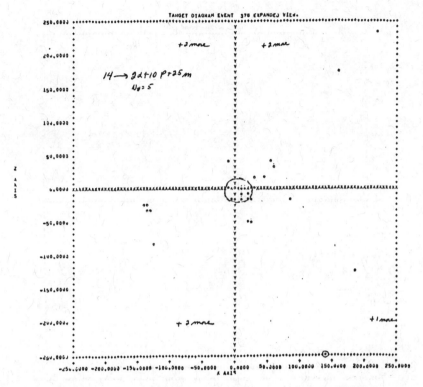

Fig. 3. A target diagram having a cluster of 4 and several pairs of mesons. The circle contains the 2 α-particles and 10 protons from the primary.

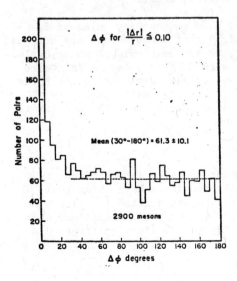

Fig. 4. Difference in azimuthal angles of mesons whose radial distance from the primary direction in the target diagram is the same within 10%.

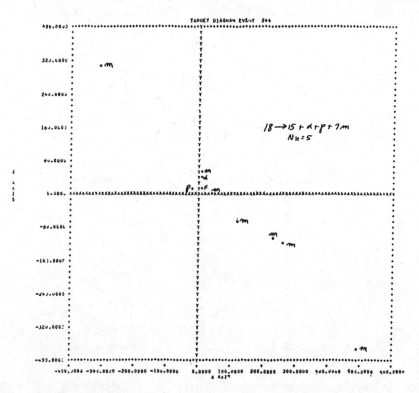

Fig. 5. The 7 mesons emitted in this collision of argon lie along a straight line. The Z axis extends from -400μ to $+400\mu$, the x axis from -400μ to $+600\mu$ and the scales are not equal.

ment surviving the interaction, and a consequent low value of f. This fact, on its own, argues strongly that these events are not just statistical fluctuations. However, we are developing a Monte-Carlo simulation that should allow us to examine this question in a more quantitative way. At present, we have no good explanation for these events, although it seems that they must represent some highly correlated behavior in the production of mesons that is peculiar to peripheral nucleus-nucleus collisions.

CONCLUSION

A careful analysis of 400 nucleus-emulsion interactions including angular measurements on all particles have shown that about six to seven percent of the mesons are emitted in pairs whose separation < 5°. From one to two percent of the interactions show 'fan-like' emission of mesons which appear as lines on target diagrams.

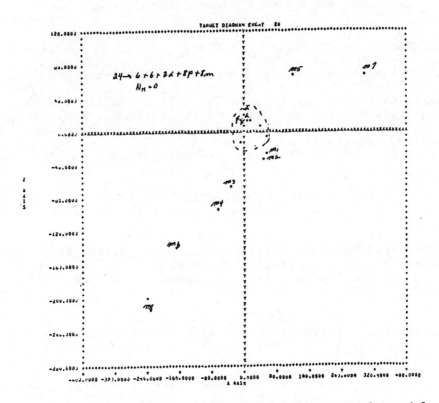

Fig. 6. A Z = 24 breaks up into 2 carbons, 2 α-particles and 8 protons shown in the circle. The 8 mesons are emitted in a line. The Z axis extends from −280μ to +120μ, the x axis from −400μ to +400μ in the target diagram.

REFERENCES

1. Heckman, H.H., H.J. Crawford, D.E. Greiner, P.J. Lindstrom, and Lance Wilson, Phys. Rev. C17, 1651 (1978).
2. Poskanzer, A., Invit. paper at Conf. on Nuclear Structure, Tokyo, Preprint LBL-6586 (1977).
3. Stock, R. and A. Poskanzer, "Comments on Nuclear and Particle Physics" VII, No. 2, p. 41 (1977).
4. Waddington, C.J., Phil. Mag. 2, 1059 (1957).
5. Winzeler, H., Nuclear Physics 69, 661 (1965).
6. Yano, F.B. and S.E. Koonin, "Determining Pion Source Parameters in Relativistic Heavy Ion Collisions", Cal Tech Preprint (1978).

Chap. 4   10-1000 TeV Interactions

MULTIPLE MESON PRODUCTION IN THE $\sum E_\gamma > 2 \times 10^{13}$ eV REGION

Brasil-Japan Emulsion Chamber Collaboration

J.Bellandi Filho, J.L.Cardoso Jr., J.A.Chinellato, C.Do-
brigkeit, C.M.G.Lattes, M.Menon, C.E.Navia O., A.M.Oliveira, W.A.Rodrigues Jr., M.B.C.Santos, E.Silva, E.H.Shibuya
K.Tanaka, A.Turtelli Jr.
Instituto de Física Gleb Wataghin-Universidade Estadual de Campinas, Campinas, São Paulo, Brasil

N.M.Amato, F.M.Oliveira Castro
Centro Brasileiro de Pesquisas Físicas, Rio de Janeiro, RJ, Brasil

H.Aoki, Y.Fujimoto, S.Hasegawa, H.Kumano, K.Sawayanagi, H.Semba, T.Tabuki, M.Tamada, S.Yamashita
Science and Engineering Research Laboratory, Waseda University, Shinjuku, Tokyo, Japan

N.Arata, T.Shibata, K.Yokoi
Department of Physics, Aoyama Gakuin University, Setagaya, Tokyo, Japan

A.Ohsawa
Institute for Cosmic Ray Research, Tokyo University, Tanashi, Tokyo, Japan

ABSTRACT

Cosmic-ray-induced nuclear interactions are observed by emulsion chambers exposed on Mt.Chacaltaya, Bolívia (5220m above sea level). We detect photons from nuclear interactions which occurred within the petroleum pitch layer (1/3 nuclear mean free path), set 150 cm above the emulsion chamber.

Photons of energy >100 GeV are detected in each event, the highest detected photon energy being 40TeV. 350 events are found to contain 4 or more photons, whose energy sums, $\sum E_\gamma$, range from 5 TeV up to 500 TeV, and are adopted for analysis.

Multiplicity-, energy-, angular-, p_T-, and other related distributions of the photons are presented, comparison is made with the accelerator data, and discussions are given.

INTRODUCTION

The idea of fireball as a basic physical entity for multiple meson production was introduced even before the π-meson and muon discovery[1]. In this picture it is assumed that in nuclear collisions of high energy there is the production of an intermediate state-fireball after the collision and before the isotropic meson production.

Our experiment detects the gamma-rays of the π^0 decay produced in the interaction of high energy cosmic particles with the nucleons of the atmosphere - "A-jets" - and with the nucleons of a carbon target - "C-jets".

One of the important results of the analysis of our data is that both, the transverse and longitudinal momentum distributions of those gamma-rays - $p_{T\gamma}$ and $E_\gamma / \sum E_\gamma$ distributions can be fitted by an exponential type function. This can be interpreted as an isotropical decay into gamma rays of a moving center, that is, the fireball produced in the hadronic interaction.

The early C-jets data were consistent with the existence of a single type of fireball, characterized by its mass and "temperature", the H-quantum[2)3)].

Through the analysis of the A-jets and of the more recent C-jets data, we found a significant number of events which have greater multiplicity and transverse momentum than the previous ones. In 1968 a new type of fireball was introduced in our model, the "S-H-quantum"[3)], with larger mass and temperature than the H-quantum.

In the same year by the study of extremely high energy events we found experimental support for another type of fireball of still larger mass and temperature, the UH-quantum[3)].

In this paper we present the latest C-jet data and its analysis based on the fireball model.

EXPERIMENTAL METHOD

A) EMULSION CHAMBER DESCRIPTION.

The emulsion chambers exposed to high energy cosmic rays at Chacaltaya Laboratory, 5220 m above sea level, have the structure illustrated in Fig.1. Each chamber is composed of the following four parts: the upper detector, the target layer, the air gap and the lower detector. Table 1 gives an outline of the four emulsion chambers used in the present work.

The detector is a multi-layered sandwich of Pb-plates, X-ray films and nuclear emulsion plates. Electrons and gamma-rays, either from outside the chamber or produced in the sandwich, generate electron showers in the detector. These electron showers leave dark spots on X-ray films and tracks in nuclear emulsion plates.

In the present work, we are concerned with the nuclear interactions occurred in the target layer of petroleum pitch, whose thickness corresponds to 1/3 nuclear mean free path, or to 0.4 radiation length. The gamma-rays pro

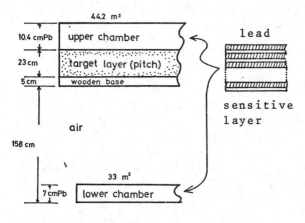

FIG.1 Structure of the emulsion chamber exposed at Chacaltaya Laborory.

duced in one nuclear interaction pass through the rest of the target layer, through the air gap and arrive at the lower detector, where they are detected as a group of electron showers concentrated within a small region of few mm. The air gap is enough to separate the cores of these electron showers, so it is possible to make the energy measurement of each individual shower. The upper detector works, in the present experiment, as a shield to prevent the arrival of atmospheric gamma-rays and electrons at the target layer and at the lower detector.

Table I Outline of emulsion chambers

Chamber number		15	16	17	18
Upper chamber	area (m^2)	44.2	44.2	44.2	44.2
	thickness(cmPb)	7.8	7.8	10.4	9.1
Lower chamber	area (m^2)	33.0	20.4	33.0	33.0
	thickness(cmPb)	6.0	15.0	7.0	7.0
Exposure (days)		295	370	567	570

B) DETECTION OF THE EVENTS.

The X-ray films of the lower detector are scanned for dark spots caused by the gamma rays produced in the nuclear interactions occurred in the target, "C-jets". The threshold energy for this scanning is around $\sim$ 3 TeV. The energy measurement of each electron shower is done by track counting on the emulsion plates observed at the microscope. The threshold energy for each shower is $\sim$ 0.1 - - 0.2 TeV. The microscopic scanning for the shower cores of each C-jet is done over a circular area of radius r $\sim$ 2.5 mm. Since the distance between the middle of the target and the detection plane is $\sim$ 170cm, the restriction on r means that we only observe the gamma-rays emitted inside a solid angle $\Omega_\gamma \sim 7 \times 10^{-6}$ st (in laboratory system). The target thickness implies an uncertainty of $\sim$ 5% in an gular measurement, while the estimated uncertainty in energy measurement is from 10% to 20%.

C) SELECTION OF EVENTS.

Up to the present time we have analysed 350 events with 4 or more gamma-rays and with $\sum E_\gamma$ (total observed energy) > 5 TeV. The highest energy observed C-jet has $\sum E_\gamma$ = 120 TeV.(One event has $\sum E_\gamma \sim$ 500 TeV, but its detailed analysis is impossible because the energy is too high). In the following sections we present the analysis of 57 events with $\sum E_\gamma$ > 20 TeV.

D) VARIABLES USED IN THE ANALYSIS OF THE EVENTS.

In the present experiment, like in most of the cosmic ray ones, it is not possible to obtain information about the incident particles that interact inside the target. The emission angle of the gamma-rays (θ_γ) of a C-jet is defined as the angle between the direction of the gamma-ray and the direction of the energy center of the C-jet. In this way, the chosen reference axis is closer to the moving fireball direction than to the incident particle direction. Instead of the incident particle energy E_o, we use the total observed energy $\sum E_\gamma$ of each C-jet. So, the variable $x_\gamma = E_\gamma/E_o$ commonly used in accelerator experiments is replaced by the fractional energy $E_\gamma/\sum E_\gamma$. In order to compare accelerator and cosmic ray results it is necessary to know the gamma-ray inelasticity, $k_\gamma =$
$= \sum E_\gamma/E_o$.

The experimental limits on E_γ > 0,1 MeV and on Ω_γ < 7×10^{-6} st, restricts our observation to gamma-rays emmited in the very forward region. In order to reduce the effects of these restrictions we select the C-jets with $\sum E_\gamma$ > 20 TeV.

GENERAL RESULTS

A) CHACALTAYA AND BALLOON RESULTS.

First we compare the energy and angular distributions of gamma-rays of C-jets with those from the balloon emulsion chamber of Sato et al. Table 2 gives some characteristics of the events detected by both experiments. In Fig.2, where the integral fractional energy distribution $f = E_\gamma/\sum E_\gamma$ is presented, it can be seen the agreement of the data of the two experiments. Fig. 3A and 3B are the angular distributions (log tan θ_γ) in the laboratory system and in the mirror system (beam system), where the

Table II Experimental conditions

Experiment	Balloon	Chacaltaya
Interaction energy $\sum E_\gamma$ (TeV)	0.6 - 5.0	20 - 120
Average $\langle \sum E_\gamma \rangle$ (TeV)	2.2	40.5
Observation range:		
γ-ray energy E_γ (TeV)	> 0.03	> 0.2
γ-ray fractional energy f	> 0.014	> 0.004
γ-ray emission angle θ_γ (radian)	< 0.01	< 0.001
Event number	15	57

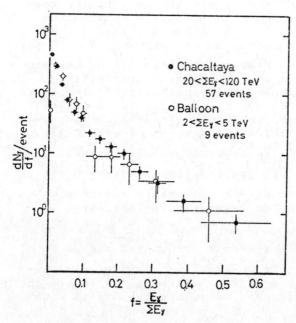

FIG.2 Fractional energy distributions of gamma-rays.

incident particle - with energy E_o = $\sum E_\gamma / k_\gamma$ is at rest (k_γ is 0.20 ∿ 0.25) The distributions are similar in shape and in height, but are shifted by a factor ∿ 20 (the ratio of the average observed energy) in Fig.3A. Fig.4 shows the (integral) transverse momentum distributions. The effect of the detection loss appears when $p_T < E_{min} \theta_{max}$ that is, when $p_T <$ 100 MeV/c for our data, and when $p_T <$ 300 MeV/c for balloon data. The distributions agree in the unaffected region $p_T >$ 300 MeV/c.15

B) COSMIC RAY AND ACCELERATOR RESULTS.

Both cosmic-ray experiments, Chacaltaya and balloon, although differing by a factor ∿ 20 in the observable energy agree on the distribution of gamma-ray fractional energy $f = E_\gamma / \sum E_\gamma$, transverse momentum $p_{T\gamma}$ and emission angle θ_γ^* in the mirror system. One might think that the

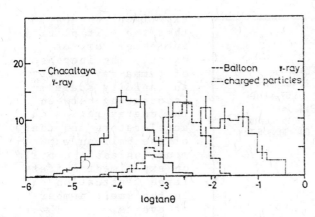

FIG. 3A Angular distributions of gamma rays in the laboratory system. Angular distribution of charged particles observed by the balloon experiment is also shown for comparison.

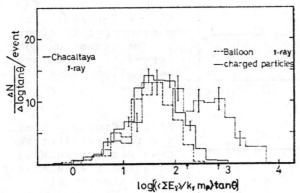

FIG. 3B The same distributions as Fig. 3A transformed into the mirror system (k_γ is assumed as 1/6)

agreement reveals persistence of the scaling rule up to the cosmic ray energy region of $\sim 10^{14}$eV. Indeed, at this meeting we learned the results of simulation calculation made by Ellsworth, Yodh and Gaisser based on the extrapolation of accelerator data to cosmic ray regions[5]. They found a good agreement on the fractional energy distribution, and the result on $p_{T\gamma}$ distribution seems to have no appreciable difference in very forward-emitted gamma-rays ($\theta_\gamma \Sigma E_\gamma < 2$GeV).

But, the comparison of the rapidity distribution of gamma-rays, as seen in Fig. 5, shows significant difference. The cosmic ray events of Chacaltaya have experiment large multiplicity 50% more, than the simulation result. The agreement of the f-spectrum is not inconsistent with this fact, because they showed that the f-spectrum is rather insensitive to changes in y-distribution.

Thus we arrive at the conclusion that the Chacaltaya events of $\sim 10^{14}$ eV are not consistent with the scaling extrapolation of accelerator data, but show an appreciable multiplicity increase. The agreement with the balloon

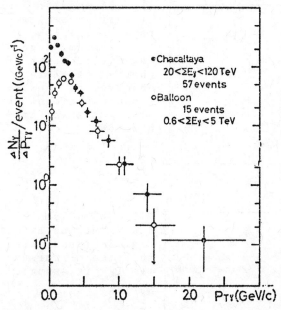

experiment suggests that the multiplicity increase-- or more - exactly the increase of gamma-ray density in rapidity scale -- is occuring between the energy region of accelerators and that of the balloon experiment, unless the balloon experiment suffers large fluctuation due to the small number of events.

FIG. 4 Transverse momentum distributions of the gamma-rays observed by Chacaltaya and balloon experiments.15

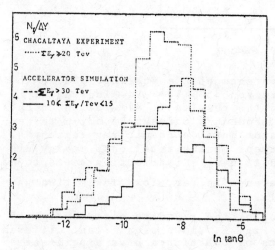

FIG. 5 Comparison of the rapidity distribution with calculated results of accelerator simulation by Ellsworth, Yodh and Gaisser[5])

ANALYSIS

A) MULTIPLICITY AND p_T

Ar first, we will try to see the existence of two types of events without using the fireball hypothesis. In Fig.6, $<p_{T\gamma}>$ means the average transverse momentum in one event and n_γ is the multiplicity of the event divided by its rapidity interval. Only events with $\Sigma E_\gamma > 20$ TeV and $N_\gamma > 4$ were considered. In computing the above two quantities, the first gamma-ray has no effect. In the figure, one sees that the experimental points scatter widely in the $<p_T>$-n_γ diagram. Scaling of accelerator data of lower energy tells us that the points will distribute around $n_\gamma \sim 2$ (one neutral pi-meson per one rapidity interval) and $<p_T> \sim 150 \text{MeV}/c$, while our gamma-ray detecting device will have a bias for neutral-pi-meson-rich events and make n_γ larger. Thus the group of C-jets with $n_\gamma = 2 \sim \sim 4$ will be the one expected from the extrapolation of accelerator informations. Then a cluster of events distributed around $n_\gamma = 6 \sim 8$ show existence of another type of C-jets with larger multiplicity and larger $<p_T>$ than the ones expected from the accelerator information.

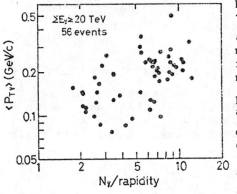

FIG. 6 $<p_{T\gamma}>$-n_γ correlation diagram for 57 C-jets with $\sum E_\gamma > 20$ TeV.

B) $p_{T\gamma}$ DENSITY IN RAPIDITY SCALE.

Fig. 7 shows the dependence of $\sum_\theta p_{T\gamma}$ on log θ for the C-jets. $\sum_\theta p_{T\gamma}$ is the sum of p_T of the gamma-ray emitted within an angle θ. In order to improve the comparison of the events, the angular scale is normalized by log $(\theta \sum_{all} E_\gamma / \sum_{all} p_T)$. Here again one sees the separation of events into two groups with different slopes. The quantity $\sum_\theta p_T$ represents the transversal momentum flow in the rapi-

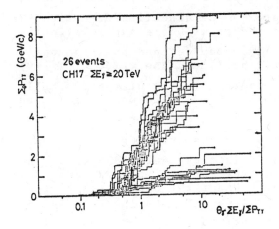

FIG. 7 Variation of $\sum_\theta p_{T\gamma}$ with log $(\theta \sum E_\gamma / \sum p_{T\gamma})$, expressing side-way flow of momenta.

dity scale, and its linear increase with log θ means that the particular type of C-jets with large p_T and large n_γ mentioned above is not just due to fluctuation.

FIREBALL ANALYSIS

A) FIREBALL ALGORITHM.

The difference between these two types of events can be better seen in terms of fireball language. For a cluster of gamma-rays from isotropic decay of a fireball we are able to estimate the rest energy either from their gamma-ray invariant mass,

$$m_\gamma c^2 = (\sum_i E_i \cdot \sum_i E_i \theta_i^2)^{1/2} \tag{1}$$

or from the sum of the gamma-ray p_T,

$$\sum p_T = \sum_i E_i \theta_i / c \tag{2}$$

because there will be a relation valid on the average,

$$m_\gamma c = \frac{4}{\pi} \sum p_T . \tag{3}$$

Now we take a cluster of gamma-rays with $\theta_i < \theta$ in a C-jet and calculate $m_\gamma(\theta)c$ and $\sum_\theta p_T$ varying θ. For small θ, $m_\gamma(\theta)c$ is smaller than $\frac{4}{\pi}\sum_\theta p_T$, showing that the cluster is of pancake-shape oblate to the direction of motion. Increasing θ, we arrive at the point where the relation(3) is satisfied. The gamma-ray group defined by(3) is consistent with an isotropic fireball, and the value of $m_\gamma(\theta)$ here is assumed to give the rest energy of the fireball in the event. For larger θ, $m_\gamma(\theta)c$ becomes larger than $\frac{4}{\pi}\sum p_T$, showing that tha shape is now of a cigar type.

There are cases where the observed angular region does not reach the point of isotropy defined by(3). In this case, we have to make the estimation by an extrapolation in the following way. Under the hypothesis of isotropic fireball, the quantities $\sum_\theta p_T$ and $\sum_\theta E_\gamma$ are the following functions of θ in their average behaviour,

$$\sum_\theta p_T = \frac{m_\gamma c}{2} \left(\tan^{-1}\Gamma\theta - \frac{(1-\Gamma^2\theta^2)}{(1+\Gamma^2\theta^2)^2} \right) \tag{4}$$

$$\Sigma_\theta E_\gamma = \Gamma \cdot \mathcal{M}_\gamma c^2 \left[1 - \frac{1}{(1+\Gamma^2\theta^2)^2} \right] \qquad (5)$$

We construct the quantities, $\Sigma_\theta P_T$ and $\Sigma_\theta E_\gamma$, with the experimental data of a C-jet, and their fit with the theoretical curves given by (4) and (5) determines the values of two parameters in the relations, i.e., the gamma-ray mass of a fireball $\mathcal{M}_\gamma$ and the Lorentz factor of its motion Γ.

Fig. 8 presents the histogram of the rest energies of the fireballs ($\mathcal{M}_\gamma c^2$) so calculated. Each blank square represents the value obtained by the first method with use of (3) and each crossed square the one by the second method with use of (4) and (5). The histogram shows two peaks, one with $\mathcal{M}_\gamma c^2 \sim 1$ GeV (H-quantum) and the other with $\mathcal{M}_\gamma c^2 = 4 \sim 7$ GeV (SH-quantum).

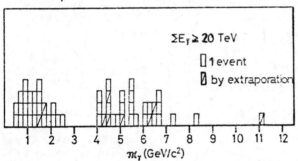

FIG. 8 Histogram of the visible rest energies $\mathcal{M}_\gamma c^2$ of the observed fireballs.

FIG. 9 shows the diagram of the rest energy $\mathcal{M}_\gamma c^2$ vs. the average energy of gamma-rays $<E_\gamma^*>$ in the fireball system. One sees that the average gamma-ray energy $<E_\gamma^*>$ is larger for the group with larger rest energy.

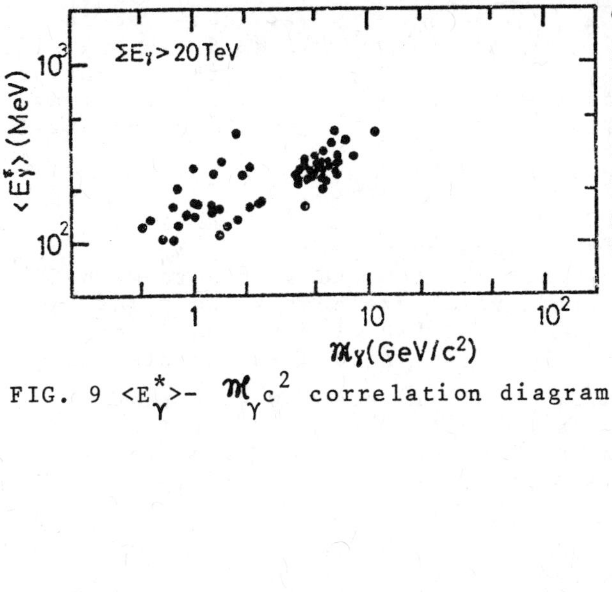

FIG. 9 $<E_\gamma^*>$ - $\mathcal{M}_\gamma c^2$ correlation diagram.

B) FIREBALL SIMULATION.

The thickness of the carbon target of the emulsion chamber is around 1/3 of the nuclear mean free path. Therefore, the probability of successive interactions of the incident particles is not negligible. The effect of these interactions on the invariant mass of the gamma-rays of each C-jet is verified through the simulation of the experimental conditions, under the fireball model.

In the simulation carried out by Santos and Turtelli[6], it is assumed the emission of a fireball with rest energy 2.8 GeV/c^2 in each nuclear interaction. In its rest system occurs the isotropic thermal emission of π-mesons with temperature KT = 110 MeV. The simulated events are analysed in the same way as the experimental ones. In Fig. 10 we present the invariant mass distribution for pure and contaminated events. Actually, due to successive interactions, we obtain invariant m_γ > 3 GeV/c^2. Notwithstanding, the fraction of simulated events with such large mass values is quite smaller than the experimental one. Thus the successive interactions cannot explain the two peaks observed in Fig. 8. In Fig. 10 we also present the invariant mass distribution obtained by Ellsworth, Yodh and Gaisser[5].

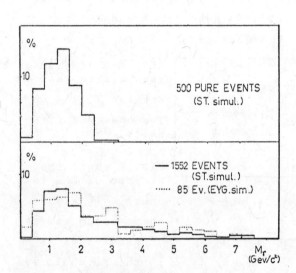

FIG. 10 Histogram of rest energies of simulated fireballs.

C) H AND SH-QUANTUM.

Among the 57 highest energy C-jets with $\sum E_\gamma$ > 20TeV, there are 26 events with H-quantum and 30 events with SH-quantum. The production rates are nearly the same.

Fig.11 gives the distribution of fractional energy f = $E_\gamma / \sum E_\gamma$ in integral form. One sees that the distribution for the SH-quantum events is steeper than that of the H-quantum events, showing the larger multiplicity of the former. The distribution for the H-quantum events is well represented by a straight line, i.e. an exponential

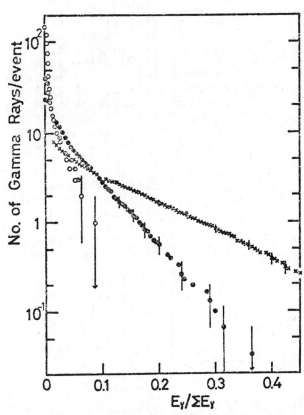

FIG. 11 Integral fractional energy distribution for H- and SH-quantum.

function of energy fraction, and its extrapolation to zero energy gives the gamma-ray multiplicity of one H-quantum ∿ 6. Smooth extrapolation to zero of the distribution for SH-quantum events gives ∿20-30 as the estimated gamma-ray multiplicity of one SH-quantum. Fig. 12 presents the p_T distribution in differential form. SH-quantum events have p_T distribution less steep, showing larger average p_T value. A drop of the distribution in small p_T region of < 100 MeV/c is due to the detection limit effect, as discussed before.

The correlation between "fireball mass $\mathcal{M}_\gamma$" and $<p_T>$ or multiplicity N_γ (or steepness of f-spectrum) was shown by Ellsworth, Yodh and Gaisser at this meeting from their acceleartor simulation calculation[5]. This seems to put in doubt the existence of the two types, H and SH-quantum. We do see such correlation in the experimental data shown in Fig. 9 in each one of the two groups. Since we have the relation $\mathcal{M}_\gamma \sim N_\gamma <p_T>$, the correlation should be expected. The existence of these two types is strongly supported by the observed correlation between $<p_T>$ and $\bar{N}_\gamma$, which is not predicted by the scaling model of accelerator simulation.

CONCLUSION

A) REMARKS ON H-QUANTUM.

It has already been stated that the H-quantum events

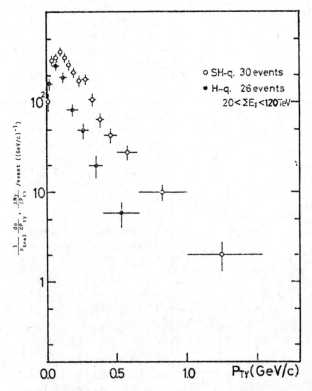

FIG. 12 Differential $p_{T\gamma}$ distribution

have gamma-ray multiplicity and transverse momentum expected from the extrapolation of accelerator experiments. In the accelerator experiments, the scaling property of the multi-hadron productions is widely observed, not only in hadronic collisions but also in collisions involving leptons and electrons.
In Fig. 13 the cosmic-ray data are compared with the results of non hadronic experiments. To compare these results it is necessary to relate the variable z = (energy of a charged hadron / (total available energy) with the fractional energy $f = E_\gamma / \sum E_\gamma$. Under the assumption of equal probability for emission of π^+, π^0 and π^- the relation is $z = (2/3)f$.

It can be seen that the distribution H-quantum events agrees well with the accelerator ones, but the cosmic-ray distribution extends into smaller fractional energy region, because the available energy for hadronic production is larger. It can be also seen that the f distribution of cosmic ray events behaves as $1/f$ for $f \ll 1$.

Our conclusion is that the multi-hadronic production in the accelerator region is dominated by the H-quantum emission, and the scaling phenomena is a consequence of the constant mass and temperature of the fireball, i.e. the H-quantum. The H-quantum emission is observed up to the energy region of few hundreds of TeV.

B) REMARKS ON SH-QUANTUM.

SH-quantum, with larger multiplicity and higher p_T, is observed only in the cosmic ray energy region and this is the reason why some features of the cosmic-ray events cannot be explained by the simple extrapolation of sca-

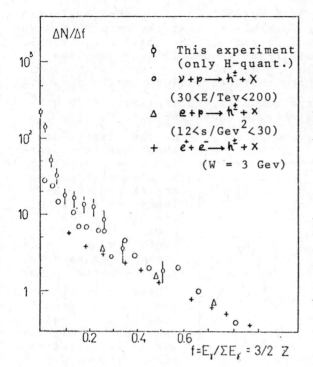

ling hypothesis. SH-quantum decays into π-mesons either directly or through the emission of H-quanta. Considerations on average multiplicities and mean transverse momenta leads to the picture that an SH-quantum decays on average, into 5 to 6 H-quanta. At the present stage, however, the observed distributions of energies, emission angles and transverse momenta of gamma-rays are consistent both with the cascade and the direct decay predictions.

FIG. 13 Z- distribution for Chacaltaya and non hadronic experiments.

C) REMARKS ON UH-QUANTUM.

The analysis of two extraordinary events - Andromeda[8] and Texas Lone Star[9] - leads us to the hypothesis of UH-quantum. Remarkable as they are, they have inherent ambiguities: the former can be a result of successive interactions in the atmosphere and the latter has some probability to be due to a heavy primary nucleus.

Presently we have two more possible examples of UH-quantum production. One is S112, a large nuclear event which occurred in the atmosphere, very near our apparatus.[10),11),12)] Examination of the cascade development of the shower cores, together with the gamma-ray pair invariant mass distribution, yields (250 ± 80) m for the most probable height of the main interaction. 117 gamma-rays and 24 local nuclear interactions are observed, the total observed energy is 878 TeV. Contamination due to ancestral and posterior atmospheric interactions is negligible.

The other is I153, a C-jet.[11),13)] Opening angle measurement excluded the possibility that the event was a superposition of plural interactions due to a narrowly collimated bundle of nuclear-active particles. 40 gamma-rays are observed with $\sum E_\gamma$ 42 TeV.

These two events are characterized by large $<p_T>$, 0.4 to 0.5 GeV/c, and large gamma-ray multiplicity, $\sim$ 70 per unit rapidity interval. Fig.14 shows the $\sum_\theta p_T$ vs. log ($\sum_\theta E_\gamma$) plots of the two events. In the same figure are shown the same plots for some typical examples of H-quantum and SH-quantum events. As is seen in the figure, the plots of the two events behave more or less in the same way p_T-flow per unit rapidity interval being 7 to 10 times that of the SH-quantum events. This leads to interpreting them as UH-quantum events. The plots of Fig.14 yield $\sim$ 70 GeV/c^2 as the best estimation for m_γ of a UH-quantum.

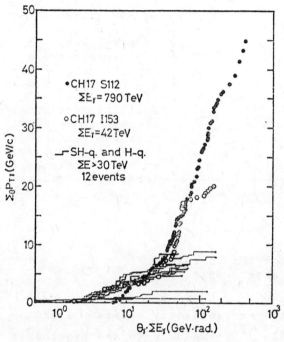

FIG. 14 $\sum_\theta p_{T\gamma} \cdot \theta_\gamma \sum E_\gamma$ correlation for H-, SH- and two UH-quanta.

Fig.15A and B illustrate the $p_x - p_y$ diagrams of the two events. Both show multi-jet-like structures in the p_T - plane. The characteristic feature of each jet agrees with that of an H-quantum event. Therefore it is inferred that a UH-quantum decays via H-quanta.

The highest energy gamma-ray pair in Event I153, indicated by circles in Fig. 15A, has its origin in the middle of the air gap between the target layer and the lower detector. This cannot be reconciled with a neutral pi-meson directly produced in the interaction, and the most plausible explanation is a production of a particle decaying with a mean life-time $\sim 10^{13}$ sec, with emission of a neutral pi-meson among its decay products[14)].

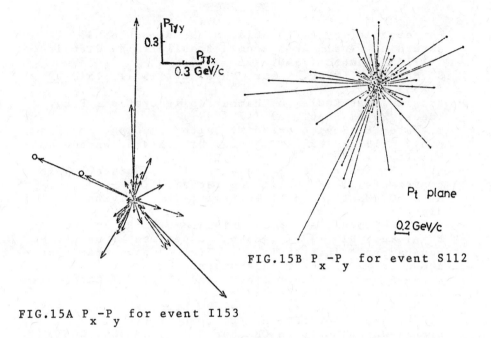

FIG.15B P_x-P_y for event S112

FIG.15A P_x-P_y for event I153

ACKNOWLEDGEMENT

The collaboration experiment is financially supported in part by Conselho Nacional para o Desenvolvimento Científico e Tecnológico and Fundação de Amparo à Pesquisa do Estado de São Paulo, in Brasil and Institute for Cosmic Ray Research, University of Tokyo, in Japan.

REFERENCES

1) For example, P. A. Pompéia, M. D. de Souza, G. Wataghin, Anais da Academia Brasileira de Ciências, T XII, setembro (1940).
2) S. Hasegawa, Prog. Theor. Phys. 26 (1961), 151; 29 (1963), 128.
3) Brasil-Japan Emulsion Chamber Collaboration, Prog. Theor. Phys. Suppl. no. 47 (1971) 1.
4) Y. Sato, H. Sugimoto and T. Saito, J. Phys. Soc. Japan 41 (1976) 1821 - Y. Sato, Dr. Thesis, Waseda University, (1978).
5) R. W. Ellsworth, G. B. Yodh and T. K. Gaisser, this conference (1978 [following paper].
6) M. B. C. Santos and A. Turtelli, Jr., private communication.
7) For references, see R. D. Field and R. P. Feynman, Phys. Rev. D15 (1977), 2590. Experimental points of non-hadronic interactions in Fig. 13 have been reproduced from Fig. 6 of this paper.
8) Japan-Brasil Collaboration, Conf. Pap., XII Cosmic Ray Conf., Hobart 7 (1971) 2775.
9) D. H. Perkins and P. H. Fowler, Proc. Roy. Soc. A278 (1964) 401. For the analysis, see ref. 3).
10) Brasil-Japan Emulsion Chamber Collaboration, Conf. Pap. XV Cosmic Ray Conf., Plovdiv 7 (1977) 195, 201.
11) Brasil-Japan Emulsion Chamber Collaboration, Conf. Pap., XIV Cosmic-Ray Conf., Munchen 12 (1975) 4297.
12) H. Semba, ICR-Report-65-78-9 (Univ. of Tokyo), (1978) submitted to Nuovo Cim.
13) N. Arata, Nuovo Cim. 43A (1978) 455.
14) K. Niu, E. Mikumo and Y. Maeda, Prog. Theor. Phys. 46 (1971) 1644. See also K. Sawayanagi, Phys. Rev. D (to appear).
15) Detection region of gamma-rays in the balloon experiment covers roughly the forward hemisphere in CMS, while that in Chacaltaya experiment covers about the forwardmost half of the forward hemisphere. Therefore, the observed multiplicities have been multiplied by factors 2 and 4, respectively, for the balloon experiment and for the Chacaltaya experiment to normalize the data in Fig. 4.

Computer Simulation of Emulsion Chamber C-Jets
Using Conventional Particle Physics

R. W. Ellsworth, G. B. Yodh[†]
University of Maryland
College Park, Maryland

T. K. Gaisser
Bartol Research Foundation
University of Delaware
Newark, Delaware

ABSTRACT

We report results of computer simulations of interactions of energetic hadrons in the Brazil-Japan emulsion chamber at Mt. Chacaltaya. This experiment provides the most direct information that currently exists on strong interactions up to about 100 Tev lab energy. We find most features of the data to be in good agreement with straightforward extrapolation of accelerator data from $E_{lab} \leq 2$ Tev using Feynman scaling. We find however that because the detector only measures energetic secondary photons, it is rather insensitive to the energy dependence of inclusive cross sections.

MOTIVATION

Emulsion Chambers have proven themselves to be powerful tools for investigating high energy interactions of the cosmic ray beam. There is now a considerable body of emulsion chamber data from interactions, both in the atmosphere and in local targets, at energies in excess of 20 Tev and as high as 1000 Tev. Some of the results suggest the onset of fundamentally new processes at these energies. Not all of the interpretations are absolutely conclusive, however; a number of biases are imposed on the data by the experimental conditions. An important question, then, is to what extent can a straightforward extrapolation of accelerator-determined particle physics predict the emulsion chamber data?

In this study we have focussed on the interactions in a local target ("C-jets"). This is because, compared to atmospheric interactions, the complications due to multiple interactions and cascading of produced γ's above the detector are greatly reduced.

The only experiment with a large amount of data from a local target, at high energies, is that of the Brazil-Japan Collaboration [1]. This work started on Mt. Chacaltaya, Bolivia, in the early

†Present address: Physics Section, National Science Foundation, Washington, D. C. 20550

sixties and is still in progress. The experiment measures the energies of produced γ rays, and their emission angles. The collaboration has reported on some observed structure in the invariant mass distribution of forward γ's, at incident primary energies above 50 Tev. The mass distributions have been interpreted as evidence for the formation of "fireballs" - new states of matter with definite masses. In this paper we address the following question: Can an extrapolation of conventional multi-particle production, together with a simulation of the experimental conditions, and the biases imposed by cuts on the data, account for the observed results? Also, what is the sensitivity of the experiment to possible scaling violations?

In the simulation we have (A) used a scaling model of multi-particle production (B) drawn incident energies from a power law spectrum (C) included multiple interactions in the target, (D) included the experimental energy and angular resolutions, and (F) treated the "data" as the Brazil-Japan group has done.

INTERACTION MODEL

In order to make use of the large amount of existing experimental data on inclusive single particle production at high energies, we have adopted an independent emission scaling model. Secondary particle center-of-mass (CM) momenta are drawn at random from probability distributions which result from Feynman scaling of the invariant inclusive cross sections. Thus the momentum of each secondary is uncorrelated with that of any other -- there are no clusters. Secondary particles are created until the available CM energy is exhausted.

We have chosen to neglect, at this stage, any effects due to the nuclear composition of the target; the target is assumed to be a proton. Recent accelerator measurements[2] of forward secondaries produced as nuclear targets show a ~ 20% increase in multiplicity of minimum ionizing secondaries for A ~ 14 targets. For the energetic forward secondaries with which this simulation is most concerned, the difference is considerably smaller.

We have also chosen to neglect the observed[3] correlation between longitudinal and transverse secondary momenta. In creating a secondary, then, the simulation program first selects, at random, a value of Feynman X ($X = 2p_{\parallel}^{*}/\sqrt{s}$) from an appropriate distribution, selects a value of $p_{\perp}$ from its distribution and then computes the CM momentum vector.

More specifically, the following set of steps is executed for each interaction:

1) The CM longitudinal momentum of the projectile and the target proton are independently chosen. The X distribution used is of the form:

$$\frac{dn}{dx} = \text{constant} \qquad (1)$$

This form is in rough agreement with accelerator data on leading particles in p-p collisions.[4] It results in a mean inelasticity of 0.5 for both π-p and p-p collisions.

2) The transverse momenta of the projectile and target are independently chosen from a distribution of the form

$$\frac{dn}{dp_\perp} \propto p_\perp \, e^{-Bp_\perp}$$

this distribution is also used for secondary particles. The constant B is 6,5,4 $(Gev/c)^{-1}$ for pions, kaons and protons respectively. A CERN, ISR experiment, that of M. Banner et al.[5], fitted their data to the same form and found the constant B to be $5.72 \pm .07$, $5.91 \pm .07$ and $4.06 \pm .25$ for π^+, π^- and protons, at $\sqrt{s}$ of 44 Gev. The results of M. G. Albrow et al.[6] support a value of B for kaons between the values for π and p. The CERN - Columbia - Rockefeller collaboration[7] found B for π^o's produced at small $p_\perp$ at the ISR to be 6.0. This form does not actually provide the best fit to all the data, but its agreement seems adequate, and it is convenient for simulation purposes. It also is obvious that this form does not take into account the observed deviation from exponential behavior at high $p_\perp$ and high CM energy. Note that only the shapes of the $p_\perp$ and X distributions are important in the Monte Carlo procedure. The number of "draws" from a distribution is determined by energy conservation, as discussed below.

3) Secondary pions and kaons are created in the ratio π/K = 88/12. Pions are randomly tagged as neutral with an average probability of 1/3. Secondary baryon production is not included. Longitudinal momenta are found from the X distribution which would result from an invariant cross section of the form

$$E \frac{d^3\sigma}{dp^3} \propto e^{-AX} \qquad (3)$$

The value of A is randomly taken to be either 5.8 or 8.3 with 50% chance of each. In Figure 1, this functional form is compared with 400 Gev p-p data obtained by J. R. Johnson et al.[8] The agreement is better for pions than kaons, but is satisfactory for our purposes, since we will use the simulation to test the sensitivity of the emulsion chamber data to gross changes in the X distributions.

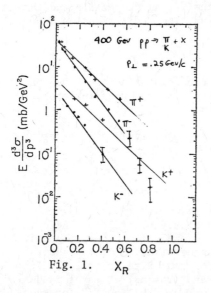

Fig. 1.

The values of A were changed, for this purpose, in subsequent runs (see below). The same X distributions are used for secondaries from p-p, π-p and K-p interactions.

4) Secondary transverse momenta are selected from $p_\perp$ distributions given above. Azimuths are chosen at random. Approximate momentum conservation is obtained by pointing the CM momentum vector into the hemisphere opposite to that of the sum of all previous momenta in the event.

5) After the momentum of each secondary is determined, the total amount of CM energy used so far is computed. If the most recently created particle makes the total energy greater than the actual CM energy, this particle is randomly included 50% of the time, and the momenta are Lorentz-transformed to the laboratory. If there is more energy available, more secondaries are produced until the available energy is used up.

The simulation procedure conserves momentum only approximately. One consequence of this is that the total energy of the outgoing particles is not exactly the same as the incident energy. Since the last particles produced are secondaries which are, for the most part, rather soft in the CM, the energy unbalance is not very severe. The fractional CM energy error is approximately 4%. The lack of exact transverse momentum conservation also induces a spread in the energy weighted center of all outgoing particles. This spread has been verified to be negligible in comparison with that of the simulated experiment, the latter being due to the selection of π^o's and to experimental resolution.

SIMULATION OF THE EXPERIMENT

The specific features of the Mt. Chacaltaya Emulsion Chamber Experiments which were included in the simulation are the following:

1) The energy spectrum of the incident hadrons. We assumed that the hadrons incident on the detector were protons, with an integral energy spectral slope of 1.70. Incident energies were generated, with this distribution, above 10 Tev. It is likely that some (perhaps 30%) of the incident hadrons are pions. Recent

Fermilab data[10] show that secondary particle X distributions in π-p collisions are somewhat different from those in p-p collisions. Effects due to these differences will not be present in our current simulation.

2) Multiple interactions in the target. A typical target had a thickness of 34 gm/cm². The first interaction point was randomly distributed over the depth of the target. The Monte Carlo program followed each created hadron and generated new interactions[9] until either

    a) the hadron left the target, or

    b) the energy of the hadron was below a nominal threshold. This threshold was taken to be 200 Gev, the stated energy detection threshold in the emulsion chamber.[11] All produced π⁰'s were allowed to decay into γ's and the γ's were "trajected" into the emulsion chamber.

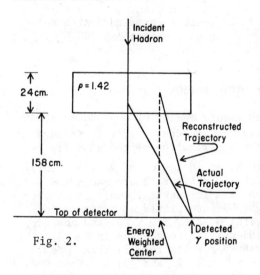

Fig. 2.

3) Energy and Angular Resolution. The main sources of experimental error are the energy resolution of the chamber for individual γ's, the lack of knowledge of the lateral position of the incident hadron, and the depth of the interaction in the target. The procedure used by the experimenters is to find the energy weighted center of the set of γ's associated with an event. This was then taken to be an estimate of the intersection point of the incident hadron trajectory with the detector. The interaction was assumed to have occurred mid-deep in the target. This permitted the reconstruction of the emission angle of each gamma. The effect of these uncertainties is shown schematically, for a vertically incident hadron in Figure 2. We have assumed that the zenith angle of the incident hadron was determined with negligible error by the emulsion chamber, so only vertically incident hadrons were simulated. The program puts in these experimental conditions through the following steps:

    a) Energy error. The energy of each γ is randomly shifted by an energy error drawn from a gaussian with 25% width.

b) Angle error. The position of the intersection of the γ with the top of the detector is assumed to be determined with negligible error. The energy weighted center of the gamma cluster is computed:

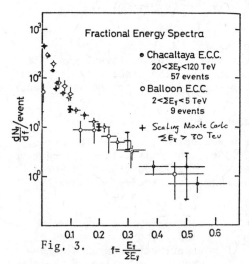

Fig. 3.

eg. $X = (\Sigma E_j)^{-1}(\Sigma_j E_j X_j)$. The reconstructed interaction point is then directly above the energy weighted center, in the middle of the target. This yields the reconstructed emission angle, θ_i, which is used in the computation of the invariant mass (see below).

4) Event selection. To be considered for further analysis, an event must have at least 4 gamma rays with energies greater than 200 Gev.

ENERGY DISTRIBUTION

At high energies the laboratory energy of a produced secondary, E_s is related to the laboratory incident energy E_i through Feynman X: $E_s \cong X E_i$. Hence the distribution of the quantity $f = E_\gamma/\Sigma E_\gamma$ might be expected to be an indicator of scaling violations in the fragmentation region. In Figure 3 are shown the f distributions for the simulation, and for several emulsion chamber experiments. The agreement might be interpreted as evidence for no scaling violations up to several hundred Tev. We find however, that the experiment is rather insensitive to changes in the X distribution. In Figure 4 are shown results for (1) the slope parameter B = 5.8 and 8.3, (2) B = 9 and 12, and (3) a cutoff for x > 0.1 in events with incident energy exceeding 50 Tev. These are compared to the Chacaltaya error envelope, shown as the shaded region. Only dramatic changes in the X distribution can be discerned by looking at the f distribution.

This insensitivity was pointed out by Wrotniak[12] et al. for A-jets, in which atmospheric cascading plays a role. We have found that even for C-jets, where cascading is not a problem, this lack of sensitivity remains. The origin of this effect is the large level of fluctuation in the amount of energy going into detected gammas. This tends to wash out the correlation between incident energy and ΣE_γ, and the gamma energies are effectively drawn at random.

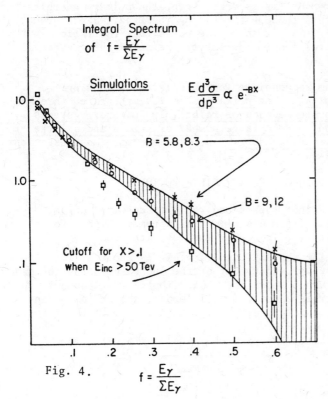

Fig. 4. $f = \dfrac{E_\gamma}{\Sigma E_\gamma}$

For random X's, the particular shape of the X distribution tends to make the f distribution insensitive to the X distribution. This is best illustrated with a simple (but incorrect) X distribution for gammas:

$$\frac{dn}{dx} = c\, e^{-Ax} \quad (4)$$

To draw X's at random from such a distribution, one chooses a random number, r, on (0,1). Then $X_j = -\ln(r_j)/A$ gives the correct distribution of X's. But it is evident that the quantity $E_j/\Sigma E_j = X_j/\Sigma X_j$ does not contain the parameter A at all, so its distribution is completely insensitive to the slope, hence to changes in it. When the more correct distribution is used (e^{-ax}/x), there is some A dependence, but it is very weak.

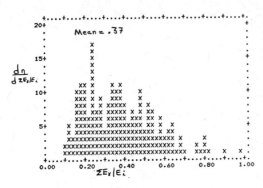

Fig. 5

Another result obtained from the simulation is the distribution of the ratio $\Sigma E_\gamma/E_{incident}$, shown in Figure 5. As mentioned above, the distribution is quite broad; the mean is .37, implying that the incident energy is on the average only 2.7 times ΣE_γ. An additional simulation, made with a single incident energy and a very low detection threshold gives a mean of .16, as expected. If our model is correct, then, the primary energies are somewhat lower than those estimated by the Brazil-Japan group, using a mean of .20 - .25. It should be noted that the ratio depends on the spectral index; the larger the index the greater is the value of the ratio.

The mechanism responsible for this is the following: Because of the steeply falling primary spectrum, the fixed threshold of 200 Gev favors selection of relatively low energy events which, by fluctuation, have a large number of gammas and a large fraction of the interaction energy. We will see more manifestations of this mechanism below.

PSEUDO-RAPIDITY DISTRIBUTIONS

We first note that the pseudo rapidity density ($dn/d\eta$) observed by the Brazil-Japan Group[13] is extremely high. For a set of events with $\Sigma E_\gamma \simeq 40$ Tev ($\bar{E}_i \simeq 100$ Tev) the near-central density is about 6. At the ISR, at 2 Tev, it is about 2.[14] The simulation shows, however, that most of the increase is due to the selection bias. Figure 6 shows the η distribution for two cuts on ΣE_γ; the peak density is about 4.6. (In model, the unbiased central η density stays constant at about 2). The role of the threshold is also evident here. For ΣE_γ in the range 10 - 15 Tev the threshold prevents the η density from reaching its maximum.

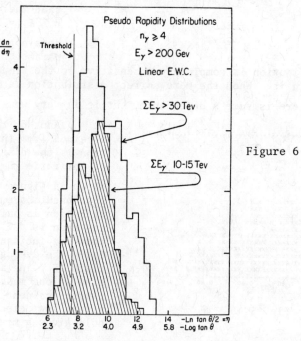

Figure 6

TRANSVERSE MOMENTUM DISTRIBUTION

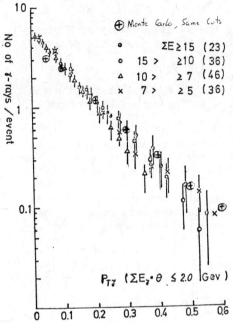

Figure 7 gives the $p_\perp$ distributions for the data and the simulation, subjected to the cut: $\theta_\gamma \cdot \Sigma E_\gamma < 2$ Gev. The agreement simply emphasizes the fact that large $p_\perp$ events are rare, even at these energies. A study of the incident energy dependence of the average $p_\perp$ might yield some new information. With more data this will be feasible. A diplot of mean $p_\perp$ vs. rapidity density is shown in Figure 8, for events with ΣE_γ greater than 20 Tev. (Unlike the Brazil-Japan group's treatment, we have not thrown out the most forward γ). As expected, there is no structure evident in this plot.

Fig. 7

INVARIANT MASS DISTRIBUTIONS

The Brazil-Japan Group has a special algorithm for finding the invariant mass of an object in the forward-most part of the rapidity distribution. It is assumed that the object decays isotropically, and that it decays partly into gammas.

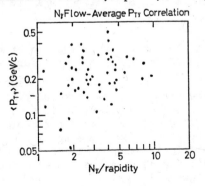

Fig. 8

The quantity
$$M_1^2 = \left(\sum_{j=1}^{N} E_j\right)\left(\sum_k E_k \theta_k^2\right) \tag{5}$$
is the small angle approximation to the invariant mass of N particles of energy E and emission angle θ. For an isotropic decay of an object with mass M_2, on the average
$$M_2 = (4/\pi) \sum_j p_{\perp j} \tag{6}$$
The algorithm is: (1) Order the gammas by angle. Start with the smallest; (2) Compute M_1 and M_2 after adding each gamma. When M_1 becomes ≥ M_2, stop. Then $M_\gamma \simeq M_1 \simeq M_2$. Various methods can be used to interpolate between the "overshooting" and the previous mass; differences among them are not very significant. If all the gammas are used up before $M_1 = M_2$, the Brazil-Japan group uses an "extrapolation method" to find M_γ. This is based on predicted average rates of increase of the mass; its current application is somewhat subjective, so we have chosen to eliminate from both the Monte Carlo and the real data those events which could require "extrapolation".

Some insight into the physical meaning of this procedure can be obtained by re-writing the mass expressions:
$$M_1^2 = \sum_{j,k} p_{\perp j} p_{\perp k} e^{-|y_j - y_k|} \tag{7}$$
where y is rapidity,
$$M_2^2 = \frac{16}{\pi^2} \sum_{j,k} p_{\perp j} p_{\perp k} \tag{8}$$

For the first gamma, $M_2 > M_1$. Now if there were a group of forward gammas in a dense cluster in y-space, then M_1 would not overtake M_2 until a large step in y were taken and so a large number of gammas contribute to M_γ. <u>Hence large M_γ's are due to clusters in rapidity.</u>

Some of the data presented as evidence for fireballs - the M_γ distribution for ΣE_j > 20 Tev, is shown in Figure 9, together with the simulation results. The gap in the distribution at 3 Gev is, of course, not reproduced by the simulation. The χ^2 probability for the hypothesis that the two distributions are drawn from one parent is about 3%. The existence of a threshold for such structure is suggested by a much lower average M_γ for events with ΣE_γ < 20 Tev. The simulation shows just the same trend. We think that large M_γ's in the simulated data are caused by accidental rapidity clusters. At lower ΣE_γ, as Figure 6 shows, the rapidity density does not become large enough to allow accidental clusters of sufficient local density to yield large M_γ's. The possible association of apparent fireballs with accidental clusters was pointed out by De Tar and Snider.(15)

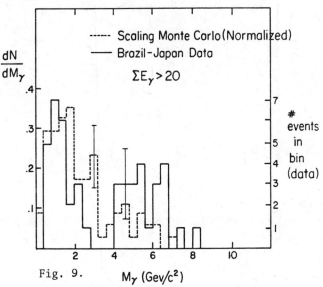

Fig. 9.

Through other simulation runs, we have found the M_γ distribution to be rather sensitive to the shape of the X distribution used. The one used here is by no means a perfect representation of the data at large X. In addition, possible incident pions were not included in this simulation. Quantitative agreement with the data, or lack thereof, then, is not highly significant. But some qualitative conclusions can be made: A scaling model of conventional particle physics <u>can</u> produce large M_γ's, with roughly the correct distribution, but the distribution <u>has no gap</u>.

We have studied the effect on the M_γ distribution of the energy resolution, multiple interactions in the target, and the angular resolution due to the finite target thickness. None is very significant. The effect of the use of the energy weighted center rather than the true incident particle trajectory (which is unknown) is not small. It will be discussed further below.

To further examine the effects of various selection biases, we have divided the simulation "data" of Figure 9 into "H quanta" and "SH quanta" by cutting at M_γ = 3 Gev, as the Brazil-Japan group has done. In Figure 10 the $p_\perp$ distribution of the γs in the two groups are shown. The two distributions are different, in a way similar to the real data. The reasons are that: (A)M_γ and $\bar{p}_\perp$ γ are necessarily correlated (see equation 8) and (B) M_γ is clearly correlated with the number of gammas used per event.

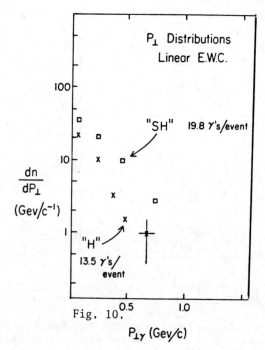

Fig. 10.

Figure 11 shows the integral spectrum of $E_\gamma/\Sigma E_\gamma$ for the two groups. Again, the same features are observed as in the real data: The SH spectrum falls more rapidly. The reason for this effect in the simulated data is as follows: H quanta correspond to events with a large gap in the forward y distribution, SH to events with no gap and more clustering. To achieve this, H quanta must have a few gammas at high energies (y's) and SH quanta must have more gammas at lower energies (y's).

Use of the Energy Weighted Center.

In the simulation, both the energy weighted center (EWC) of the gammas, and the intersection of the incident proton with the detector plane are known. We have computed distribution of the angle between true "beam axis" and a line from the interaction to the EWC. This angle is the deviation of the reconstructed beam axis from the true beam axis. With the EWC determined as described above, the mean value of this angle is 3×10^{-5} radians. Gammas in this region of pseudo rapidity (log tan θ < -4.5) have large fractional errors in $p_\perp$ and θ. As a test of the sensitivity of the M_γ distribution to this effect, we re-analyzed the simulated data using a "quadratic" EWC, i.e.

$$\bar{X} = \left(\sum_j E_j^2\right)^{-1} \sum_j X_j E_j^2 \qquad (9)$$

The new M_γ distribution was dramatically different; <u>the mean for $\Sigma E_\gamma > 20$ Tev, dropped from 2.36 to 1.23</u>. Both definitions of the EWC have been used: The Bristol-Bombay emulsion chamber experiment [16], at lower energies, used the quadratic formula for the EWC; the Brazil-Japan group uses the linear one. The large effect on the M_γ distribution made by simply changing methods of finding the EWC suggests that the M_γ distribution obtained in experiments of this type may have limited physical significance.

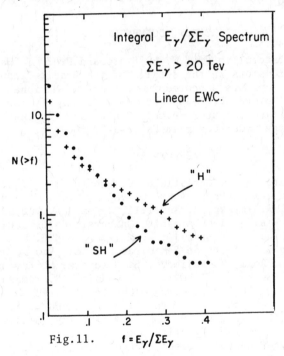

Fig.11. $f = E_\gamma / \Sigma E_\gamma$

CONCLUSIONS

We summarize here the conclusions we have drawn from this study.

(A) There is no evidence for scaling violations for incident energies up to several hundred Tev. But the sensitivity of experiment is low; large-scale changes in the X distribution would be necessary to cause an observable effect.

(B) The falling energy spectrum and fixed threshold favor selection of high multiplicity events. The mean primary energy is about 2.7 times ΣE_γ.

(C) There are no dramatic changes in the transverse momentum distribution at these energies.

(D) Except for the gap, the simulated M_γ distribution agrees roughly, but not exactly with the data. The qualitative differences between H and SH quanta seen in the $p_\perp$ and f distributions are mainly due to biases imposed by the selection criteria. Whether remaining quantitative differences are significant remains to be investigated.

ACKNOWLEDGEMENTS

We are grateful to Prof.Y. Fujimoto and Prof. S. Hasegawa for extensive discussions on the details of the Brazil-Japan experiment. This work was suppored in part by the National Science Foundation Grants No. PHY-77-01438 and PHY-77-08661, by the University of Maryland Computer Science Center and by the U. S. Department of Energy under Contract ER-78-5-02-5007.

REFERENCES

1. J. A. Chinellato, P. L. Christiano, C. Dobrigkeit, J. B. Filho, C. M. G. Lattes, M. Luksys, A. M. Oliveira, M. B. C. Santos, E. H. Shibuya, K. Tanaka, A. Turtelli, Jr. and A. Verlade (Instituto de Fisica Gleb Wataghin, Universidade Estadual de Campinas, Campinas, SP, Brazil); N. M. Amato and F. M. Oliveira Castro (Centro Brasileiro de Pesquisas Fisicas, Rio de Janeiro, RJ, Brazil); H. Aoki, Y. Fujimoto, S. Hasegawa, H. Kumano, K. Sawayanagi, H. Semba, T. Tabuki, M. Tamada and S. Yamashita (Science and Engineering Research Laboratory, Waseda University Shinjuku, Tokyo, Japan); N. Arata, T. Shibata and K. Yokoi (Department of Physics, Aoyama Gakuin University, Setagaya, Tokyo, Japah); A. Ohsawa (Institute for Cosmic Ray Research, The University of Tokyo, Tanashi, Tokyo, Japan), preceding paper and references therein.
2. W. Busza et al., Phys. Rev. Lett. 34, 383 (1975).
3. See, for example, F. E. Taylor et al., Phys. Rev. D14, 1217 (1976).
4. See, for example, Figure 5 in T. K. Gaisser, R. J. Protheroe and K. E. Turver, and T. J. L. McComb, Rev. Mod. Phys. 50, 864 (1978).

References - continued

5. M. Banner et al., Phys. Lett. 41B, 547 (1972).

6. M. G. Albrow et al., Phys. Lett. 42B, 279 (1972).

7. F. W. Busser et al., Phys. Lett 46B, 471 (1973).

8. J. R. Johnson et al., Fermilab Pub. 77/98 - Exp. 7100.284. The cross section is actually plotted against $X_R = 2E^*/\sqrt{s}$, but $X_R = X$ for $X >> 2\sqrt{\frac{p^2 + m^2}{s}}$.

9. The interaction lengths used were 86 gm/cm^2 for protons, and 134 gm/cm^2 for pions and kaons.

10. D. Cutts et al., Phys. Rev. Lett. 40, 141 (1978) and W. W. Toy, Ph. D. Thesis, M.I.T., 1978.

11. Brazil-Japan collaboration, Proc. 13th ICRC, Denver, 1973, p. 2210.

12. A. Krys, A. Tomszewski, J. A. Wrotniak, Proc. 15th ICRC, Plovdiv, 1977, Vol. 7, p. 254; A. Tomaszewski and J. A. Srotniak, Ibid, p. 255.

13. Brazil-Japan collaboration, paper contributed to International High Energy Physics Conference, Tokyo, 1978.

14. W. Thome et al., Nucl. Phys. B129, 365 (1977).

15. C. DeTar and D. Snider, Phys. Rev. Lett. 25, 410 (1970).

16. P. K. Malhotra et al., Nuovo Cimento 60, 404 (1965).

CHARACTERISTICS OF NUCLEAR INTERACTIONS AROUND 20 TEV
THROUGH THE ANALYSIS OF THE MONTE-CARLO SIMULATION

Hisahiko Sugimoto
The Sagami Institute of Technology, Fujisawa

Yoshihiro Sato
Faculty of Education, Utsunomiya University, Utsunomiya

ABSTRACT

Features of high energy nuclear interactions around 20 TeV are studied by means of a balloon-borne emulsion chamber with jet producer. Experimental results are compared with the monte-carlo simulation assuming the inclusive distributions of the parent π^0 (η^0) mesons and the observation conditions of secondary γ-rays detected with the chamber.

The quantities concerning the mean multiplicity, the angular distribution, the transverse momentum spectrum and the fractional energy spectrum of the produced particles were derived through the monte-carlo simulation. The results were compared with those of accelerator experiments. They show that the mean transverse momentum of π^0-mesons grows slowly with incident energy. They also indicate the appreciable increase of multiplicity and the scaling breakdown faster than logS in the central region, resulting the gradual steeping of the fractional energy spectrum in the same region. The distribution of the invariant mass of two γ-rays, $M_{\gamma\gamma}$, also indicate the appreciable production of η^0 mesons compared with the simulation.

1) INTRODUCTION

The recent direct measurements of charged multiplicity distribution for P-P collisions by means of a streamer chamber detector at CERN ISR energies by Darriulat et.al. show that particle densities are observed to rise in the central region as $\sqrt{s}$ increases. The multiplicity distribution in this region deviate from a poisson law, thus giving evidence for correlations of clustering of produced particles. In this situation it is important to investigate how the various properties of inelastic collisions vary at upper energies around 20 TeV. It is the main purpose of this experiment to study the gross features of nucleon-nucleus collisions around 20 TeV.

The emulsion chamber was flown at an atmospheric depth of 10 g/cm^2 for about 10 hours, 1974, from Sanriku Balloon Center, the Institute of Space and Aeronautical Science, University of Tokyo.

In this report, 15 events are selected for the present analysis. The results of the analysis were compared with the monte-carlo simulation assuming the inclusive distribution of parent π^0 mesons and the detection condition of secondary γ-rays observed with the emulsion chamber. We also assume that the observed γ-rays come from the decay of π^0 (η^0) mesons.

2) METHOD OF SIMULATION

The events used for the present analysis are all nucleon-nucleus

(target;acrylics) interactions in the producer layers of the chamber, and we must take into account the target effect for the simulation, although not so effective. We simulated the nucleon-nucleus(acrylics) interactions around 20 TeV as follows.
1) The observed secondary γ-rays are assumed to come from the decays of π^o and η^o mesons. The ratio of π^o to η^o mesons is assumed to be about 20 %. Consequently the contribution of $\eta^o - 2\gamma$ decay to all the secondary γ-rays is estimated to be 8 %, considering the branching ratio of $\eta^o \to 2\gamma$ decay mode.
2) The multiplicity of secondary π^o mesons was sampled successively until the multiplicity of simulated γ-rays through the filter of detection bias mentioned below reaches the multiplicity of the secondary γ-rays of the individual event observed in the experiment. In special case KNO type multiplicity distribution of mesons was adopted in order to investigate the event selection bias and to evaluate the target effect.
3) The inclusive distributions of π^o mesons are assumed as follows. The angular distribution of π^o mesons in form of $\log\tan\theta_{\pi^o}$ is
$$dN/d\log\tan\theta_{\pi^o} \propto \exp\left[-(M-M_0)^2/2\alpha\right]$$
where $M = \log\tan\theta_{\pi^o}$, $M_0 = 1.0 \times 10^{-2}$ rad., $\alpha = 3.0$
The transverse momentum distribution of π^o mesons is
$$dN/dP_{t\pi^o} \propto P_{t\pi^o}\exp(-P_{t\pi^o}/P_0)$$
where $P_0 = 0.22$, $\langle P_{t\pi^o}\rangle = 2 P_0 = 0.44$ GeV/c
The fractional energy spectrum of $\pi^o(\eta^o)$ mesons is
$$f\,dN/df \propto \exp(-f/f_0)$$
where $f = E_{\pi^o}/\Sigma E_{\pi^o}$, $f_0 = 0.25$
5) Obsevation condition of secondary γ-rays with emulsion chamber is very complicated, and it is assumed as follows.
$E_\gamma > 30$ GeV, $\theta_\gamma > 1\times 10^{-2}$ rad., $P_{t\gamma} > 0.05$ GeV/c

3) RESULT OF THE ANALYSIS AND COMPARISON WITH THE SIMULATION

3-1) On the Experimental Data

Events for the present analysis were selected by scanning dark spots on X-ray films in the lower chamber with naked eyes. Scanning efficiency for nuclear interactions with $\Sigma E_\gamma > 2.0$ TeV is considered to be close to unity. Nine events of $\Sigma E_\gamma > 2$ TeV were used for the analysis.

Energies of secondary γ-rays are estimated by counting the electron numbers of cascade showers in the radius of 50 μm. Energies are calibrated by exposing the same type of chamber to the electron beam of Fermi laboratory at the energies of 300 GeV and 100 GeV. The result of the energy calibration experiment is shown in fig. 1.

Detailed descriptions on the design of chamber and experimental procedure are shown in refs. 2).

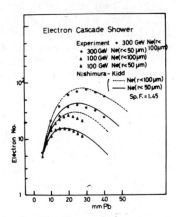

Fig.1 The transition curve of the cascade shower electrons. Fitting lines are by the caluculation of Nishimura and Kidd. Data are results of Fermi-lab. calibratio exp.

3-2) Multiplicity

The average charged multiplicity $\langle n_s \rangle$ and the dispersion $\langle n_s \rangle /D$ were found to be $\langle n_s \rangle = 31 \pm 9$ and $\langle n_s \rangle /D = 1.4 \pm 0.5$ at level of confidence of 75%. The effect of target nucleus on the average multiplicity is estimated using the multiplicity ratio $R_A = \langle n_s \rangle_{PA} / \langle n_s \rangle_{PP}$ for acrylics, where $\langle n_s \rangle_{PP}$ and $\langle n_s \rangle_{PA}$ are average charged multiplicity in P-P and P-A collisions. R_A was estimated to be 1.4, which is consistent with the result reported by Ogata et.al.[4]. Furthermore, the selection bias in the collection procedure of the events is estimated to be about 10%. Consequently the corrected average charged multiplicity in nucleon-nucleon collision is $\langle n_s \rangle = 19 \pm 5$ at the mean energy of about 20 TeV. Energy dependence of mean charged multiplicity is faster than logS, and it can be fitted with $(\log S)^2$ or more rapid dependence.

Mean charged multiplicity in central region of pseudo-rapidity interval of $|\eta| < 1.5$ is estimated to be $\langle n_s \rangle = 10.2 \pm 1.4$. Nuclear effect of target material was also removed by estimating the R ratio to be 1.6 in the central region of $|\eta| < 1.5$. Energy dependence of that is shown in fig.3. Data of 400 GeV/c proton beam with the emulsion chamber are consistent with the result of ISR experiment with streamer chamber. Data of balloon experiment indicate that the mean charged multiplicity in central region of $|\eta| < 1.5$ grows faster than logS. Although the error bar is large, it indicates the appreciable breakdown of scaling in central region.

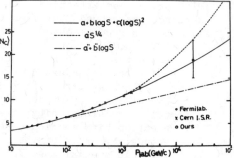

Fig. 2 The mean charged multiplicity as a function of primary energy. Dark spots are the data of Fermi-lab. Crosses are the data of ISR by Darriulat et.al.

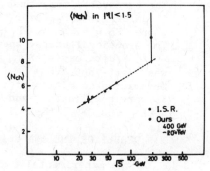

Fig. 3 Mean charged multiplicity in central region of $|\eta| < 1.5$ as a function of energy. Dark spots are the result of ISR by Darriulat et al. Circles are the result of ours with emulsion chamber experiments at the energy of 400 GeV/c proton beam and around 20 TeV with balloon chamber.

3-3) Angular Distribution

The $\log \tan \theta$ distribution of charged particles is shown in fig.4 and represented as the histogram in the figure. Full line represents the simulated $\log \tan \theta$ distribution, which is fitted with the Gaussian distribution of $\alpha = 3.0$.

The $\log \tan \theta$ distribution of γ-rays of the present experiment is shown in fig.5 and represented as the histogram in the figure.

Full line represents the result of simulation, assuming the same type of angular distribution of π^0 mesons as that of charged particles. The detection condition of γ-rays is also simulated as mentioned in 2). The data are well fitted with the simulation.

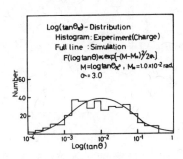
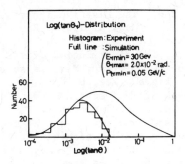

Fig.4 Histogram represents the logtanθ distribution of charged particles. Full line represents the simulation.

Fig.5 The logtan distribution of secondary γ-rays. The histogram represents the exp. data. The full line represents the simulation.

3-4) Transverse Momentum Distribution

Transverse momentum distribution of $\pi^0(\eta^0)$ mesons was simulated as $dN/dP_{t\pi^0} \propto P_{t\pi^0}\exp(-P_{t\pi^0}/P_0)$, where $P_0 = 0.22$ GeV/c. The transverse momentum distribution of γ-rays in integral form is shown in fig.6. The data of the present experiment are well represented by the simulation. The mean value of $P_{t\pi^0}$ is estimated to be $\langle P_{t\pi^0}\rangle = 2P_0 = 0.44$ GeV/c.

3-5) Fractional Energy Spectrum

The fractional energy spectrum of γ-rays, $E_\gamma / \Sigma E_\gamma$, is shown in fig.7 in integral form. Full line represents the result of simulation, assuming the inclusive distribution of parent $\pi^0(\eta)$ mesons, as $f(dN/dF) \propto \exp(-f/f_0)$ where $f = E_{\pi^0}/\Sigma E_{\pi^0}$, $f_0 = 0.25$. The event collection bias of $\Sigma E_\gamma > 2$ TeV as well as the detection condition of secondary γ-rays is included in the simulation as described in section 2. The experimental data are well fitted with the simulation parameter of $f_0 \approx 0.25$. If we define the quantity X_1 as $X_1 = E_{\pi^0}/E_0$, where E_0 means the primary energy, the X_1-distribution can be derived using the relation $X_1 = k_\gamma E_{\pi^0}/\Sigma E_{\pi^0} = k_\gamma f$. If we also assume that the mean value of the inelasticity released to γ-rays, $\langle k_\gamma \rangle$, is

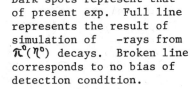

Fig.6 The integral transverse spectrum of γ-rays. Dark spots represent that of present exp. Full line represents the result of simulation of -rays from $\pi^0(\eta^0)$ decays. Broken line corresponds to no bias of detection condition.

around $0.2 < \langle k_T \rangle < 0.4$ according to the event collection bias, the inclenation parameter of the X_1-spectrum corresponds to $0.05 < X_0 < 0.1$ in the central region.

3-6) The Distribution of the Invariant Mass of Two γ-Rays

The distribution of the invariant mass of two γ-rays, $M_{\gamma\gamma}$, is shown as histogram in fig.8. The peak of π^0 mesons and slight peak of η^0 mesons are observed. Probability of observing both two γ-rays of η^0 mesons is smaller than that of π^0 mesons, due to the 38% branching ratio of $\eta^0 \to 2\gamma$ decay mode and the heavy mass of η^0 mesons. Background distribution of $M_{\gamma\gamma}$ is more fitted to the simulation of η^0 mixing than that of π^0 mesons only included.

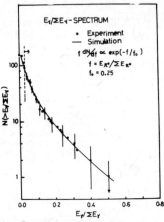

Fig.7 Fractional energy spectrum of γ-rays in integral form. Full line represents the simulation.

4) DISCUSSION AND SUMMARY

The two component model[5] of multi-particle interactions was confirmed by the many results of accelerator experiments such as PISA STONY BROOK[6]. The model is composed of the diffractive part and non-diffractive part. The diffractive part has a slow increase of multiplicity with energy and forward or backward assymetric angular distribution in the fragmentation region, while for the non-diffractive component, the multiplicity increases faster than logS and the the secondary particles are produced in the central region.

In the present analysis, the average multiplicity of charged particles in the central region of $|\eta| < 1.5$

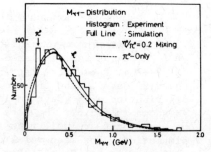

Fig.8 The distribution of $M_{\gamma\gamma}$. Full line represents the result of simulation including the contribution of $\eta - 2\gamma$ decay about 8%. Broken line represents that of only π^0 mesons.

was shown to be increasing as logS or faster already from the ISR energies, as in fig.3. It means the appreciable increase of multiplicity and the scaling breakdown in the central region. It corresponds to the gradual steepning of the X-spectrum in the central region, although the X-spectrum is not so sensitive as the increase of multiplicity. We derived the fractional energy spectrum in the range of $0.03 < E_\gamma/\Sigma E_\gamma < 0.3$, which corresponds to the X_1 range of $0.01 < X_1 < 0.1$ in the central region. The comparison with accelerator experiments of $P + P \to \gamma + X$[7)8)9] is shown in table 1, together with the result of transverse momentum. In the table 1, X-spectra of accelerator experiments were fitted with the distribution, $x(dN/dx) \propto \exp(-x/x_0)$ in the central region of $x < 0.1$, not in the fragmentation region. The inclenation of x-spectra in the table seem to be steeper than those

TABLE 1

Comparison with other experiments of P+P→γ+X

Expt.	Detector	Incident Energy (GeV)	$P_o, <P_{t\pi^o}>$ in GeV/c $dN/dP_t \propto P_t \exp(-P_t/P_o)$	X_o $dN/dX \propto \exp(-X/X_o)$
Fermilab Ref. 8	Bubble Chamber	205	$P_o = 0.20 \pm 0.02$ $<P_{t\pi^o}> = 0.40 \pm 0.04$	0.08
Fermilab Ref. 9	Bubble Chamber	303		0.075
ISR Ref. 7	lead-glass	1500	$<P_{t\pi^o}> = 0.322 \pm 0.016$	0.083
Ours	emulsion chamber	20,000	$P_o = 0.22 \pm 0.01$ $<P_{t\pi^o}> = 0.44 \pm 0.02$	0.075 ± 0.025

of many other experiments, which insist on the scaling behavior of x-spectrum. It is due to the fitted range of x-spectrum, and also due to the multiplicity increase in the central region. In our case we observed fractional energy spectrum in the central region, not in the whole region, because of the event selection bias. In the fragmentation region x-spectrum will be flatter than the observed spectrum. In the two component model, it means that the non-diffractive component is increasing considerably with energy, while the diffractive component is constant or decreasing with energy. In fact, there are some experimental results[10] which insist on the slight decrease of diffractive component with energy.

Thus, the conclusions are as follows from the analysis of nuclear interactions at the energy around 20 TeV.
1) The mean multiplicity of charged particles in nucleon-nucleon collisions in the central region of $|\eta|<1.5$ is estimated to be $<n_S> = 10.2 \pm 1.4$, which together with the result of ISR experiment indicates the appreciable increase of multiplicity and scaling breakdown faster than log S in the central region.

2) The transverse momentum spectrum of π° mesons is represented as $dN/dP_{t\pi^\circ} \propto P_{t\pi^\circ} \exp(-P_{t\pi^\circ}/P_0)$, where where $P_0=0.22$ and $<P_{t\pi^\circ}>=2P_0=0.44\pm0.02$ GeV/c. The result shows the slow increase of the transverse momentum of π° mesons with the incident energy.
3) The fractional energy spectrum of γ-rays, $E_\gamma/\Sigma E_\gamma$, in the central region of $0.03<E_\gamma/\Sigma E_\gamma<0.3$ is well fitted with the distribution that the fractional energy spectrum of parent π° mesons is $f(dN/df) \exp(-f/f_0)$, where $f=E_{\pi^\circ}/\Sigma E_{\pi^\circ}$, $f_0=0.25$. The range of f corresponds to the range of X_1, $0.01<X_1<0.1$, in the central region. The spectrum seems to be steeper compared with that of central region as well as that of fragmentation region at lower energies. It corresponds to the multiplicity increase in the central region.
4) The distribution of the invariant mass of two γ-rays, $M_{\gamma\gamma}$, was compared with the simulation and the fair amount of production of η° mesons was indicated. It is consistent with the correlation of clustering of produced particles reported by accelerator exp.

ACKNOWLEDGMENTS

The authors are grateful to Professor J. Nishimura for helpful discussions and also to all the members of Sanriku Balloon Center, University of Tokyo, for the successful flight. Our thanks are due to Drs. T. Saito, M. Noma and T. Matsubayashi for their helpful contribution on the experiment. The simulation calculation was made with use of a computer, TOSBAC 3400, of the Institute for Nuclear Study, University of Tokyo.

REFERENCES

1) P. Darriulat et al., Nuclear Phys. B129 (1977) 365.
2) Y. Sato et al., J. Phys. Soc. Japan vol. 41 no. 6 (1976) 1821.
3) J. Nishimura, Prog. theor. Phys. Suppl. 32 (1964) 72.
4) F. Fumuro et al., Proc. Int. Cosmic Ray Conf., Plovdiv (1977) HE59.
5) K. Wilson, Cornell Preprint, CLNS-131 (1970).
6) L. Foa et al., Phys. Letters 44B (1973) 119.
7) G. Neuhofer et al., Phys. Letters 37B (1971) 438.
8) G. Charlton et al., Phys. Rev. Letters 29 (1972) 1759.
9) F. T. Dao et al., Phys. Rev. Letters 30 (1973) 1151.
10) Y. Takahashi et al., ICR-Report 67-78-11, Tokyo.

SOME 10-100 TeV EVENTS

F. Fumuro
Faculty of Science, Kwansei Gakuin University,
R. Ihara
Faculty of Science, Konan University,
J. Iwai
Science and Engineering Research Laboratory Waseda University,
T. Ogata
Institute for Cosmic Ray Research University of Tokyo,
I. Ohta and Y. Sato
The Faculty of Education, Utsunomiya University,
H. Sugimoto
The Sagami Institute of Technology,
Y. Takahashi
Faculty of Engineering Science, Osaka University.

ABSTRACT

The emulsion chambers with producing layers were exposed to the cosmic rays at aeroplane altitude. A total of 12 events produced in the producing layers with the energy 10-100 TeV were analyzed. An analysis was made for production of energetic particles and unstable particles in high energy hadron nucleus interactions.

INTRODUCTION

Recently the study of high energy hadron nucleus interaction has attracted special interest, because it may yield direct information on the space time development of the hadronic production process. A number of experimental results on hadron nucleus interactions at accelerator energies have been obtained and analyzed by many authors.
On the other hand, the experimental results at cosmic ray energies are not yet well established at present. To make more systematic studies of hadron nucleus interactions at energies 10-100 TeV, emulsion chamber experiments at aeroplane altitude are continued since 1972[1]. In this paper, the results of analysis of hadron nucleus interactions observed in the emulsion chambers are presented. A report is made also on the evidence for the existence of new unstable particles produced by super high energy (>100 TeV) hadron nucleus interactions.

EXPERIMENTAL PROCEDURE AND RESULTS

The emulsion chamber used in this experiment consists of two parts, upper chamber with producing layers in which jet showers are produced and lower chamber with shower detector in which cascade showers are developed.

The lower chamber has 8 radiation lengths of lead. In the producing layers, three kinds of metal plates, such as aluminum, copper and lead are inserted between the nuclear emulsion plates so that nuclear interactions take place in known elements. The total thickness of the producing layers is 0.12 nuclear interaction lengths. An emulsion chamber of this type has the dimensions, 20 cm x 25 cm x 10 cm and weighs about 25 kg.

The exposures were performed using a Jet Cargo aeroplane of Japan Air Lines, at altitude of 260 gr/cm^2. In a total exposure of 480 m^2 hours, 12 clean jet showers originating in the producing layers were found. Of the twelve jets, five occurred in the glass base on which is coated nuclear emulsion, one in the copper plate and six in the lead plate.

In all these jet showers, not only charged secondary particles but also many cascade cores and electron pairs produced by gamma rays from the decay of π° mesons were observed well separately from one another. Measurements were made on the energies of gamma rays and emission angles of both gamma rays and charged particles. The results are listed in Table 1.

Their individual gamma ray energy distributions are presented in Fig. 1.

One can see from Fig. 1 that characteristic events with an energetic gamma ray having the energy about one order greater than the rest are observed in hadron light nucleus interactions.

Table 1

Event	incident particle	target	n_s	n_γ	ΣE_γ (TeV)
J 2		Pb	6	3	0.13
J 8		Pb	21	7	1.14
KG 1	P	Glass	16	7	1.80
KG 3	neutral	Glass	16	8	1.32
KG 4		Cu	13	2	0.74
KG 7	P	Glass	35	10	30
KG 8	neutral	Glass	9	4	0.95
KG 9		P	28	6	1.69
j 1		P	54	9	2.53
j 2		P	10	4	1.05
k 1	P	Glass	9	2	1.43
l 1		P	43	7	2.47

The Japan Brazil collaboration reported the observation of same type events, which they named "Torpedo".[2] The physical meaning of the torpedo, however, is not yet clear.

Torpedo type events.

Since the torpedo type events occur more frequently than is expected from the thermodynamical model, it is difficult to consider that such energetic gamma rays among the secondary particles are emitted thermodynamically from a fire ball. To make the production process of torpedo more clear, we examined the fractional energy including the charged secondary particles, $E\pi_1^°/E_{rad}$, where

$E\pi_1^o$ is the highest energy among the secondary π^o mesons and Erad the total radiated energy. Erad was derived by using $E\gamma$ and $E\pi^{\pm}$ which were calculated under the assumption of constancy of transverse momentum, (0.4 GeV/c). The fractional energy distributions are shown in Fig. 2, where open circles are data for hadron light nucleus (A<30) interactions and black points for hadron heavy nucleus (A=207) interactions. Even in this small sample of events, one sees that there are different types of nuclear interaction. In some cases a single π^o meson carries off more than half of the total radiated energy in an interaction.

Those torpedo type events are observed frequently in the hadron light nucleus interactions, but very rarely in hadron heavy nucleus interactions. In other cases, many π^o mesons of comparable energy are emitted.

Assume that the production process of the energetic particles is as follows: The primary nucleon is ex-

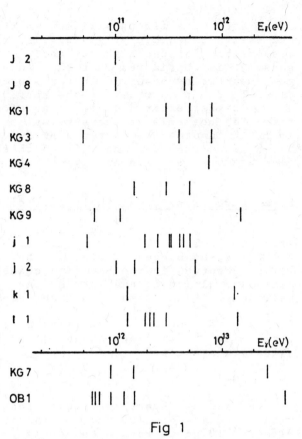

Fig 1

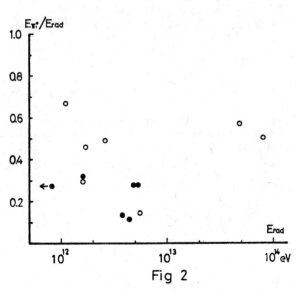

Fig 2

cited after the first collision in the target nucleus, and decays immediately, giving rise to pions which are responsible for the highest energy part of the $\pi°$ mesons. It is considered that the features of the energetic particles observed in torpedo type events reflect directly the properties of the survival nucleon, which may be a higher excited state. Therefore, very useful information not only on the survival nucleon itself, but also on the fundamental particle production process could be obtained from the analysis of those energetic particles produced by high energy nucleon nucleus interactions.

Successive decay of unstable particles.

Recently, two superhigh energy jet showers with energy $\Sigma E_\gamma > 30$ TeV, which is equivalent to primary energy > 100 TeV, have been obtained. One of these, Event KG-7, was observed in the emulsion chamber exposed at an aeroplane altitude, and another one, Event OB-1, at a balloon altitude. Among the many secondary particles produced in jet showers, a number of unstable decay particles were found. The detailed analysis on these particles is now in progress and so here preliminary results are presented.

Event KG-7.

Jet shower, Event KG-7, was found in the glass base with thickness of 1100 μm. 35 charged secondary particles and 10 gamma rays with energy greater than 0.2 TeV were observed. The total energy of the observed gamma rays was about 30 TeV. X, Y and Z projections of characteristic tracks A,B,C,D and E are shown in Fig. 3.

Tracks A,B and C.

As seen from the three dimensional view, tracks A,B and C are closely correlated with one another. The convergent point of these tracks was examined by measuring the relative distances among the tracks. The distance of the convergent point from the vertex "O", where the primary particle interacts, is determined to be 410 μm. Fig. 4 shows a schematic view of the event near the vertex. Momenta of the particles A,B and C were measured by the relative scattering method, and the results are given in Table 2 together with these emission angles and transverse momenta.

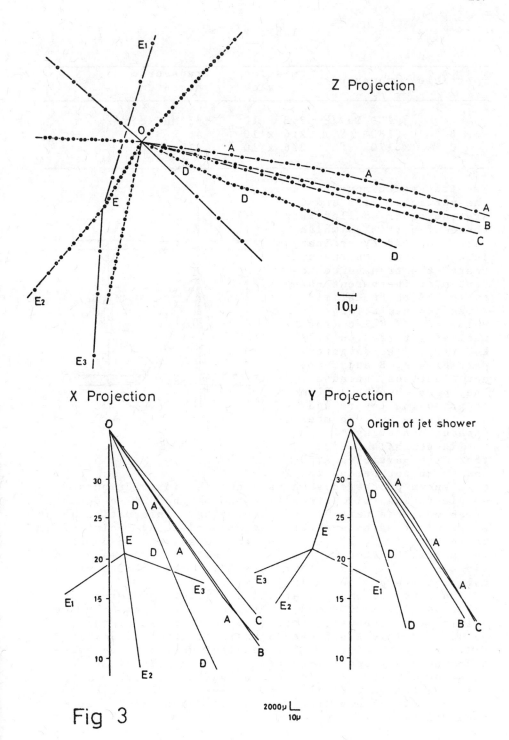

Fig 3

Table 2

particle	momentum	emission angle	transverse momentum
A	230 GeV/c	2.8×10^{-4}	65 MeV/c
B	180 "	2.6×10^{-4}	46 "
C	170 "	3.6×10^{-4}	61 "

The transverse momenta of the particles A, B and C are one order smaller than those of particles which were produced by ordinary interactions. On the contrary, the transverse momentum of the parent particle "X" to the primary direction has a large value of 2.7 GeV/c (for nuclear interaction 2.7/K_γGeV/c). The daughter particles A, B and C are not electrons, because they pass through 4.8 cm in producing layers and 4.5 c.u in lower chamber without cascading.

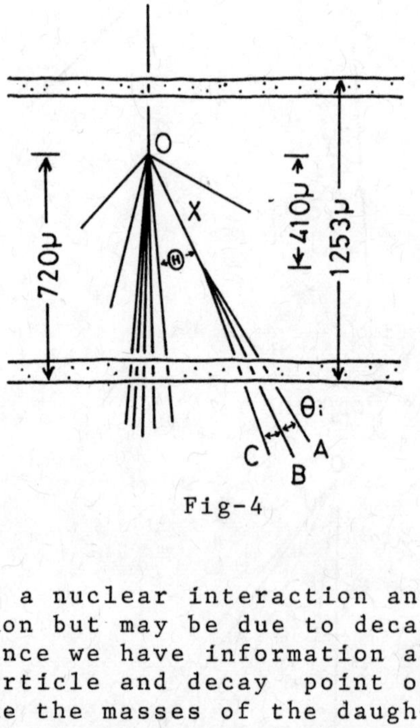

Fig-4

These facts indicate that the particles A, B and C can not be due to well known processes such as a nuclear interaction and an electromagnetic interaction but may be due to decay of an unstable particle. Since we have information about momentum of each daughter particle and decay point of parent particle, if we assume the masses of the daughter particles then we are able to estimate the mass and lifetime of the parent particle "X". The results of calculations are shown in Table 3.

One should remark that the particle A suffered deflection, 4.9×10^{-4} rad, at the point 2.05 cm from the origin of jet shower, after 1.45 cm flight from this point, particle A suffered again deflection, 3.9×10^{-4} rad. In spite of careful scanning, no particles emitted from these scattering points were observed. Determinations of the lifetime of the particles A and A' by assuming the mass of these particles to be 3.6 GeV and 2.5 GeV, give the values of 1×10^{-12} sec and 5×10^{-13} respectively.

Table 3

daughter particle assumed mass			parent particle	
A	B	C	mass GeV	lifetime 10^{-14} sec
$\pi(0.14)$	$\pi(0.14)$	$\pi(0.14)$	0.5	0.1
$c_2^*(3.6)$	$P(0.94)$	$\pi(0.14)$	6.0	1.4
$c_2^*(3.6)$	$\Sigma(1.19)$	$\Sigma(1.19)$	6.5	1.5

Track D

The transverse momentum of particle D is estimated from its emission angle and momentum as 2.6 GeV/c, which is as large as the parent particle "X" has. As seen from Fig. 3, this particle also suffered double deflections at the point 2.10 cm from the origin of jet and 0.96 cm away from the first scattering point. The probability that the double deflections were caused by hadron nucleus elastic scattering is as small as 3×10^{-4}, so it is possible to attribute these deflections to successive decay of an unstable particle. Assuming the mass of particles D and D' also to be 3.6 GeV and 2.5 GeV, the lifetime of these particles are estimated as 3×10^{-13} sec and 9×10^{-14} sec respectively.

Track E

Particle E splits into three charged particles E1, E2 and E3 at a point in the lead plate with thickness 500 μm, 2.59 cm from the origin of jet. The probability that a nuclear interaction occurred in this distance is $\sim 10^{-1}$. The transverse momenta of the particles E1 and E3 have normal values 250 MeV/c and 230 MeV/c. It is therefore plausible to consider that these charged particles were created by hadron lead nucleus interaction.

Generally in high energy hadron heavy nucleus interactions, however, a large number of secondary particles involving heavy evaporation tracks are emitted within a wide angle. In this case, except for these three charged particles, neither any gamma rays nor heavy evaporation tracks coming from the vertex are observed, so there is some possibility that they are decay products of the unstable charged particle "E".

New kind of successive decays and multibody decays were observed in event KG-7. As mentioned above, these phenomena cannot be due to nuclear interactions and electromagnetic interactions, so these deflections observed in the producing layers may be due to decay of new short lived particles with higher quantum number.

Event OB-1.

One of jet shower event OB-1 has been observed in an emulsion chamber exposed by a balloon which was launched from Mildura Balloon Center for the collaboration experiment of Japan Australian emulsion group. This event has 98 charged secondary particles and 31 gamma rays with a total energy of ΣE_γ about 40 TeV. The energy spectrum and logtanθ plot are shown in Fig. 5, for both gamma rays and charged particles not including the leading particle.

Three dimensional views of the characteristic particles are shown in Fig. 6a, b and c, where the dotted lines indicate the flight path of the gamma rays and the chain line the primary direction. The features of this event are as follows:

1) The primary energy of this event is estimated >100 TeV.
2) Energy spectrum of secondary particles shows a typical torpedo type.
3) A large number of secondary particles are created by proton light nucleus interaction.
4) Three Vee Tracks (AB, DF and JK) are observed among the secondary particles.

This event is of particular interest because of enormous multiplicity and typical torpedo characteristic in a superhigh energy nucleon light nucleus interaction.

Vee Events

Tracks A and B

In the sensitive layer of nuclear emulsion plate No. 37, there appeared Vee tracks A and B with relatively large opening angle 5.1×10^{-3} rad, and these tracks were emitted at large angles 1.0×10^{-2} rad with respect to the primary direction. The vertex "a" of these tracks was estimated to be 2.17 cm downward from the origin of jet. From this vertex a gamma ray with energy 500 GeV was emitted.

Tracks F and L

The vertex "c" of this event was found in the same layer in which Vee tracks A and B appeared, and the distance of this vertex from the origin of jet was estimated to be 2.15 cm. Opening angle between the tracks was relatively large, 4.4×10^{-3} rad. As seen from the three dimensional view, one of the Vee tracks F suffered deflection, 1.4×10^{-3} rad, at the point "d" traveling 3.63

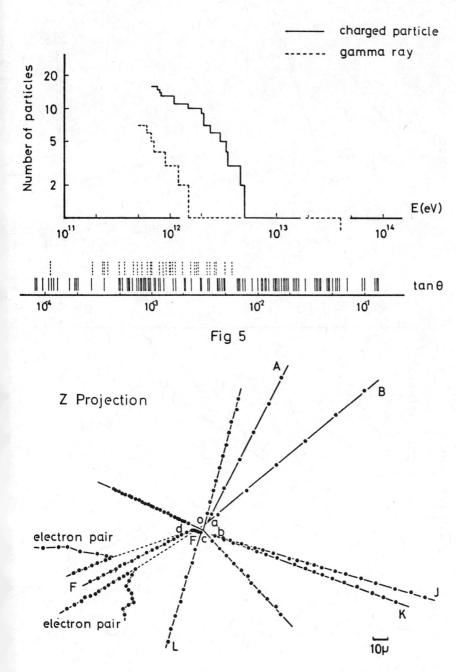

Fig 5

Fig 6c

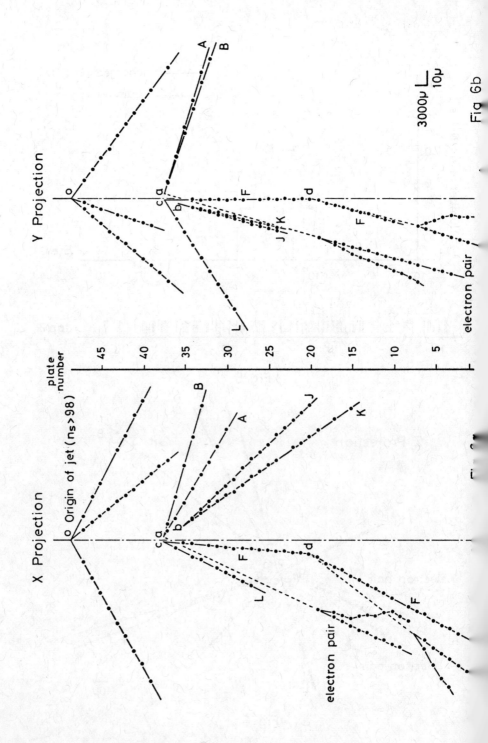

Fig 6b

cm from the vertex c and from this point a gamma ray with energy about 100 GeV was emitted. For reasons of momentum conservation, at least one more neutral particle must be emitted from the scattering point d.

Tracks J and K

This Vee event consists of two tracks J and K starting from a vertex "b" in the metaacryl base of No. 35 nuclear emulsion plate. The opening angle of Vee tracks was measured to be 7.6×10^{-4} rad and the emission angles of these tracks were 2.9×10^{-3} rad and 2.1×10^{-3} rad respectively. The distance of the vertex from the origin of jet was 2.79 cm. Though, each track was followed 5.8 cm in length in producing layer, neither electron cascades nor electron pairs due to bremsstrahlung along its path were found.

Our preliminary measurements appear to indicate that three Vee events observed in the producing layers are due to the decay of unstable neutral particles, which may be also decay products of a massive neutral particle.

DISCUSSIONS

As seen above, in superhigh energy nuclear interactions a number of deflected tracks, Vee tracks and multisplit tracks were observed among the many emitted particles. Some example of such complicated phenomena were already reported by Niu et al.[3] Though it is difficult to make a definite interpretation for those phenomena, the results of preliminary analysis seem to suggest that most of those are due to successive decays and multibody decays of the unstable particles which may be charmed baryons with higher charm quantum number or new kind of hadrons with heavy quarks such as bottom quark and top quark. Because of large amount of available energy, it is expected that the new higher mass states could reveal themselves at the superhigh energy 100 TeV region.

ACKNOWLEDGMENTS

We would like to thank the members of Japan Air Lines for making exposures at aeroplane altitude, and the staff of Mildura Balloon Launching Station who carried out the balloon flight.

REFERENCES

1. F. Fumuro, R. Ihara, T. Ogata and Y. Yukimasa, Proc. of Int. Cosmic Ray Sympo. on High Energy Phenomena, Tokyo, 1974, 179.

2. Japan-Brasil collaboration, Proc. of 12th Int. Cosmic Ray Conf., Hobart, 7 (1971) 2786.
3. S. Kuramata, K. Niu, K. Niwa, K. Hoshino, E. Mikumo, Y. Maeda, T. Nagai and S. Tasaka, Proc. of 13th Int. Cosmic Ray Conf., Denver, 3 (1973) 2239. See also the paper of K. Niu in this volume.

BARYON PAIR PRODUCTION WITH LARGE DECAY Q-VALUE

Brasil-Japan Emulsion Chamber Collaboration

J.Bellandi Filho, J.L.Cardoso Jr., J.A.Chinellato, C.Dobrigkeit, C.M.G.Lattes, M.Menon, C.E.Navia O., A.M.Oliveira, W.A.Rodrigues Jr., M.B.C.Santos, E.Silva, E.H.Shibuya K.Tanaka, A.Turtelli Jr.
Instituto de Física Gleb Wataghin - Universidade Estadual de Campinas, Campinas, São Paulo, Brasil

N.M.Amato, F.M.Oliveira Castro
Centro Brasileiro de Pesquisas Físicas, Rio de Janeiro, RJ, Brasil

H.Aoki, Y.Fujimoto, S.Hasegawa, H.Kumano, K.Sawayanagi, H.Semba, T.Tabuki, M.Tamada, S.Yamashita
Science and Engineering Research Laboratory, Waseda University, Shinjuku, Tokyo, Japan

N.Arata, T.Shibata, K.Yokoi
Department of Physics, Aoyama Gakuin University, Setagaya Tokyo, Japan

A.Ohsawa
Institute for Cosmic-Ray Research, University of Tokyo, Tanashi, Tokyo, Japan

ABSTRACT

Events of a new type have been observed by the Chacaltaya Emulsion Chamber Experiment[1,2] which can consistently be interpreted as "baryon" pair decay of a heavy intermediate state of rest mass 20-30 GeV/c^2.

ATMOSPHERIC EVENTS OF "BINOCULAR" TYPE

Tentatively we give the morphological name of "binocular events", to the analyzed events, since each consists of two, and only two, γ-ray (and/ or electron) groups with parallel incidence, making in an emulsion chamber two clusters of electron showers separated by a large distance.

The clusters have a lateral spread of order 1 cm and their appearance suggests that each has a common origin, being the result of a nuclear interaction in the atmosphere above the chamber. The pair separation distance is large compared to the cluster size, sometimes as much as nearly 20 cm.

Table I summarizes the characteristics of "binocu-

lar" events we have analysed. Column 1 (C-1) gives the
event name; C-2 and C-3 the visible energies E_1 and E_2

Table I. Characteristics of Binocular Events

event	E_1 (TeV)	E_2 (TeV)	$<r_1>$ (cm)	$<r_2>$ (cm)	R_{12} (cm)	X_{12} (TeV.cm)
16-s180	160.2	40.5	1.13	0.86	28.7	2312
17-s204	104.1	42.7	0.53	0.39	27.0	1801
18-A261	55.1	33.8	0.62	3.99	26.6	1150
18-A230F2	92.4	19.4	0.43	2.9	19.4	815.3
18-A234	220.6	60.0	0.68	*	6.2	710.0
18-s197F1	31.0 (45.0)	19.6	1.03	1.73	25.2	621.2
18-s112F1	77.0	27.0 (20.0)	0.68	1.02	12.8	583.6
18-s124	32.7	15.3	1.35	0.11	20.8	465.2
18-A217	33.0	22.0	*	*	17.1	460.8
18-s112F2	89.8	27.3 (4.3)	0.19	0.77	8.43	417.4
15-s27	20.1	18.8	0.52	0.59	20.3	394.4
18-A241F2	59.9	39.6	0.71	0.22	5.42	264.0
18-A310	37.7	20.4	0.14	0.07	9.38	260.2
18-A228F2	83.5	25.5	0.11	0.43	2.80	109.0

*   Single
()   Energy of Pb-jet contained in a cluster

of the constituent clusters; C-4 and C-5 the mean lateral spread of the clusters, $<r_1>$ and $<r_2>$, defined as

$$<r_i> = \sum E_{\gamma j} r_j / \sum E_{\gamma j} \qquad (1)$$

where r_j is the distance of the j-th shower measured from the energy weighted center of the cluster i to which it belongs. C-6 gives the distance R_{12} between the respective energy-weighted center of the two clusters.

The events analysed have been selected under the criteria $R_{12} / <r>_{sup} > 5$ and $X_{12} = (E_1 E_2)^{1/2} \cdot R_{12} > 100$ TeV cm, where $<r>_{sup}$ means the larger of $<r_1>$ and $<r_2>$.

It seems now that no successful interpretation of this type of event can be made within the present accepted knowledge of nuclear and electromagnetic interactions unless an accumulation of extreme fluctuations are assumed in every step of shower formation, which seems unreasonable. The expected frequency of chance coincidence in arrival of two independent clusters with almost parallel incidence was estimated to be negligibly small compared to the observed number of events.

GEMINION HYPOTHESIS

We discuss now the nature of the pair particles which gave rise to the clusters. Each cluster has a structure similar to that of a usual γ-ray family of the same energy region. Usual γ-ray families are due to atmospheric nuclear interactions and we have analysed several hundreds events of this kind. The comparison of the clusters analyzed with the usual γ-ray families shows that their production heights are not very large ($\gtrsim$ 500 m). This allows one to trace back each cluster to its original atmospheric interaction.

Let us now assume a hypothetical particle "geminion" with mass M_{gem} which decays into two "baryons" within very short life time. When a "geminion" is produced high in the atmosphere and the two baryons of its decay product travel over a long distance until making their secondary interactions, we expect to observe atmospheric events of such "binocular" type.

Under the geminion hypothesis, we have the following kinematical relation for the production height H of a geminion and the distance R_{12} between the two clusters.

$$H M_{gem} c^2 = R_{12} (E_{N1} E_{N2})^{1/2} \qquad (2)$$

where E_{N1} and E_{N2} are the energy of two baryons. We now replace E_{N1} and E_{N2} with the observed cluster energy E_1 and E_2, and construct the quantity χ_{12} defined as,

$$\chi_{12} = R_{12}(E_1 E_2)^{1/2} = H k_\gamma M_{gem} c^2 \qquad (3)$$

k_γ is the γ-ray inelasticity. C-7 of Table I shows the value of χ_{12} for the analysed events.

The expected distribution of χ_{12} can be derived from the altitude variation of production rate of "geminion" and its probability to produce "binocular" events. Since the height of secondary interaction of "baryons" is confined near the chamber, we have the approximate expression,

$$f(\chi_{12}) = \text{const.} \exp(-\chi_{12}/\lambda k_\gamma M_{gem} c^2) \qquad (4)$$

where λ is the nuclear mean free path. Fig. 1 gives

the observed distribution of χ_{12} which agrees with the expected one (eq.(4)). Taking notice of the selection criterion of $\chi_{12} > 100$ TeV cm, we are able to estimate the mass of geminion from the average value $<\chi_{12}>$ as,

$$k_\gamma M_{gem} c^2 = (<\chi_{12}> - 100 \text{ TeV cm})/\lambda \quad (5)$$
$$= 5.2 \pm 1.4 \text{ GeV}$$

If we put $k_\gamma = 0.2$, we have the geminion mass as, $M_{gem} \sim 26$ GeV.

We now show that the hypothesis that all "binocular" events are produced through the same intermediate state with constant mass $k_\gamma M_{gem} \sim 5$ GeV/c^2 is compatible with our data. Indeed, the energy and angular distribution of the decay products in the Geminion rest frame is

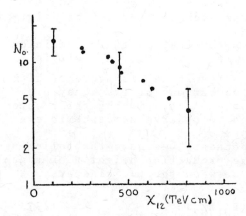

FIG. 1 χ_{12} Integral distribution.

FIG. 2 Fractionary energy distribution.

$$f(E'^*, \theta^*) = \frac{1}{4\pi} \cdot$$
$$\cdot \delta(E'^* - M_{gem}) \quad (6)$$

since it is a reasonable assumption to suppose (in our sample of events) that Geminion is produced unpolarized. Then the energy distribution in the laboratory frame will be

$$g(E') dE' = dE'/E_{tot} \quad (7)$$

where $E_{tot} = \gamma M_{gem}$ is the Geminion total energy in the laboratory frame. Now we don't measure E' and E_{tot}, but E and $\sum E_\gamma$, the "visible energy". Supposing that k_γ

the mean inelasticity is constant we end with

$$\frac{dN}{N_o} (E / \sum E_\gamma) = dx \qquad (8)$$

where $x = E / \sum E_\gamma$, $\sum E_\gamma = (E_1 + E_2)$.

Our data for this distribution is shown in Fig.2, and it is consistent with the expected distribution (eq. (8)). We observe a cut-off at small and large x, due to the detection threshold.

THE EVENT "CASTOR-POLLUX"

In the course of hadron bundle scanning looking for "Centauro" type events, we found a high energy double spot event on X-ray films in the lower part of the emulsion chamber. The distance between the two spots is about 1.6mm, and the microscopic observation of both in nuclear plates shows core structure consistent with the occurence of nuclear interactions within the chamber.

One of the showers, "Castor", has a multiple core structure and becomes observable only below 6 c.u. of lead and therefore is due to a local nuclear interaction in the lead plate of the lower chamber.

"Pollux", on the other hand, comes into observation from the out set, at 3 c.u. of lead onward, in the lower chamber. It contains a few associated small showers separated by about 100 μm from the main core, and, furthermore, a small Pb-jet shower is observed near the main core under 12 c.u. of lead. From these characteristics, it is identified as result of a nuclear interaction close to the bottom of the upper chamber.

The visible energies of "Castor" and "Pollux" are determined as 11.0 ± 1.0 TeV and 8.0 ± 1.0 TeV respectively.

The relative opening angle between the showers is determined comparing the relative distance on the nuclear plates under 6, 8 and 10 c.u.. The result is shown in Fig 3, where the abscissa gives the geometrical depth of the nuclear plate measured from the top of the lower chamber and the ordinate the relative distance between the showers. The straight line is the least-square fit to the measurements and yields the height of their common origin as $H = 3.0 \pm 0.2$ m above the chamber. Incidentally, the chamber's roof of small thickness and its support wood-beam is located at that height. The dotted line in Fig.3 represents the "would-be" rate of change with depth of the relative distances if the showers were to come from the central level of the target layer, and it is in ob-

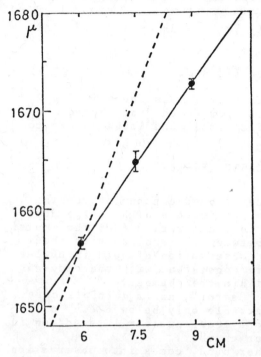

Fig. 3 Distance between the two showers at different depths

vious disagreement with the data.

Fig.4 illustrates the overall view of the event thus reconstructed. We may summarize the main characteristics of the "Castor-Pollux" event as follows: at least two "baryons" are produced at 3.0 m above the lower chamber. One of them makes a nuclear collision near the bottom of the upper chamber and the other in the lower chamber. The total material thickness of the whole chamber is about 1.4 nuclear mean free path. The transverse momentum of each shower with respect to the direction of the energy weithted center is 2.56 ± 0.40 GeV/c. Assuming the γ-ray inelasticity, k_γ, of both interactions as 0.2, we arrive at the conclusion that the "baryons" are emitted with transverse momenta of (12-13) GeV/c.

If we assume the two "baryons" here are from decay of a "geminion", we have $k_\gamma M_{gem} c^2$ = 5.0 ± 0.8 GeV in this "Castor-Pollux" case.

CONCLUSION

The "geminion" hypothesis becomes now plausible after seeing striking morphological resemblance of atmospheric "binocular" events and the event "Castor-Pollux", and both giving almost the same estimated mass value, $k_\gamma M_{gem}$ ~ 5 GeV/c^2. With k_γ about 0.2, the "geminion" has mass of (20-30) GeV/c^2 and goes into two "baryons". With the term "baryon", we mean a hadron which does not undergo rapid γ-decays. We believe that nucleons and anti-nucleons would be most probable candidate for "baryons".

The rest mass of the "geminion", (20-30) GeV/c^2, stands for the characteristic unit of energy liberation at violent nuclear collision, because we already know the existence of such an energy unit called "SH-quantum" in the Multiple Meson Production through observation of both C-jets and atmospheric jets with the Chacaltaya emulsion

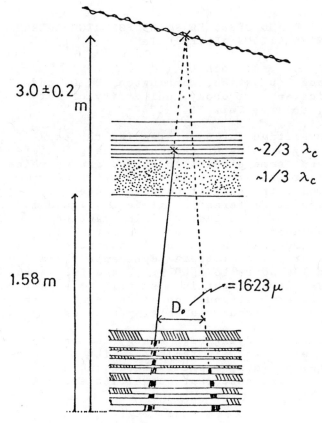

chambers[1]. In these jets, an SH-quantum appears as an isotropically emitting fire ball with mass (20-30) GeV/c^2.

If this is the case, we shall come to an understanding of the significant abundance in nature of this energy quantum. Thus we know also the existence of a "Mini-Centauro" fire-ball with the same rest mass which decays into about 15 "baryons"[2]. Now the decay mode into a "baryon" pair has come into observation with a sizable channel width, not as a fluctuation tail of the

Fig. 4 Schematic view of the event "Mini-Centauro" mode.

ACKNOWLEDGEMENT

The collaboration experiment is financially supported in part by Conselho Nacional para o Desenvolvimento Cientifico e Tecnologico, Fundação de Amparo à Pesquisa do Estado de São Paulo, in Brasil and Institute for Cosmic Ray Research, University of Tokyo, in Japan.

REFERENCES

1. Brasil-Japan Collaboration, "Multiple Meson Production in the $\Sigma E_\gamma > 2 \times 10^{13}$ eV Region" presented at this conference.
2. Brasil-Japan Collaboration, "A new type of Nuclear Interactions in the $\Sigma E_\gamma > 10^{14}$ eV Region" presented at this conference.

A SUMMARY OF EXPECTED AND OBSERVED RATES FOR HIGH ENERGY INTERACTIONS ABOVE 10 TeV

G. B. Yodh
National Science Foundation, Washington, D. C. and
Dept. of Physics and Astronomy, University of Maryland, College Park MD 20742

To set the perspective for understanding the capabilities of cosmic ray studies of very high energy interactions I will present some simple tables and graphs which will enable the reader, interested in doing a cosmic ray experiment, to estimate the size of detectors needed for getting desired rates at different altitudes. Figure 1 gives the size and weight of detector (2 interaction lengths in thickness) needed to get one interaction in the detector of energy $\sim$1000 TeV in one year's exposure. The horizontal bar for aeroplane altitude represents variation of flight time from 100 hours to 1000 hrs., and for satellites it represents 7 days flight in a shuttle to a year's exposure in a free flyer. Attenuation of the beam in the atmosphere is taken into account to obtain these estimates.[1]

The techniques for studying high energy interactions require large area detectors. In table I I list the techniques for satellite, balloon, mountain level and air shower experiments and indicate what quantities are meas-

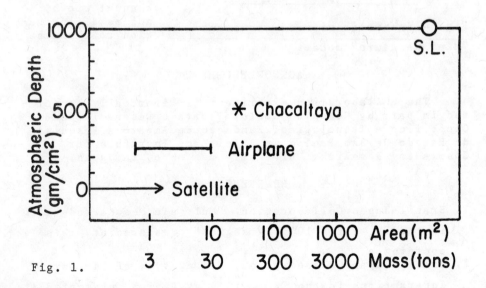

Fig. 1.

ISSN 0094-243X/79/49152-12 $1.50

ured, as well as those not determined, along with a comment on how well one can measure those that are measurable.

TABLE I

Study of Individual High E Events

Low Flux needs large collecting area

Different particles need diff. instruments (e.g., $-\gamma$, h, μ, e)

Exp. Tech.	Quantity Meas.	How well?
Nucl. Emulsion chambers Balloon, Satellite (Accelerators) $\sim$1m^2 Days. <100 TeV	Charged ptl Energy γ-energy Location Primary	$\pm$ 20% M.S. $\pm$ 10% Cascade $\sim$ 1μm seen
X-ray, Nucl. Em. Chambers Mt. Level. $\sim$100$\sim$1000m^2 yr Up to $\sim$1000 TeV	hadron-jets γ-jets Location Primary	$\pm$ 20% $\pm$ 20% 1 to 10 μm not seen
Calorimeters Up to 1000 TeV.	Total energy of h Direction Location Primary	$\pm$ 10% $\pm$ 1° $\sim$cm Sometimes.

Other methods:
Shower cores $\longrightarrow$ Ne_i, r_{ij}, Ne^T, N_μ^t ....
Multiple μs' $\longrightarrow$ E_μ^{th}, r position-angle...

Next I present a naive summary of event rates for emulsion chambers exposed at various altitudes based upon discussions with Japanese colleagues[2] and a summary of reported events from the Pamir experiment.[3] This is shown in Table II.

TABLE II

A Naive* Summary of Emulsion Chamber Event Rates with $\Sigma E_\gamma > 100$ TeV

Experiment	Depth g/cm²	Exposure factor m², Sr, sec	No. of Events with $\Sigma E_\gamma > 100$ TeV	Comments
1. Chacaltaya				
CH-14	550	1×10^9	25	Only upper chamber exposed and completely scanned.
CH-15-16-17	550	4×10^9	100 (estimated)	Two Storey chamber with nuclear emulsion $E_\gamma > 200$ GeV
2. Mt.Fuji	650	1.7×10^{10}	80	Thin single layer EC with X-ray film only $E_\gamma > 1.5$ TeV
3. Airplanes	260	9×10^6	9	Two Storey E.C. with nuclear Em. Complete scan. $E_\gamma > 100$-300 GeV
4. Pamirs	~600	3×10^{10} / 10^{10}	83 (reported)	Γ-block only $E_\gamma > 2$ TeV X-ray films only Γ+H and Γ+3H $E_h > 5$ TeV

*Caution: Event selection criteria + steep spectrum = severe biases.

The event classification for Mt. Chacaltaya experiments for $\Sigma E_\gamma \geq 200$ TeV is given next:[4]

Total # ~ 40   Centauros reported: 5
                Minicentauros: 6

The 40 events have a number of γ-ray jets varying from about 10 to 100, the spatial size varies from 0.5 cm to about 15 cm. Ordinary families (40-11=29) have only a few jets in the lower chamber because of the higher threshold needed for detecting hadrons.

A summary of double and multicore event rates can be presented as follows:[5] A multicore event at Mt. Chacaltaya is defined by the following procedure illustrated in the accompanying sketch

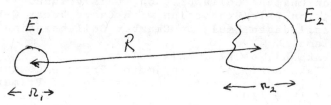

One requires $R/<r> > 5$, $\sqrt{E_1 E_2} \cdot R > 100$ TeV-cm. Number of observed double cores with $\Sigma E_\gamma > 100$ TeV is about 10; and these events do not have background of "low energy γ" due to atmospheric cascading.

On Mt. Fuji[6], with similar selection criteria, for $\Sigma E_\gamma > 100$ TeV 2 double cores and 2 triple core events were observed out of a total of 80 events. At energies between 50 and 100 TeV the rate was 2 double core events out of 40 total.

Airplane exposures[7] had a different criterion, $\sqrt{E_1 E_2} \, R \gtrsim 10$ TeV-mm. They have observed 4 multicore events out of 9 events total.

Finally, I give a summary[8] of best estimates of spectra of primary cosmic rays: these are differential spectra, where E is energy per nucleon and units are m^{-2}, Sr^{-1}, sec^{-1} GeV/c^{-1}.

Protons: $1.5 \times 10^4 \, E^{-2.71 \pm .06}$

Alphas : $1.42 \times 10^3 \, E^{-2.75 \pm .05}$

CNO : $1.23 \times 10^2 \, E^{-2.75 \pm .05}$

Iron : $1.27 \, E^{-2.36 \pm .06}$

These apply up to total energies of the order of 10^{15} eV.

REFERENCES

1. See the appendix of "Unusual Interactions Above 100 TeV: A Review of Cosmic Ray Experiments with Emulsion Chambers", G. B. Yodh in <u>Prospects for Strong Interaction Physics at ISABELLE</u> (D. P. Sidhu and T. L. Trueman, editors) 1977.
2. Y. Fujimoto, S. Hasegawa, T. Yuda, H. Sugimoto, Y. Takahashi, private communication at the conference.
3. Pamir Emulsion Chamber Collaboration (S. A. Slavatinsky, et al.) Paper HE-53, Proc. 15th Inter-

national Cosmic Ray Conference (Plovdiv) 1977, p. 226 and Proc. 14th International Cosmic Ray Conference (Munich) 1975, p. 2370.
4. "New Interactions $\sim 10^{15}$ eV", Brasil and Japan Emulsion Chamber Collaboration, this volume.
5. "Baryon Pair Production with Large Decay Q-Value," Brasil-Japan Emulsion Chamber Collaboration, this volume.
6. M. Akashi, et al., this volume.
7. Y. Takahashi and H. Sugimoto, Private Communication.
8. J. A. Goodman, etal., "Composition of Primary Cosmic Rays Above 10^{13} eV from the Study of Time Distribution of Energetic Hadrons Near Air Shower Cores," this volume.

How fast does the multiplicity of particle production in p-p collisions really increase with primary energy?

by

Erwin M. Friedlander

Lawrence Berkeley Laboratory, Berkeley, California 94720

It has been well-known for a long time that the dependence of the mean multiplicity, m, of particles produced in high energy p-p collisions on the primary laboratory energy E_o is much weaker than $\sim E_o^{1/2}$, the fastest rise allowed kinematically or implied e.g by Heisenberg's original theory.[1] At the same time the rise is faster than $\sim \ln E_o$ as predicted by scaling models [2] and even somewhat faster than the $E_o^{1/4}$ dependence predicted by thermodynamical-hydrodynamical considerations [3], [4] implying a Lorentz contracted production volume.

This paper is concerned with the problem of predicting the evolution of m with E_o at super-high energies (say, beyond 10 TeV), not just from an extrapolation of fits to values of m observed in the energy range covered by present accelerators, but by using regularities observed in the structure of multiplicity distributions (abbreviated hereafter as MD), too. It turns out that—provided these regularities continue to hold at very high E_o—a very fast increase, $\sim E_o^{1/2}$ is asymptotically expected.

In order to emphasize the Poisson-like shape of MD's[†] it is preferable to use instead of the probability W(k) for observing k secondaries the quantity

$$Y(k) \equiv \ln[k! \, W(k)] \qquad (1)$$

Obviously, if W(k) is a Poisson distribution (PD) of mean, say, a, Y is linear in k:

[†] Hereafter we will be concerned only with negative multiplicities $k = (n_{ch}-2)/2$ where n_{ch} is the total multiplicity of charged secondaries; k refers to created particles only.

$$Y = -a + k \ln a \qquad (2)$$

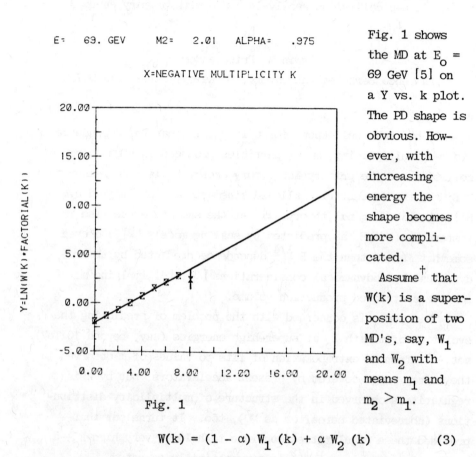

Fig. 1

Fig. 1 shows the MD at E_o = 69 GeV [5] on a Y vs. k plot. The PD shape is obvious. However, with increasing energy the shape becomes more complicated.

Assume[†] that $W(k)$ is a superposition of two MD's, say, W_1 and W_2 with means m_1 and $m_2 > m_1$.

$$W(k) = (1 - \alpha) W_1(k) + \alpha W_2(k) \qquad (3)$$

where one component, say W_2, is Poisson and all we know about W_1 is that it practically dies out beyond some value k_c of k. Then, beyond k_c,

$$Y \approx - m_2 + \ln \alpha + k \ln m_2 \; ; \qquad (4)$$

i.e., we get again a straight line with slope $\ln m_2$ but with an intercept depending on both m_2 <u>and</u> the relative weight α of the Poisson component.

[†] As has been done in Ref. [6] for mathematical expediency, but without presenting compelling evidence for Poisson components!

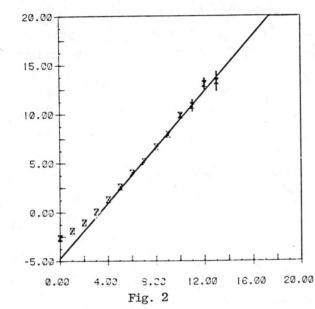

Fig. 2

This is well seen in fig. 2 which shows the MD at 300 GeV, [7] the statistically best measured sample to date (10^4 events).

The fact that this pattern is consistently repeated over the whole range of accelerator energies is shown in fig. 3, which displays the scaling variable Z, defined as

$$Z = \frac{m + \bar{y}}{m \ln m} , \qquad (5)$$

as a function of

$$X \equiv k/m , \qquad (6)$$

for all available data from 50 to 2100 GeV. On such a plot PD would be a straight line of slope 1 passing through the origin.

As can be seen, beyond x ~ 1.5 the points (drawn from both HBC exposures between 50 and 405 GeV and ISR results between 500 and 2100 GeV equivalent laboratory energy) do indeed cluster along a straight line confirming up to the highest available energy the presence of a PD component (W_2). As to the residue (W_1), it can be shown that, in spite of the apparently high statistics gathered to this day, the accuracy is insufficient to define its shape. A PD is not excluded, although a more complicated structure

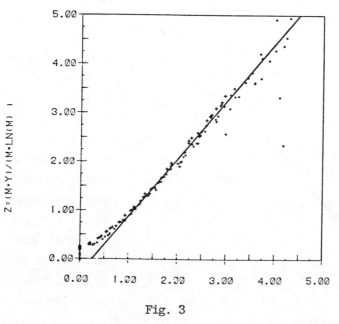

Fig. 3

may be indicated by the HBC-data.[†] However, the accuracy is just sufficient to estimate the mean values m_2 and m_1 (the Poisson component and the residue).

These values are plotted against E_o (di-log plot) in fig. 4, together with the overall mean m of k.

As can be seen, the energy dependence of both m_1 and m_2 can be well parameterized as $\sim E_o^\delta$ with δ close to 1/2 (the fitted values are $\delta = 0.54 \pm 0.03$ for m_2 and—understandably with lower accuracy— $\delta = 0.57 \pm 0.13$ for m_1).

It thus appears that p-p collisions can be regarded as a mixture of (at least) two types of events, each of which produces particles with a multiplicity law $\sim E_o^{1/2}$.

This can be reconciled with the relatively slow variation of the mean of the mixture, namely m (indeed, for m, $\delta = 0.35 \pm 0.01$, if it is parameterized in the same way as m_1 and m_2) only if α (the relative weight of W_2) decreases with E_o. Within the statistical

---

[†] The largest uncertainty resides in $W_1(0)$ because of the systematical effects connected with estimation of the elastic contribution.

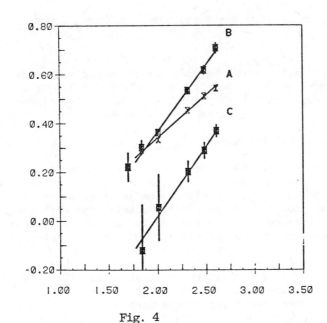

Fig. 4

and systematic uncertainties of the data, α appears to decrease like $E_o^{-(0.3 \pm 0.1)}$. If this trend continues at very high energies, W_1 should become dominant beyond, say, 50 TeV (at ~ 400 GeV W_1 and W_2 have comparable weights) and there one might expect m to increase like $E_o^{1/2}$. This may be essential in understanding the development of extensive air showers in the early stages of their evolution.

References

1. W. Heisenberg, Kosmische Strahlung, Springer, Berlin (1953) p. 148 ff.

2. R. P. Feynman, Phys. Rev. Lett. 23, 1415 (1969).

3. E. Fermi, Progr. Theor. Phys. 5, 270 (1950).

4. L. Landau, Izv. A. N. SSSR, s. fiz. 17, 1 (1953).

5. V. Ammosov et al, Phys. Rev. Lett. 42B, 519 (1972).

6. J. Lach and E. Malamud, Phys. Lett. 44B, 474 (1973).

7. A Firestone et al, Phys. Rev. D10, 2080 (1974).

HIGH ENERGY INTERACTIONS
AND
COSMIC RAY COMPONENTS IN THE ATMOSPHERE

K.Kasahara
Institute for Cosmic Ray Research, Univ. of Tokyo, Japan

ABSTRACT

A brief summary is given of the analysis of cosmic ray muons, hadrons and electron-photon components ranging from 1 to 100TeV. In particular, the connection with possible rising cross-sections and violation of scaling in hadron-hadron collisions is discussed.

INTRODUCTION

In a previous paper[1], we have shown that the reproducibility of the high energy cosmic ray spectra* in the atmosphere by the hadronic scaling hypothesis (with constant inelastic cross-sections) and the conventional proton-dominant primary cosmic rays with power spectral index ~1.7 gets worse and worse as the interaction energy involved goes higher: phenomena reflecting ~10TeV region interactions (muon spectrum in 1- 10TeV, gamma rays[@] and hadrons at airplane or balloon altitudes in 1-10TeV) are reproduced within a feasible margin (factor ~1.5), while those relating to 100TeV or higher energy region (gamma rays at mountain altitudes) are far from permitting the conventional interpretation.
The situation is summarized in Table I.

Table I Reproducibility of the spectra[#]

Component/place	Energy range (TeV)		Description by scaling and primary, $E_0^{-1.7}$
	Observed	Corresp'ing primary	
Muon/sea level	1-10	7-70	Fairly good
Gamma/balloon	0.5-5	<30	Fairly good
Gamma/airplane	1-10	10-100	Poorer than muon, but not so bad
Gamma/mountain	1-100	100-10000	Bad. Observed flux is 5-20 times lower than expected
Hadron/mountain	5-100	50-1000	Bad but better than gamma

*The spectrum here means the uncorrelated one which is constructed by collecting all observed particles disregarding time and spatial correlations.
@By gamma rays we imply the electro-magnetic component.
#Data employed are emulsion chamber data (see ref. 1).

WHAT CAN SAVE THE SITUATION?

The possible error in energy determination or some experimental biases cannot be a cause of the descrepancy, as one sees in Table I that the observed energy ranges overlap in different observation points. There is no realistic combination of the interaction parameters to reproduce the whole phenomena, provided that the cross-section is more-or-less constant. Possible ways for saving the situation are to assume the gradual steepening of the (equivalent proton) primary sepctrum and/or to incorpolate rising cross-section and break of scaling. A quantitative method for treating such effect is given in reference[1,2].

The models we assumed to incorpolate the rising cross-section are the following three which may cover the wide variety of possible changes of the nuclear interaction.

To start with, we note that many of the predicted rises of $\sigma(pp)$ are converted to $\sigma(p\text{-air})$ by the Glauber theory to give the energy dependence of $\sigma(\text{inel}) = \sigma_0 E^\delta$ with δ ranging from 0.02 to 0.14 in the energy interval of 10^{12}eV to 10^{19}eV.* The most promising value of δ is ~0.04 which is consistent with accerelator experiments as well as a value reduced from cosmic ray experiments up to ~20 TeV[3].

The first model which we call M0 assumes the complete scaling of one particle distribution, i.e., $x d\sigma/dx/\sigma_{\text{inel}} = \phi(x)$ is constant. By virtue of high energy, we can define x to be the ratio of secondary particle energy E to the incident energy E_0 in the laboratory system ($x = E/E_0$). Many papers dealing with the rising cross-section seem to use the same assumption. This is, however, perhaps not the case; change of the cross-section should lead to some change in one particle distribution.

Model M1 is introduced so as to be a counterpart of M0. There are two physical processes: one is constant cross-section part, $x d\sigma/dx/\sigma_c = \phi(x)$, and the other is the rising cross-section part, $x d\sigma/dx/\sigma_i$ where $\sigma(\text{inel}) = \sigma_c + \sigma_i$. The latter is assumed to be responsible for particles at $x \bar{=} 0$ which can be neglected when we consider the cosmic ray spectra. The model is a realization of the rising cross-section connected to the scaling violation in the central region as observed in ISR experiments. The one particle distribution $x d\sigma/dx/\sigma_{\text{inel}}$ shrinks with energy by $E_0^{-\delta}$. The value of δ is small so that scaling behavior in the fragmentation region does not explicitly break down.

Model M2 takes into account the explicit break of scaling in the fragmentation region. Namely, we assume

$$\frac{1}{\sigma_{\text{inel}}} \frac{d\sigma}{dx} = aE_0^{2\alpha} \cdot e^{-bE_0^\alpha x}$$

The fragmentation region shrinks with energy by $E_0^{-\alpha}$: $\langle x \rangle \propto E_0^{-\alpha}$ and $\langle n \rangle \propto E_0^\alpha$. If we take $\alpha = 1/4$ and $\sigma_{\text{inel}} = \text{const}$, M2 is nothing but the origianl CKP model[4].

The calculations under these assumptions are compared to the

*For a fixed model, δ is of course constant.

experimental data. The results are summarized in Table II.

Tabel II Values of δ needed to interpret the spectra*

Model	Altitude variation		Atten. length		Overall spec. at 650g/cm^2		Slope of spec.		Muon at sea level
	$\gamma(1)$	h(10)	$\gamma(1)$	h(10)	γ	h	$\gamma(10)$	h(20)	
M0	>0.1	~0.06	>0.1	~0.06	>0.1	~0.05	~0.1	~0.04	10% increase of flux for δ=0.1
M1	~0.1	~0.06	~0.1	~0.06	~0.1	~0.05	≥0.1	~0.035	10% decrease of flux for δ=0.1
M2 α=1/4	~0.04	≲0.04	~0.04	0.03	~0.03	0.03	≤0.03	0.03	α=1/4 is too large

*For the case of the primary, $E_0^{-1.7}$. Figures in the parentheses indicate approximate energy in TeV. γ indicates the gamma-rays and h the hadrons.

Note that the experimental values of the attenuation length and slope of the spectra can be obtained without knowing the absolute value of energy. The values of δ's in those entries are consistent with those requiring absolute value of energy scale.

It is clear that models M0 and M1 do not give a consistent δ for gamma-rays and hadrons while model M2 (α=1/4) gives a more-or-less unique value, δ=0.03~0.04 excepting that it fails in describing *low* energy phenomena. This fact indicates that α=1/4 is too large at least in the 10 TeV region.

The equivalent-proton primary might not be a single powered one. One might think it difficult to treat such case. We can, however, overcome the problem by drawing primaries required by a specific model in order to obtain a consistent picture with observed data. If the expected primaries thus obtained cannot be connected smoothly, the model should be rejected.

CONCLUSION

a) If the scaling in the fragmentation region should hold with σ ~ const, the equivalent-proton should steepen at ~30TeV rather abruptly. This leads to the assumption that among the primary composition, heavy elements which are as heavy as Fe, should replace protons; at 10^{17}eV there must be the Fe-component almost exclusively.

b) For Models M0 and M1 which assume rising σ and no explicit break of scaling in the fragmentation region, the following is required: for gamma data $\delta \gtrsim 0.1$ and for hadron data $\delta \sim 0.05$, if the equivalent-proton primary is $E_0^{-1.7}$. Thus the models are rejected. If we take into account the possibility of other type of primary, it is difficult to find a solution except the case of $\delta \sim 0$, which is nothing but the case a) above. This fact tells that the upper bound of δ is ~ 0.05 which may be compared with the value $\delta \sim 0.04$ quoted previously.

c) For Model M2, which introduces explicit break of scaling in the fragmentation region, $\delta=0$ cannot give a consistent picture but $\delta=0.03 \sim 0.04$ gives an overall agreement with data except at low energies. The value, $\alpha=1/4$, is too large for low energy phenomena ($\sim$10TeV region): the Chacaltaya C-jet experiment[5] ($<E_0> \sim$*10TeV) gives $\exp(-8x)$ for π^0 in the fragmentation region while ISR results can be approximated by $\exp(-6x)$, which leads to $\alpha \sim 0.12$. Moreover, it is difficult to take $\alpha=1/4$ at 1TeV region where accelerator data are available. There is, however, some indication in ISR experiment that α could be taken to be $\sim 0.1^2$.

There are some gamma ray family events showing the property of large multiplicity ($\alpha \gtrsim 1/4$)[6]. These seem to tell that model M2 is close to the reality. In that case we must assume that α is changing with energy:

$\alpha < 0.15 \quad E \ll 100\text{TeV}$
$\alpha \gtrsim 0.25 \quad E \gg 100\text{TeV}$
$\delta \sim 0.03 \sim 0.04$ at least up to $\sim$ 100TeV region.

The values of α and δ in $\gtrsim$100TeV region depends on the nature of the primary.

In a word, the present status of the spectrum analysis favours the large multiplicity model rather than the Fe-component dominance + scaling model, though the latter case is not completely rejected. Then, we can say, at least, that there is some extraordinary phenomena in the regime of cosmic rays.

ACKNOWLEDGMENT

The author is indebted to the members of Mt.Fuji group in many respects through the series of works which he has been making on the analysis of the cosmic ray spectra.

REFERENCE

1. K.Kasahara, Nuovo Cim. A3, 333(1978)
2. Y.Takahashi, in this volume.
3. G.B. Yodh et al, Phys. Rev. Lett. 28 1005(1972)
4. G.Cocconi, Nucl. Phys. B28, 341(1971)
5. Brazil-Japan Emulsion Chamber Collaboration, in this volume.
6. M.Akashi et al, in this volume

HIGH ENERGY INTERACTIONS AND COSMIC-RAY COMPONENTS IN THE ATMOSPHERE

Y. Takahashi
Faculty of Engineering Sciences, Osaka University, Toyonaka 560

ABSTRACT

The global features of hadronic interactions at very high energy (10^{12} eV - 10^{15} eV) are investigated from the observations of cosmic-ray spectra, specifically, of electro-magnetic and hadronic components. Data are compared with calculations, considering the scaling breakdown, rising cross-section and the bending primary hypothesis. From the analysis of cosmic-ray propagation in the atmosphere, it is concluded that the energy spectra of high energy cosmic-rays are most indicative of the scaling violation in the fragmentation region with $x/\sigma_{in} \cdot d\sigma/dx \propto s^{2\alpha} \exp(-B \cdot s^{\alpha} \cdot x)$, $\alpha \approx 0.20 \pm 0.05$, and the slowly rising cross-section with $\sigma \approx \sigma_0 \cdot s^{\delta}$, $\delta \approx 0.030 \pm 0.015$, provided that the primary spectrum is not dominated by anomalous bending at around 30 TeV.

1. INTRODUCTION

During the past decade, the knowledges on multiple production of hadrons have been much refined by both cosmic-ray [1] and big accelerator experiments [2]. For instance, the energy distribution of mesons in proton-proton interaction is widely examined, and at accelerator energies (10^9-10^{12} eV), the Feynman Scaling Law or the Hypothesis of the Limiting Fragmentation [3,4] is sometimes said to be valid within the measured accuracy of 10 % or less [5]. On the other hand, however, it is widely accepted that with the global scaling behavior, there exist some exotic characters in multiple production, which are suggested by the discoveries of large P_T phenomena [6], new particle productions [7], the rising cross-section [8] and the increase of the particle density in the central region [9].

Although these new phenomena take place with rather small probabilities, they are observed to be dependent on the incident energy. Therefore, it is well expected that at cosmic-ray energies, they might turn out to be dominant terms in the mechanism of particle production, giving rise to the breakdown of simple scaling pictures.

It is well-known that cosmic-ray spectra in the atmosphere reflect the integral quantities, specifically, the inelastic cross-section and the inclusive cross-section ($x \geq 0.1$). These spectra of muons, gamma-rays and hadrons so far measured with emulsion chambers show substantial differences from scaling predictions[16]. However, most of the measured spectra are of complicated convolution of many parameters that it is desirable to get them separated by providing the particle spectra at shallower atmospheric depth. This paper deals with the kernels of convolution, guided by the newest spectra measurement on the airplane and the calculation of the generalized diffusion equation [18]. From the analysis of these spectra, one can infer to the validity of the hadronic scaling law at very high energies. The conclusions are reconsidered in conjunction of the bubble chamber data and recent streamer chamber experiment at CERN-ISR.

2. SPECTRA MEASUREMENT ON THE AIRPLANE [17]

We shall describe here the newest spectrum measurement performed on the aircraft, which observe the energy spectra of atmospheric gamma-rays and hadrons in the energy region of 40 GeV - 40 TeV (gamma-rays) and 1 TeV - 40 TeV (hadrons). The apparatus adopted in this experiment is the emulsion chamber [17] by the reason that it holds high efficiency and the good resolution in observing high energy ($E \geq$ TeV) particles. The composition of the chamber is illustrated in Fig.1. It is easily understood that the apparatus is of the same design with those applied in the observations of cosmic-rays at mountain elevations [11] and of muons underground. The choice of the same design enables one to have a set of various cosmic-ray spectra which are almost free from relative systematic errors in making mutual comparisons.

The emulsion chambers are flown on a jet airplane (JAL CARGO JA-8018). Ordinarily, the ship has taken a trans-pacific round course, keeping the level flight duration of 8-10 hours. Before the film processing, most chambers continue to fly successively, until they get the integral exposure time of 500 - 600 hours. The average flight height of them is given by flight recorders. The effective atmospheric depth in average, $<H>$, is 260 g/cm^2, which is obtained by taking the altitude variation of cosmic-rays into account as,

$$<H> = \int H(g/cm^2) \cdot e^{-H(t)/\Lambda} \cdot dt \Big/ \int e^{-H(t)/\Lambda} \cdot dt, \quad (1)$$

where, $H(t)$ and Λ denote the flight curve and the attenuation length of cosmic-rays in the atmosphere, respectively.

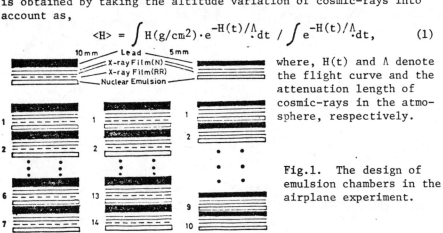

Fig.1. The design of emulsion chambers in the airplane experiment.

Cascade showers in the emulsion chamber are initiated either by gamma-rays or hadrons*. By utilizing the difference in the shower development, it is possible to separate gamma-ray spectrum from that of hadrons statistically. The prominent difference between these two components is characterized by the path length before the shower development (Δt). Let t_0 and Λ_{pb} be the mean length of cascade starting and the nuclear mean free path, then all observed showers having the value Δt in $d(\Delta t)$ are given as follows:

$$I(\Delta t) = I_\gamma \int d(\frac{t}{t_0}) e^{-\frac{t}{t_0}} \cdot \frac{1}{\sqrt{2\pi}\sigma} \exp(-\frac{(\Delta t - t)}{2\sigma^2}) + I_{pb} \int d(\frac{t}{\Lambda_{pb}}) e^{-\frac{t}{\Lambda_{pb}}} \cdot \frac{1}{\sqrt{2\pi}\sigma} \cdots \quad (2)$$

where, I_γ and I_{pb} are the flux of gamma-rays and hadronic showers. The dispersion, σ, of the Δt-measurement is 1.0 ± 0.5 c.u.. By adopting the selection criterion at $\Delta t = 4.0$ c.u. (pb), the observed events are divided into two groups. The contamination of hadrons into gamma-rays is 19 %, and vice versa, 25 %.

The zenith angle distribution of gamma-rays is inspected and it is confirmed that the observation is well reproduced by the theoretical curve with $\Lambda = 100 \pm 10$ g/cm2 (Air) for both gamma-rays and hadrons at energies exceeding 1 TeV. This value is too short to accord with the expected value for Feynman Scaling with constant cross-section 16. [$\Lambda(200 \sim 300 \text{g/cm}^2) \simeq 120 \sim 130$ g/cm2 and $\Lambda(\sim 600 \text{g/cm}^2) \simeq 135 \sim 145$ g/cm2 in scaling prediction.]

The gamma-ray spectrum (integral) in a wide energy range is constructed by combining all data from 40 GeV to 40 TeV. It is obviously seen that there is a little steepening at around a few hundred GeV from $\beta_\gamma \simeq 1.65$ into $\beta_\gamma \simeq 1.95$. The spectrum after the steepening is written as,

$$I_\gamma(\geq E_\gamma) = (1.92 \pm 0.07) 10^{-8} (\frac{E_\gamma}{1.5 \text{TeV}})^{-1.95 \pm 0.10} \text{ cm}^{-2} \cdot \text{sec}^{-1} \cdot \text{str}^{-1}. \quad (3)$$

On the other hand, the integral spectrum of hadronic showers is observed to have less steep power than gamma-rays in the same energy regime ($\Sigma E_\gamma = 1$ TeV ~ 40 TeV). It is expressed in the form of $(\Sigma E_\gamma)^{-\beta_H}$,

$$I_{pb}(\geq \Sigma E_\gamma) = (5.5 \pm 0.5) 10^{-8} (\frac{\Sigma E_\gamma}{1.0 \text{TeV}})^{-1.73^{+0.15}_{-0.10}} \text{ cm}^{-2} \cdot \text{sec}^{-1} \cdot \text{str}^{-1}. \quad (4)$$

where ΣE_γ is the transferred energy into photons.

---

* To avoid the misunderstandings, we shall define here the special terminology used in the spectrum works of cosmic-rays. "Gamma-ray" component is defined by electro-magnetic component, namely, γ-ray, electron and positron. "Hadrons" are nucleons and charged mesons.

The relation of the incident energy of hadrons (E_H) to the ΣE_γ is given by

$$\Sigma E_\gamma = <E_\gamma> \cdot E_H, \qquad (5)$$

and the conversion factor, $<k_\gamma>$, is defined by

$$<k_\gamma> \equiv [\int_0^1 f(k_\gamma) \cdot k_\gamma^\beta \cdot dk_\gamma]^{1/\beta} \simeq 0.23 \pm 0.02. \qquad (6)$$

This factor is calculated by using the inclusive distribution of survival nucleon and the charge symmetric distribution of secondary pions. It is experimentally confirmed by both cosmic-rays (Japan-Brazil Collaboration [13]) and FNAL 400 GeV proton beams (S·Dake et al. [14]), and there are Monte-Carlo simulations[10] which showed the model independence of the energy determination of hadronic showers by this method. Note that the above quantity is the conversion factor of the spectrum, which is different from the effective average of the interaction, $\bar{k}_\gamma \simeq 0.4$, as is discussed in the C-jet studies. The observed energy range of hadrons is thus converted as 4 TeV $\leq E_H \leq$ 180 TeV in (4).

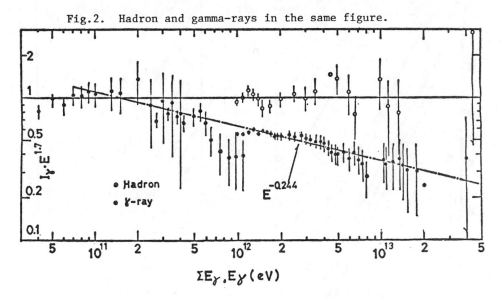

Fig.2. Hadron and gamma-rays in the same figure.

In Fig.2, observed spectra are drawn by dividing them by the scaling prediction with the ordinary primary spectrum ($\beta=1.7$). It should be noticed that hadron component will keep constant line in the figure irrespective of the scaling or its violation, on account of the nucleon dominance in hadron component at such depths as airplane altitude, and the inelastic cross-section will affect it only little that it reflects primary spectrum very well. However, one might obviously understand that gamma-rays are decreasing with increasing energy even from a few hundred GeV.

The early steepening of the gamma-ray spectrum suggests the substantial violation of the Hypothesis of the Limiting Fragmentation, because rising cross-section, bending primary and others can hardly reproduce the Fig.2 without introducing any breaks in the inclusive cross-section. The tendency of $\beta_\gamma > \beta_H$ has been observed at mountain heights [11,12] and known as the possible evidence of either the scaling breakdown or the effect of the bending primary (see K·Kasahara and Y·Takahashi [16]). This two-fold solution might be separated, because the bending effect of the primary spectrum is much smaller at such a shallow depth as 260 g/cm^2, that the steepening point of gamma-ray spectrum is hard to be understood from the existing primaries with the bending point at around 10^{15} eV. It is required to have an anomalously bending primary with $E_B \simeq 30$ TeV or less, if one tries to save the scaling of π^0 production, however, it will fail to accord with the observed hadron spectrum. Fig.3a and 3b are the data compilation at various depths and the scaling prediction in case of the anomalous bending primary with $E_B \simeq 30$ TeV. Although the global features seem to be satisfactory in such drawings, the detail of the individual spectrum will not agree well as will be discussed later.

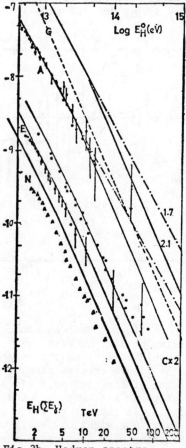

Fig.3a. Gamma-ray spectra.   Fig.3b. Hadron spectra.

3. COMPARISONS OF VARIOUS SPECTRA OBTAINED WITH EC —— SLOWLY RISING CROSS-SECTION AND SCALING BREAKDOWN

Because of the failure to describe cosmic-ray spectra within the framework of simple scaling model, one must develop the analysis by generalizing the diffusion equation. The propagation of cosmic-rays in the atmosphere is a difficult problem when one introduce the energy dependent cross-section and inclusive cross-section. It is solved analytically by iteration method [18]. The rising cross-section is approximated by a power function of the incident energy, as

$$\sigma_{in} = \sigma_0 \cdot s^\delta, \tag{7}$$

where s denotes the total energy measured in TeV ($2 \cdot E_{lab}$). The scaling violation is introduced according to the treatment by T·Shibata[25],

$$\frac{1}{\sigma_{in}} \cdot \frac{d\sigma}{dx} = A \cdot s^{2\alpha} \cdot e^{-B \cdot s^\alpha \cdot x} \cdot g(x), \quad g(x) \simeq \frac{1}{x} \tag{8}$$

where x is the fractional energy of the produced meson, and the constant parameters, A, B, and σ_0 are determined by accelerator data [16].

With use of the numerical results, let's inspect the altitude dependence of the exponent at very high energies (s≳TeV). Although the calculation is so complicated, the behavior of the solution can be simply summarized for the altitude dependence of the power exponent as follows:

$$\begin{cases} \beta_\mu - 1 \simeq \dfrac{\beta - \alpha}{1 - \alpha}, & \text{(for muons),} \tag{9a} \\[6pt] \beta_\gamma \simeq \dfrac{\beta - \alpha}{1 - \alpha} + <n_\gamma(z)> \cdot \dfrac{\delta}{1 - \alpha}, & \text{(for gamma-rays),} \tag{9b} \\[6pt] \beta_N \simeq \beta + <n_N(z)> \cdot \delta, & \text{(for nucleons),} \tag{9c} \\[6pt] \beta_\pi \simeq \dfrac{\beta - \alpha}{1 - \alpha} + <n_\pi(z)> \cdot \delta \cdot K, \text{ with } K = \begin{cases} 1, & z<z_0 \\ \dfrac{1}{1-\alpha}, & z>z_0 \end{cases}, & \text{(for pions),} \tag{9d} \end{cases}$$

The parameters, $<n_i(z)>$, are almost linear functions of $z/\Lambda(1\text{ TeV})$, the collision times. The z denotes the atmospheric depth.

Fig.4 shows how parameters (7) and (8) affect the power exponent of hadron spectrum. The data drawn in the figure are those obtained with emulsion chambers. It can be concluded that the stronger rise of inelastic cross-section like $\ln^2 s$ (corresponding δ being ~0.1) is ruled out by the existing hadron spectra. It is also understandable that the scaling violation does not affect $\beta_H(z)$ so much.

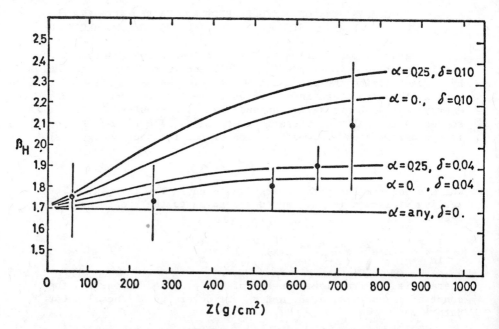

Fig.4. The altitude dependence of exponents of hadronic spectra in case of the rising cross-section and the scaling violation.

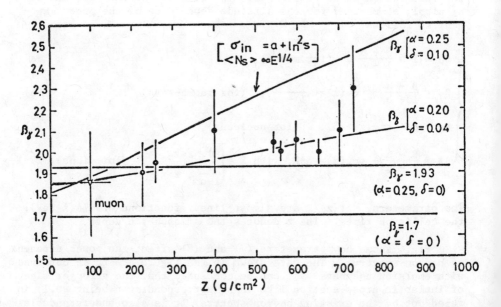

Fig.5. The altitude dependence of exponents of gamma-ray spectra in case of the rising cross-section and the scaling violation

Similar comparison is made for gamma-rays in Fig.5. The data are taken from the compilation of the past EC experiments as is listed in the end of this paper. Both hadrons and gamma-rays show a small increase in exponents with increasing atmospheric depth, that is well understood by the slow rise of the inelastic cross-section. It is however, possible to understand as the effect of the hypothetical bending primary with $E_B \simeq 30$ TeV. Nevertheless, the high exponent value of gamma-rays seems universal at every depth, that cannot be attributed to the bending primary anyway. Thus, we are forced to have a conclusion that the scaling violation is quite appreciable as is introduced in the airplane experiment.

Now, in order to treat the degree of scaling breakdown, α, and the rising cross-section, δ, let's confine ourselves to the data measured by almost the same apparatus, the same method and the same persons, namely, the airplane data and that obtained at Mt.Fuji[12]. From the mutual comparisons of them, we have obtained the best guess for α and δ. Fig.6 gives the experimental constraints for these parameters.

$$\begin{cases} \alpha \simeq 0.20 \pm 0.05, & (10a) \\ \delta \simeq 0.030 \pm 0.015, & (10b) \end{cases}$$

Fig.6. The experimental constraints for the amount of the rising cross-section(δ) and the scaling breakdown(α).

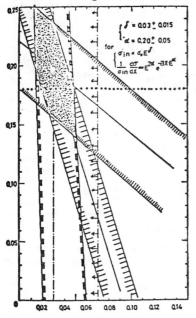

The above results sound contradictory to the scaling claim shown by muon people in the past. However, we are dealing only with EC data, and EC experiments of muon burst seem to be consistent with above discussions. In addition, the newest magnet spectrograph made by MUTRON shows that muon spectrum in TeV-10TeV regime[15] has steeper power than those believed in the past from indirect measurements. To settle the muon problems, it is desirable to wait still for the confirmation of the steepness. By this reason, (10a) is better taken as the scaling violation for "π^0-meson".

The threshold energy of these characters is estimated as less than 1 TeV at most, from the bending point of gamma-rays at airplane altitude. It is also confirmed by the supplementary measurement of low energy (300 GeV -2 TeV) gamma-rays performed at Mt. Fuji.

For the reader's convenience, we shall cite the compilation of Cosmic-ray spectra, obtained with emulsion or emulsion chambers.

Table 1.

Upper Atmosphere [Primary($*g/cm^2$)]	Observed Energy Range(bare)	Power	Observer (Ref.)
1. Satellites*)	10 GeV - 10 TeV	$\beta_p = 1.60 (\leq 1\ TeV)$ $= 2.10 (>1\ TeV)$	Grigorov (a) -1971
	10 GeV - $5 \cdot 10^{15}$ eV	$\beta_{TES} = 1.60$	Grigorov (b)
2. Balloon**)	50 GeV - 2 TeV	$\beta_p = 1.75 \pm 0.03$	Ryan et al (c) -1972
[Secondary(g/cm^2)]			
3. Balloon at 60 g /cm^2	2 TeV - 12 TeV 2 TeV - 15 TeV	$\beta_H = 1.56 \pm 0.17$ $\beta_\gamma = 2.20 \pm 0.20$	Amineva et al (d) -1975
4. Balloon***) at 26 g/cm^2	100 GeV - 2 TeV	$\beta_\gamma = 1.95^{+0.3}_{-0.2}$	Kidd et al. (e) -1962
5. Balloon at 37 g /cm^2	300 GeV - 5 TeV 300 GeV - 10 TeV	$\beta_\gamma = 1.75 \pm 0.20$ $\beta_H = 2.04 \pm 0.22$	P.H.Fowler (f) -1963
6. Balloon at 57 g /cm^2	100 GeV - 2 TeV	$\beta_\gamma = 2.0 \pm 0.2$	(g) Minakawa et al. -1958
7. Airplane#) at 220 g/cm^2	300 GeV - 2 TeV 2 TeV - 10 TeV 300 GeV - 2 TeV 2 TeV - 10 TeV	$\beta_H = 2.5 \pm 0.2^{(°)}$ $\beta_H = 3.5 \pm 0.3^{(°)}$ $\beta_\gamma = 2.3 \pm 0.2^{(°)}$ $\beta_\gamma = 2.8 \pm 0.3^{(°)}$	(h) J.Duthie et al. -1962
8. Airplane at 200 g/cm^2	200 GeV - 20 TeV	$\beta_\gamma = 1.7 - 1.9$	(i) Smorodin et al. -1968
9. Airplane at 225 g/cm^2	2 TeV - 50 TeV	$\beta_\gamma = 1.9 \pm 0.15$	(j) Smorodin et al. -1977

*). Calorimeter experiment, not EC, on board Satellites.

**). Calorimeter experiment, not EC, on board balloons.

#). Emulsion Stack with similarly designed Lead absorber as EC, but not exactly same with EC. Flown on Comet airplanes. It is understood there might be some systematical errors in the energy measurement.

10. Airplane at 260 g/cm²	40 GeV - 500 GeV 500 GeV - 40 TeV (40 GeV - 40 TeV) 1 TeV - 40 TeV	$\beta_\gamma=1.75\pm0.13$ $\beta_\gamma=1.95\pm0.10$ $\beta_\gamma=1.90\pm0.05$ $\beta_H=1.73^{+0.15}_{-0.10}$	Present Data (k) Takahashi et al. -1975 -1977
11. Mt. Lenin Peak at 400 g/cm²	2.5 TeV - 25 TeV	$\beta_\gamma=2.1\pm0.2$	(l) Chedyntseva & Nikolskii -1976
12. Mt. Chacaltaya at 540 g/cm²	200 GeV - 100 TeV 2 TeV - 100 TeV	$\beta_\gamma=2.05\pm0.05$ $\beta_H=1.80\pm0.10$	(m) Japan-Brazil -1973
13. Pamir at 556 g/cm² and 596 g/cm²	3 TeV - 30 TeV 2.5 TeV - 30 TeV	$\beta_\gamma=2.00\pm0.06$ $\beta_\gamma=2.05\pm0.04$	(n) Pamir Collab. -1975
14. Mt. Fuji at 650 g/cm²	250 GeV - 100 TeV 2 TeV - 100 TeV	$\beta_\gamma=2.0\pm0.05$ $\beta_H=1.9\pm0/10$	(o) M. Akashi et al. -1977
15. a mountain at 700 g/cm²	2 TeV - 30 TeV	$\beta_\gamma=2.1\pm0.14$	(p) Smorodin & Baradzei -1977
16. Mt. Norikura at 735 g/cm²	350 GeV - 1.0 TeV 1 TeV - 10 TeV 2 TeV - 23 TeV	$\beta_\gamma=2.0\pm0.2$ $\beta_\gamma=2.3\pm0.2$ $\beta_H=2.1\pm0.3$	(q) M. Akashi et al -1964
17. Mt. Norikura at 735 g/cm²	600 GeV - 4 TeV 3 TeV - 20 TeV 3 TeV - 20 TeV	$\beta_\gamma=1.94\pm0.17$ $\beta_H=2.3\pm0.04(?)$ $\beta_\gamma=2.3\pm0.27$	(r) S. Dake et al. 1977 (s) Hazama & Nishikawa -1976
18. Muons Underground. (1030 + x) g/cm²	500 GeV - 2 TeV 2 TeV - 12 TeV 2 TeV - 5 TeV 5 TeV - 20 TeV	$\beta_\mu=2.70\pm0.10$ $\beta_\mu=2.85\pm0.25$ $\beta_\mu=2.60\pm0.12$ $\beta_\mu=2.85\pm0.30$	(t) M. Akashi et al. -1975 (u) Amineva et al. -1973

It should be noticed that the above results are valid so long as the primary spectrum keeps its single power form upto 10^{15} eV with unchanging composition. If it steepens at around 30 TeV, the values in (7,8), especially, the parameter (δ), will eventually become smaller than (10). The amount of the bending power of the speculative primary might be around ~ 0.1, if we consider the hadronic spectra seriously. There are only little room for extraordinary primaries, considering various cosmic-ray indications, including the hadron spectrum on the airplane, family spectrum of gamma-rays at various depths and Air Shower observations.

4. RECONSIDERATION OF ACCELERATOR RESULTS

To incorporate the cosmic results shown above, let's review accelerator data. The break of the limiting fragmentation, i.e., $\alpha = 0.1 \sim 0.2$, will result in the decrease of particle density in the fragmentation region, of which one might remind of the work by T·Ferbel [19], who had inspected the bubble chamber data upto FNAL energies. In Fig.7, Inclusive cross-sections of $\pi^{\pm}$ at $y^* \leq 0$ ($x \gtrsim 0.15$) are displayed. As Ferbel's remark[19,20], the decrease of the fragmentation mesons is appreciable. According to the formula(8), the number of the fragmentation particles would decrease almost proportional to $s^{-\alpha}$, and the bubble chamber data are more or less consistent with $\alpha = 0.1 \sim 0.2$.

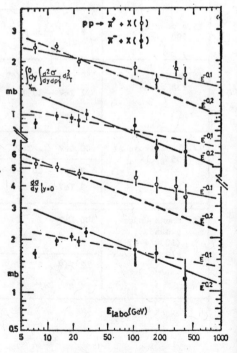

Fig.7. Bubble chamber data showing the early break of the scaling in the fragmentation.

This should be enhanced at ISR region, because there is rising cross-section consistent with (10b), and the normalized invariant cross-section, (8), per collision would be more reduced than in Fig.7 at ISR regime. Despite that the cosmic-ray results are in agreement with bubble chamber data, it is very hard to reconcile the present analysis with the data at CERN-ISR, e.g., J·N·Allaby et al.[21] and M·G·Albrow et al.[22] and with the scaling functions[23] so far proposed. Whatsoever is the accuracy of the previous ISR data, we had better pay attention to the latest ISR experiment with streamer chamber [24]. W·Thomé et al. have shown that in addition to the increase of the central region with $E \cdot d^3\sigma/dp^3]_{y=0} \propto \sqrt{s}$, there is the small decreasing tendency in the fragmentation region. They have reported that the equi-density point in the rapidity space, $y' \equiv Y - y(1/\sigma \cdot d\sigma/dy=0.8)$, is a increasing function of total energy, which we can see in Fig.8. As is shown by the straight line in the figure, it is consistent with formula(8) by adopting α = 0.1.

All these arguments might be rather heretical for the current understandings of accerelator data, but it is worth to note, for, these quantities are obtained by the inclusive spectrum constructed from all-particle-observation in an event(exclusive construction of inclusive spectrum), that the systematics of the measurement might have some advantages to the previous ISR data.

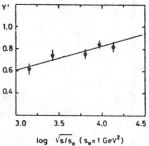

Fig.8. Streamer chamber data at ISR, indicating the violation of HLF.

5. CONCLUDING REMARKS•

As for the validity of the scaling at cosmic-ray energies, there are some critical discussions on C-jet measurement of Japan-Brazil Collaboration. (see papers contributed by Fujimoto et al and Ellsworth et al, in this volume). Apart from these problems, we can summarize the study of cosmic-ray components in the atmosphere as follows:

-1-. Hypothesis of the Limiting Fragmentation is not affirmative over the wide energy range, 10^{11} eV - 10^{15} eV. The decrease in the fragmentation region is consistent with α = 0.1~0.2.

-2-. The total cross-section is most likely continue to rise but less faster than $\ln^2 s$. Provided that the primary cosmic-ray continue to be a single power in this regime, the amount of the rise is estimated as,

$$\sigma_{in} \simeq \sigma_0 \cdot s^{0.03 \pm 0.015}, \text{ s in TeV}. \tag{11}$$

The particle multiplicity, $<N_S>$, is closely related to the formula of the scaling violation. By integrating (8), one gets $<N_S> \propto s^\alpha$. To compare the results, we have drawn the estimate by dotted lines in G•B•Yodh's drawings(NASA report, 1977). The dependence is much similar to the $\ln^2 s$ formula as in Fig.9.

The break of the scaling conjecture will give us complicated picture, and many quantities are not simple functions of a few parameters; thus it is required to improve the non-scaling models. In doing so, the factorization should be given up *.

---

* By bubble chamber study, it is drastically shown that the factorization by x and P_T is strongly violated. (Bromberg et al, see Fig. 10.)

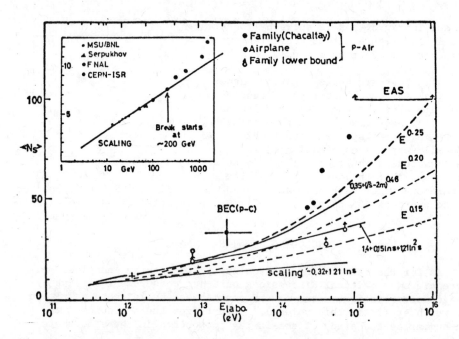

Fig.9. Multiplicity estimate from the cosmic-ray measurements of various spectra. Dotted lines are the region of the present estimation normalized to p-p interaction at 1 TeV. Points are other estimates for p-Air collision.

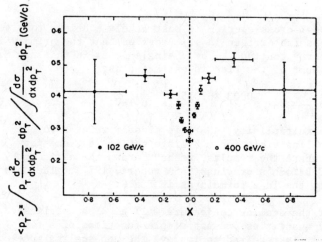

Fig.10. Bubble chamber data on the correlation of $<P_T>$-x which clearly show the inadequecy of Feynman scaling and the factorization. Figure is taken from ref.25.

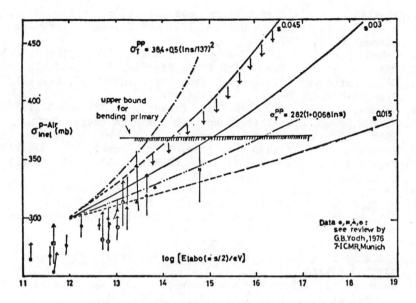

Fig.11. Estimate of the rising cross-section. The domain covered by $s^{0.045}$ and $s^{0.015}$ is the case for the single power primary. Shaded bar designates the upper bound for the bending primary.

6. ACKNOWLEDGEMENTS.

The main part of this summary is based upon the collaboration experiment on the airplane with Drs. J·Iwai and I·Ohta. The author is thankful to Profs.T.Ferbel and P·Darriulat for the informations and discussions. Discussions with Prof. A·M·Hillas and Dr. K·Kasahara are helpful in the mutual comparisons of various spectra.

7. REFERENCES·

1. Brazil-Japan Collaboration, Proc. 15th ICRC, Plovdiv,7, 195,201 and 208 (1977).
2. Data compilations are seen with relevant physics in the followings, J·Whitmore, Physics Reports, 10, 273 (1974).
R·Slansky, Physics Reports, 11c, 101 (1974).
M·Jacob, TH.1683-CERN(1973), and ISR WORKSHOP-2-1(1977).
F·E·Taylor et al., Phys.Rev., D5, 1217(1976).
3. R·P·Feynman, Phys.Rev.Lett.,23, 1415(1969).
4. J·Benecke, T·T·Chou, C·N·Yang and E·Yen, Phys.Rev., 188, 2159(1969).
5. M·Jacob, see Ref.2.
6. B·Alper et al., Phys.Lett., 44B, 521 and 527(1973).
7. K·Niu, E·Mikumo and Y·Maeda, Prog. Theor. Phys., 46,1644(1971).
8. V·V·Akimov et al., Proc. 11th ICCR, Budapest, 3,211(1969).
G·B·Yodh, Y·Pal and J·S·Trefil, Phys.Rev.Lett., 28,1005(1972).
U·Amaldi, Proc. 2nd Int. Aix Conf., CI-241,J.de Phys. Tome 34(1973) and others referred therein.

9. K•Guettler et al, Phys.Letters, 64B, 111(1976).
10. S•Dake, private communication, 1977 and M•Shibata, private communication, 1977.
11. C•M•G•Lattes et al., CKJ-Report-13, University of Tokyo,(1974).
12. M•Akashi et al., Proc. 15th ICRC, Plovdiv, 7,430 and 424(1977).
13. C•M•G•Lattes et al., Supple. Prog.Theor.Phys., 47, 1(1971), Proc.13th ICCR, Denver, 3, 2210(1973).
14. S•Dake et al., Proc. 15th ICRC, Provdiv, 7, 321 and 322(1977).
15. Mutron Group, contributed paper to the High Energy Conf., Tokyo, 1978.
16. K•Kasahara and Y•Takahashi, Prog. Theor. Phys., 55, 1896(1976).
17. Y•Takahashi, J•Iwai and I•Ohta, ICR-Report-67-78-11, (1978).
18. Y•Takahashi, CRL-Report-58-78-2, (1978).
19. T•Ferbel, UR-500, c00-3065-91, University of Rochester, and Proc. of SLAC Summer Institute on Particle Physics(1974).
20. C•Bromberg, T•Ferbel, P•Slattery, A•A•Seidel and J•C•Vander Velde, Nucl.Phys., B107, 82(1976).
21. J•N•Allaby et al, CERN 70-12, (1971).
22. M.G.Albrow et al., Nucl.Phys., B56, 333(1973), B54, 6(1973), and B73, 40,(19740.
23. many empirical formulae have been proposed, see details in ref.16.
24. W•Thome et al., Nucl. Phys., B129, 365(1977).
25. T•Shibata et al., Prog. Theor. Phys., 56, 1845(1976).

---

a. V•V•Akimov et al.,Proc. 11th ICCR, Budapest, 1,517(1969).
b. N•L•Grigorov et al., Proc. 12th ICCR, Hobart, 1, 1746(1971).
c. M•J•Ryan et al., Phys.Rev.Lett., 28, 985(1972'.
d. T•P•Amineva et al., Proc. 14th ICRC, Munich, 7, 2501(1975).
e. J•M•Kidd, Nuovo Cimento, 27, 57(1962).
   F•Abraham et al., Nuovo Cimento 28, 221(1963).
f. P•H•Fowler, Proc. 8th ICCR, Jaipur, 5, 182(1963).
g. O•Minakawa et al., Nuovo Cimento Supple., 8, 761(1958).
h. J•Duthie et al, Nuovo Cimento, 24, 122(1962'
i. A•V•Apanasenko et al., Can.J.Phys., 46, s700(1968).
j. E•A•Kanevskaya et al., Proc. 15th ICRC, Plovdiv, 7, 453(1977).
k. Y•Takahashi et al., ibid 408.
l. K•V•Cherdyntseva and S•I•Nikolsky, Sov.J•Nucl.Phys.,23, 652(1976).
m. C•M•G•Lattes et al., Proc. 13th ICRC, Denver, 3, 2219(1973).
n. V•K•Budilov et al., Proc. 14th ICRC, Munich, 7,2365(1975).
o. M.Akashi et al., Proc. 15th ICRC, Plovdiv, 7,424 and 430(1977).
p. E•A•Kanevskaya et al., ibid 453.
q. M•Akashi et al., Supple. Prog. Theor. Phys., 32,1(1964).
r. S•Dake et al., Nuovo Cimento, 41B, 55(1977).
s. M•Hazama et al., Proc. Int. C-R Symp., Univ. of Tokyo,(1974).
t. M.Akashi et al., Proc. 14th ICRC, Munich, 6, 2037(1975).
u. T•P•Amineva et al., Proc. 13th ICCR, Denver, 3, 1788(1973).

(ICRC is an abbreviation of "International Cosmic Ray Conference" and ICCR denotes "International Conference on Cosmic Rays".)

Chap. 5 Evidence for New Particles

COSMIC RAY EVIDENCES FOR SHORT-LIVED PARTICLES.

Presented by

K. Niu
Nagoya University, Nagoya, 464 Japan.

ABSTRACT

Summary of evidences from cosmic rays on production of short-lived particles, -the X particles-, is given. All of those have been obtained by means of emulsion chamber techniques. Estimated life time of the X particle is of the order of $7 \sim 8 \times 10^{-13}$ sec. Production rate of the X particle is roughly estimated as $1 \sim 5\%$ at 10 TeV range and still higher percentage is indicated at 100 TeV range. Opening of new channels at higher energy may give us proper understanding of this increase.

INTRODUCTION

Several years in advance the commencement of the "charmed" age in high energy accelerator physics, a pioneering work on short-lived particle observation had been carried out in the cosmic ray field. In the year 1971, one of Japanese group discovered one event showing a pair creation and decay of short-lived particles with life time around 10^{-13} sec and mass of $2 \sim 3$ GeV[1]. The particle was named by them as X particle.

The super high energy jet shower in which a pair of the X particle was observed was induced by a neutral primary particle with energy of 10 TeV range. A sketch of the event is shown in Figure 1. It was characterized by high multiplicity of secondary particles, $(19+70)_n$, and by a so called torpedo π^0 meson with distinctly higher energy than other's, as is shown in Figure 2. Essential points of this event are shown in Figure 3. Those are existence of a pair of kinks, 1.4cm and 4.9cm away from the origin and a coplanarity relation satisfied by the charged secondary and tertiary of the first kink and the tropedo π^0. This was interpreted as a pair decay of short-lived particles. Mass and life time of the particle was estimated assuming identity of the charged daughter as

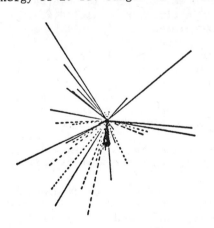

Fig. 1. Sketch of Event 6B-23 $(19 + 70)_n$

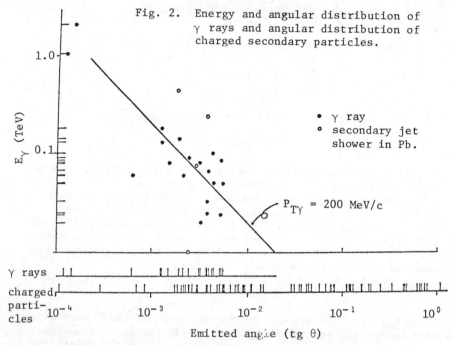

Fig. 2. Energy and angular distribution of γ rays and angular distribution of charged secondary particles.

indicated in the figure.

Just after its discovery, the X particle was pointed out to be a possible charmed particle by a Japanese theoretical group[2]. Being stimulated by the theoretical works, hunting of X particles has been carried out exposing emulsion chambers to cosmic rays and to accelerator beams.

Because of high spatial resolving power, nuclear emulsion is, at present, the only technique by which we can directly observe the decay of short-lived particles with life time arround 10^{-13} sec. Much effort has, therefore, been concentrated on the emulsion experiment to observe decays of such particles.

Further evidences have been accumulated not only from our own exposures but also from reanalysis of certain anomalous events published by others[3]. Another remarkable event with clear pair decay of the X particles was reported in 1975 by Sugimoto et al[4]. Sketch and description of the event is given later in this report.

In this paper a report is made summarizing all data from cosmic rays on production of the X particles. The summary is extended on the recent observation at several tens to hundreds of TeV region carried out by BEC group in I.C.R., University of Tokyo[5] and Japan-Brasil callaboration group[6].

To observe short-lived particles produced in super high energy nuclear interactions our Japanese groups have been utilizing detectors so called emulsion chambers. Most of accelerator physicists are not familiar with the emulsion chamber, it may assist them to make a brief explanation before summarizing the observed data.

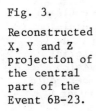

Fig. 3.
Reconstructed X, Y and Z projection of the central part of the Event 6B-23.

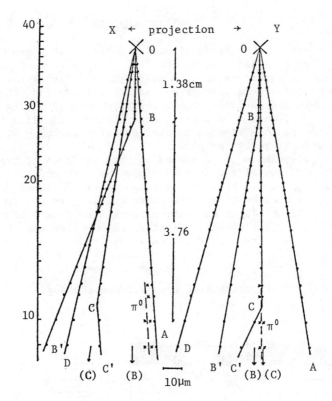

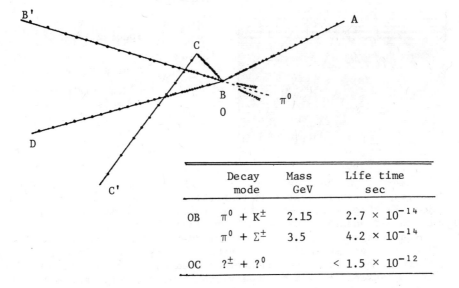

	Decay mode	Mass GeV	Life time sec
OB	$\pi^0 + K^\pm$	2.15	2.7×10^{-14}
	$\pi^0 + \Sigma^\pm$	3.5	4.2×10^{-14}
OC	$?^\pm + ?^0$		$< 1.5 \times 10^{-12}$

EMULSION CHAMBER

Emulsion chamber is a complex detector with nuclear emulsion and other material plates. Unlike the case of pure emulsion stack, nuclear emulsions are used prinicipally as the position detector of tracks of charged particles traversing the material plates in the emulsion chamber. Therefore, if a pure emulsion stack is likened to a micro bubble chamber, emulsion chamber could be compared to a complex counter detector assembly with super high spatial resolving power.

The emulsion plate coated on both surfaces of transparent substrate is utilized as a two-fold emulsion counter. Spatial resolving power of the emulsion counter is extremely high and of the order of submicron. Emulsion counters are combined with thin plates made of light or heavy materials to form a complex detector assembly consisting of a target and a momentum energy analyser. Typical configuration of the emulsion chamber is shown in Figure 4.

Fig. 4. Configuration of the emulsion chamber.

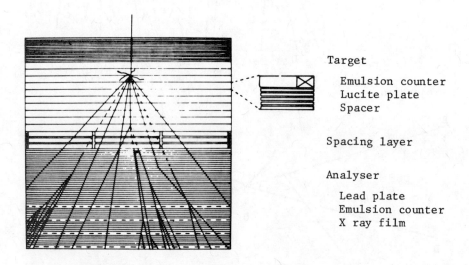

Target
  Emulsion counter
  Lucite plate
  Spacer

Spacing layer

Analyser
  Lead plate
  Emulsion counter
  X ray film

The target on the upper part of the chamber is a pile of two-fold emulsion counters, bare lucite plates and some amounts of spacers. In this part, it is possible to observe cross-sectional view of the secondary tracks each ∿1mm along the shower axis. Charged secondary tracks could be examined for sudden changes in direction several centimeters from production vertices. Special care taken in assembling the chamber enabled us to detect the relative angle between charged secondary tracks with error, in favorable cases, less than 10 micro-radians.

The analyser at the lower part of the chamber is a pile of lead plates, two-fold emulsion counters and some sets of x ray films. A track of charged particle or a group of cascade electrons could be observed at each 0.2 radiation length. An electron or a γ ray is clearly discriminated from other components by inspecting a cascade shower it induced in the lead emulsion pile. Energy and momentum of an electron or a γ ray could be estimated analysing each cascade shower. For charged particles, relative scattering method is applied to estimate their momentum. High precision in measurement of the relative distance between secondary tracks, and high scattering signal due to lead nuclei allow us to estimate momentum up to TeV/c region. Technical details and results of calibration of the method for energy momentum estimation are described in a separate paper of this issue[7].

Emulsion chambers exposed to cosmic rays at airplane or balloon altitude are examined for super high energy jet showers after processing of emulsion counters and x ray films. At first, x ray films inserted in the lower part of the chamber are scanned by naked eyes for dark spots due to cascade showers induced by γ rays. They are traced back to vertices plate by plate up into the target layer. After detecting vertices of jet showers, all secondary charged particles are followed down to the down stream end of the chamber to be examined for sudden changes in direction or other secondary phenomena. Any vee produced by neutral hadrons in the inspection volume is easily detected in the course of this analysis.

Emission and deflection angles, energy and momentum of secondary and tertiary charged particles or γ rays are measured and analysed by means of the methods described in a separate paper[7]. Tracing back of cascade showers induced by γ rays to vertices several centimeters away, and detection and analysis of kinks and vees up to several centimeters from production vertices are carried out with ease and with sufficient accuracy in the emulsion chamber. This type of emulsion chamber enabled us to extend the power of pure emulsions to study very high energy interactions to tens of TeV, and it opened a new window for search of short-lived particles.

SUMMARY OF THE X PARTICLES.

By means of the emulsion chamber techniques several cosmic ray events have been accumulated up to now in which associated decays of the X particles are observed. In those cases the background level simulating the true decay of the short-lived particles is guaranteed to be of the order of 10^{-4} as analysed by Gaisser et al.[8] Table I is the list of events obtained not only from our own exposures but also from reanalysis of old events published by others.

The event labelled 6B-23 is the first event described at the beginning of this paper, and the event BEC-II was observed by another Japanese group, Sugimoto et al.[4]. The events 11c-34 and 6b-19L are from our own exposures. The T star and ST-2 are ones digged out of reanalysis[9] of events published by others. T star[10] was observed by the Rochester Group in 1951 in their first experiment applying emulsion cloud chamber to the study of super high

Table I Event list in which an associated decay of the X particles are observed.

	Charge of primary particle	E_0 TeV	n_s	ΣE_γ TeV	E_{π^0} or γ TeV
6B-23	0	10	70	4.5	3.2
T	1	20	36	6	2.4, 1.6
ST-2	?	25	51	2.1	1.0
11c-34	0	20	70	8	3.0, 1.2, 0.6
6a-19L	1	20	20	2.0	
BEC-II	1	10	27	2.7	

energy phenomena. The jet shower ST-2[11]) was observed also in the first balloon experiment of the Japanese Emulsion Chamber Group in 1956. It is exceedingly impressive that since old days they have already observed such events very similar to ours by means of the emulsion chamber techniques. Following are the short descriptions of these events.

BEC-II ; Figure 5a is a sketch of the event BEC-II. Among 27 charged products of the jet shower were two prongs no.2 and no.20. After 3.04cm, particle no.2 scattered through 3.6×10^{-3} radians, and 3mm from the scattering point a pair of electrons materialised from a γ ray named H. Another γ ray, B, produced an electron cascade in the lower lead emulsion sandwich. These γ rays, H and B, were found to be daughters of η^0 emitted from the scattering point. The second particle no.20 travelled 6.34cm before scattering through 1.2×10^{-2} radians. Two γ rays F, and G, daughters of a π^0 meson emitted at the scattering point produced electron pairs at 4.1mm and 8.2mm down stream respectively.

The estimated invariant masses of the two X particles depend on the nature of the charged daughter. They are given in the table attached to the figure with estimated life times.

T-star; Figure 5b shows the essential feature of the T star. In this old event, were observed two π^0 mesons with much higher energy than others, 2.4 TeV and 1.6 TeV, which decayed into two γ rays at the point 2.5cm and 7cm respectively from the origin of the event. Kaplon et al. reported in their paper[10]) that they found 'delayed decay' of π^0 mesons of the order of 10^{-14} second. Kuramata et al.[9]) reinterpreted the event as a pair production of the X particles each decaying into π^0 meson and other partner(s).

ST-2; Figure 5c shows the sketch of central part of the event ST-2. Reanalysis on this event revealed following facts. One of secondary charged particles, t, suffered a multibody decay into one charged daughter one η^0 and other missing neutral daughter(s) at the point 7.63cm down stream from the primary vertex. The charged daughter, t', suffered a sudden change in direction at the point

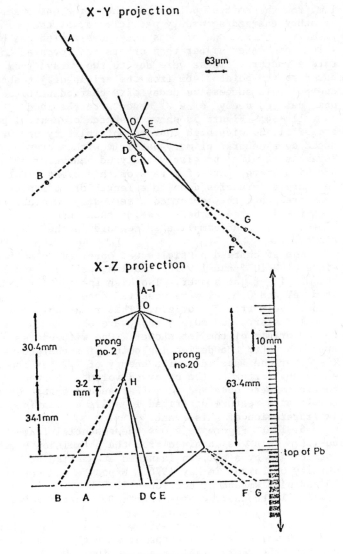

Fig. 5a

BEC-II
Sugimoto et al.[4)]

	Decay mode	Mass GeV	Life time sec
no.2	$\eta^0 + K^\pm$	1.66	5.1×10^{-13}
	$\eta^0 + \Sigma^\pm$	2.23	
no.20	$\pi^0 + K^\pm$	1.74	3.4×10^{-12}
	$\pi^0 + \Sigma^\pm$	2.36	

1cm from the emitted point. Another kink was observed on the track of other charged secondary particle, d, at the point 8.9cm from the primary vertex. One of cascade showers with the highest energy, 1 TeV, one order higher than others was revealed to show a double core structure. These were due to two γ rays from a π^0 meson produced at the point 8.4cm from the origin of jet shower. In total, one possible successive decay of a charged particle, one kink and one 'delayed decay' of a π^0 meson were observed.

11c-34; Figure 5d shows a sketch of central part of the jet shower 11c-34 with high charged multiplicity of 70. This was induced by a neutral primary particle with energy of ∼20 TeV. Kuramata et al.[9] precisely analysed behaviour of charged particles and high energy part of γ rays of this event. Following tracks of 26 charged particles down to a length of 0.25 nuclear interaction mean free path they observed 6 secondary interactions which was just the expected number. Beside these usual secondary interactions, they found complicated feature in the central part of this event.

One of charged particles was observed to suffer successive kinks of 2×10^{-3} and 1.9×10^{-3} radians at 1.1cm and 1.1+0.063cm from the origin of jet shower. Two high energy π^0 mesons with energy of 3 TeV and 0.6 TeV were observed decaying at 6.14cm and 1.1cm respectively from the origin. This remarkable feature is just like the T star. Two more γ rays were observed not to be traced back to the origin of the jet shower. One with energy of 55 GeV was traced back to the deflecting point of a charged particle "1". The other one labeled m with energy of 1.2 TeV could not be connected to any vertex in the observed volume. Further two more charged particles were observed to suffer sudden change of their direction. Complicated feature described is still preventing them to fix final interpretation of this event.

6a-19L; Figure 5e shows the essential feature of the event 6a-19L. This was obtained from the 3 emulsion chambers exposed to cosmic rays by a balloon launched at Sanriku Balloon Center, University of Tokyo, in May 1976. Exposure was carried out at a height of 31km and it continued 25.5 hours. Fuchi et al.[12] followed all secondary charged particles in the forward cone of 27 jet showers induced by singly charged primary particles. Out of 16m of total track length followed, 15 secondary phenomena were observed which is to be compared with the expected number of 12.0 assuming the geometrical cross-section.

The event 6a-19L was found in the course of this analysis. Among 20 charged secondaries, particle "20" suffered a multibody decay into 3 charged hadrons and missing neutral(s) at the point 0.79cm from the origin. One of tertiary charged particles "0-3" of a secondary jet shower induced by a neutral particle "0" at 6.3cm from the primary vertex suffered a sudden change in direction. Deflecting point was in a polystyrene film of a two-fold emulsion counter with thickness of 150μm at 0.26cm down from the secondary interaction vertex. No signal of evaporating track in the emulsion counter suggest them possibility of a decay of short-lived particle.

Fig. 5 b, c, d, e.

b) T-star; Kaplon et al.[3] [10]

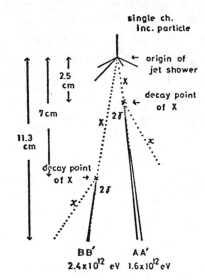

c) ST-2; K.Nishikawa[3] [11]

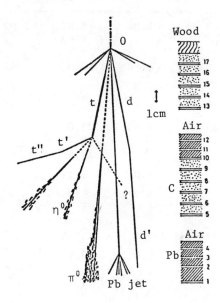

d) 11C-34; Kuramata et al.[9]

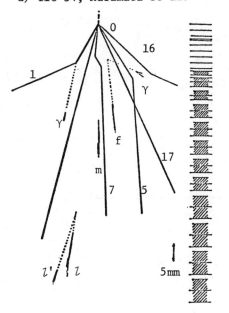

e) 6a-19L; H.Fuchi et al.[12]

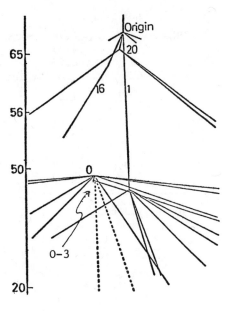

Table II Summary of the X particles.

	L cm	θ'/θ		M GeV	$\tau \times 10^{-13}$ sec	
6B-23	1.38	8.92	$\pi^0 + K^\pm$	2.15	0.27	
		1.63	or $\pi^0 + \Sigma^\pm$	3.5	0.42	pair
	4.88	10.3	$x^\pm + x^0$		~15	
T	7.3		$\pi^0 + ?$		0.1~1	pair
	2.5		$\pi^0 + ?$		0.2~3	
ST-2	7.63	21.0	$\eta^0 + x^\pm + x^0$	2~3	~20	
		9.2				
		19.6				cascade
	1.0	0.21	$x^\pm + x^0$		< 56	
	8.9	1.3	$x^\pm + x^0$		~7	multi
	8.4		$\pi^0 + x^0$		6	
11c-34	1.1	3.36	$x^\pm + x^0$		0.4	
	0.063	0.94	$x^\pm + x^0$	> 2	0.05	cascade
	1.1		$\pi^0 + x^0$		~1.2	
	6.14		$\pi^0 + x^0$		~1.4	
	1.18		$\gamma + x^\pm + x^0$		~5	multi
	1.6		$x^\pm + x^0$		~10	
	1.6		$x^\pm + x^0$		~10	
6a-19L	0.79	4.76	$x^\pm + x^\pm + x^\mp + x^0$	>1.5	~5	
		4.37				
		8.73				pair
	6.3		interaction		~18	
	0.27	1.07	$\rightarrow x^\pm + x^0$		~1	
BEC-II	3.04	2.08	$\eta^0 + K^\pm$	1.66	5.1	
		2.46	or $\eta^0 + \Sigma^\pm$	2.23		pair
	6.34	3.49	$\pi^0 + K^\pm$	1.74	34	
		3.12	or $\pi^0 + \Sigma^\pm$	2.36		

$\Sigma L = 71.3$ cm

They interpreted the event as follows. The neutral particle "0" is a neutral X particle associatingly produced with the charged X particle "20" at the original interaction. The neutral X particle "0" induced a jet shower in a lead plate suffering a charge exchange. The regenerated charged X particle "0-3" itself takes its turn to decay after 0.26cm flight from the regenerated point.
In this case, flight times of the neutral particle "0" and the charged particle "0-3" are estimated as 1.8×10^{-12} sec and 10^{-13} sec respectively.

Table II shows the summary of the X particles distingushed from produced particles of those events listed in Table I. Flight length, ratio of deflection to emission angle, estimated mass and life time of the X particles are summarized. In the observed total track length of the X particles, ∿70cm, one possible secondary interaction of the X particle was observed. This is not inconsistent with the expectation if the X particle is a strongly interacting particle as the conventional hadrons.

In Table III they are classified according to the decay type.

Table III Decay type of the X particles.

	$X^{\pm}$			X^0		
	I	II	III	IV	V	VI
	2 body	?	many body	2 body	?	many body
6B-23	$\pi^0 + x^{\pm}$					
		$x^0 + x^{\pm}$				
T					$\pi^0 + ?$	
					$\pi^0 + ?$	
ST-2			$\eta^0 + x^0 + X^{\pm}$			
		$\hookrightarrow x^0 + x^{\pm}$				
		$x^0 + x^{\pm}$			$\pi^0 + x^0$	
11c-34		$x^0 + X^{\pm}$				
		$\hookrightarrow x^0 + x^{\pm}$				
					$\pi^0 + x^0$	
					$\pi^0 + x^0$	
			$\gamma + x^0 + x^{\pm}$			
		$x^0 + x^{\pm}$				
		$x^0 + x^{\pm}$				
6a-19L			$x^{\pm}+x^{\pm}+\overline{x^{\pm}}+x^0$			
		$x^0 + x^{\pm}$				
BEC-II	$\eta^0 + x^{\pm}$					
	$\pi^0 + x^{\pm}$					
	3	8	3	0	5	0
		14			5	

Number of clear two-body decay is only 3 in 19 and this indicates the dominance of multibody decay of the X particles. Raw charge neutral ratio of observed X particles is 14:5 including detection bias. About the mass of X particles, only sample of type I in Table III are available. Assuming that the charged daughter of the X particle is a K meson, mean value of 1.85 ± 0.4 GeV is calculated from 3 samples.

Figure 6 shows a distribution of life time of the X particles estimated assuming mass value as 2 GeV. Distorting effect of the finite size of the emulsion chamber on obtained life time distribution was estimated not to be serious. Statistics being still poor, it seems unlikely to fit it with a single exponent. Roughly saying, the mean life time may be of the order of $7\sim8\times10^{-13}$ sec, but two kind of life times, $3\sim5\times10^{-13}$ sec for neutral particles and $1\sim2\times10^{-12}$ sec for charged particles, are also considerable.

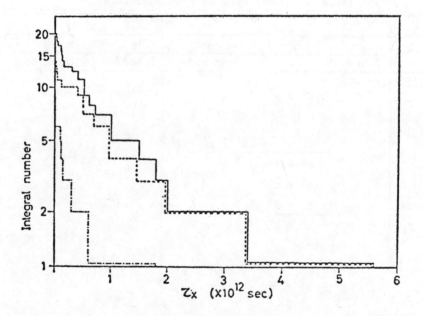

Fig. 6. Life time distribution of the X particles
—— total, ····· charged, — · — · neutral

For about the production rate of the X particles in the cosmic ray region, it is difficult to estimate it accurately, because of the detection biases. The rate of one event per $20\sim40$ observed jet showers of 10 TeV range is, however, roughly estimated as is shown in Table IV.

Table IV Rate of pair production event.

Experiment	Number of pair production event	Number of observed jet showers
AEC	2	∽ 70
BEC ('56)	1	∽ 20
BEC II	1	∽ 20
Kaplon et al.	1	?
BEC-6	1	27

1 Event/20 ∽ 40 obs. jet shower

OPENNIG OF NEW CHANNELS AT SUPER HIGH ENERGIES ?

As indicated in Table II and III, possible new kind of cascade decays and multiple production of the X particles were observed in two of the jet showers, 11C-34 and ST-2, listed in Table I.
Sources of those cascade decays are as follows.
1) Successive decay of a charmed baryon with higher charm, Ξ_c or Ω_c.

2) Decay of new heavy hadron with bottom quark to a charmed hadron followed by a decay of the latter[13].

Though a definite explanation is still difficult at present, those observations seem to suggest an opening of new channels at super high energy region.

Recently, Ogata et al.[5] reported some events showing possible multiple production of successively decaying particles.

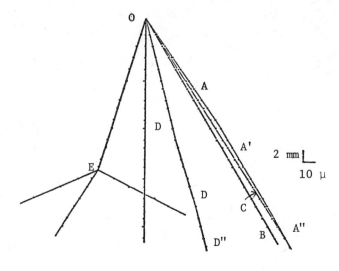

Fig. 7. KG-7.

n_{ch} = 35
n_γ = 10
ΣE_γ = 29.5 TeV
torpedo

They exposed a small emulsion chamber on board of J.A.L. freight-plane at a height of 260 gr/cm^2 for about 600 hours. A jet shower essential part of which is shown in Figure 7 was observed to be produced by singly charged particle in a glass base of an emulsion counter. Number of charged secondary particles was 35. Energy sum of 10 observed γ rays amounts to 29.5 TeV which is the highest value ever observed at airplane altitude in their series of observations.

Among charged particles produced they found three not converging to the primary vertex but to the point 410 μm down stream of it inside the glass base. This indicates that these 3 particles are tertiaries produced by a secondary particle. One of tertiary, A, observed to suffer successive deflection of the order of 10^{-4} radians after 1.85cm and 1.38cm flight. The other secondary particle, D, also observed to suffer successive deflection of the same order as the A after 2.10cm and 0.96cm flight. One more charged secondary, E, producing a three track event in a lead plate 2.59cm down stream of the primary vertex was also observed. All of those particles attracting their notice were confirmed to be hadrons because of the observed behaviour in the lead emulsion sandwich. They tried to interpret these phenomena by conservative assumption, nuclear or electromagnetic interactions, but without success.

Figure 8 is showing a sketch of another complicated event they obtained from an exposure of an emulsion chamber at balloon altitude. Number of secondary charged particles was as high as 93. Though their analysis of this event is not yet finished, the figure seems to suggest the variety of phenomena at tens of TeV range as well as dificulty of explanation at present.

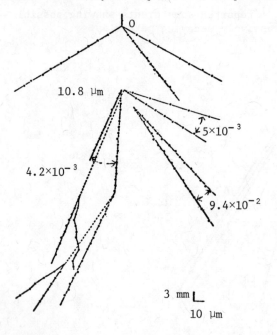

Fig. 8. OB-1.

n_{ch} = 93
n_γ = 31
ΣE_γ = 40 TeV
torpedo

HOW ABOUT AT 100 TEV RANGE ?

How about the production rate of new particles at much higher energy region above 100 TeV range ? At this energy region, Japan-Brasil emulsion chamber group has been studying energy and angular distribution of γ rays produced in the local target by nuclear active component arriving to Mt.Chacaltaya, Bolivia, 5200m above sea level.

The detector they are utilizing is a huge emulsion chamber of two story structure with interleaved local target.

High energy γ rays emited in the forward cone of a jet shower produced at the target are observed in the lower chamber as a group of cascade showers with a spread of several mm.

Daughter γ rays of short-lived particle produced in association with the jet shower and decaying in the air gap may arrive the lower chamber with flight direction not converging to the common vertex. Sawayanagi[6] tried to find such γ rays by geometrical method. A small variation of the relative distance of any two cascade cores with depth in the lower chamber gives a relative angle between arrival directions of the two initiating γ rays, as is shown in Figure 9.

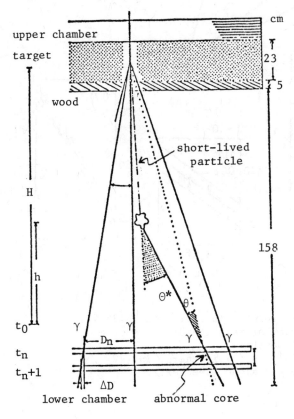

Fig. 9.

Structure of Chacaltaya emulsion chamber and idea of detecting short-lived particles.

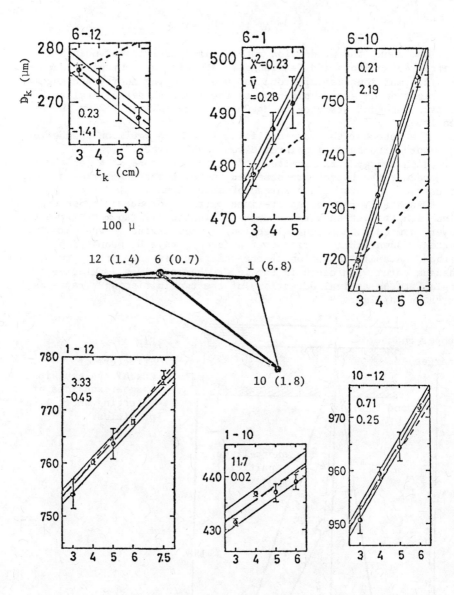

Fig. 10. Map of most forward part of Event B135-4 and variation of relative distance of cascade cores with depth.
Numbers given in ( ) is energy in TeV.
----- expected slope (see text).

The search for γ rays with abnormal arrival direction is made over 11 jet showers out of 42 samples with $\Sigma E_\gamma \leq 15$ TeV.

Out of 11 jets, 3 are found to accompany abnormal cores.

For example Figure 10 shows the result of geometrical measurement between core No.6 and other cascade cores in the Event B153-4. Dotted lines show divergence expected if both incident γ rays are produced in the target layer. One recognizes the abnormal arrival direction of the γ ray initiating the core No.6, while normal arrival directions of other γ rays are confirmed.

Another method to detect short-lived particle decaying in the air gap is to find isolated close pair of γ rays which could not be the daughter of π^0 meson produced in the target. In the Event B154-1, two such close pairs were observed in connection with abnormal cores. Estimated production heights of two π^0 mesons are 42 and 64cm above the lower chamber. In this case, emission angles θ^* and emission transverse momentum P_T^* could be estimated from the measured deflection angle θ and the energy E of the abnormal cores. These are summarized in Table V with data on two single abnormal cores and one delayed π^0 meson observed in other event[14]).

Table V Result of the cores with abnormal arrival directions, and of the delayed π^0 mesons.

Event	Energy of abnormal core (TeV)	D (μm)	h (cm)	θ^* (10^{-4} rad)	P_T^* (GeV/c)	l=H-h (cm)	τ (10^{-13} sec)
B135-4	0.7	–	–	7.0±3.0 ∼ 35±15	0.5±0.2 ∼ 2.5±1.0	30∼140	∼ 1
B138-8	3.3	–	–	3.2±1.8 ∼ 16± 9	1.0±0.6 ∼ 5.0±3.0	30∼140	∼ 1
B154-1	2.5+2.0	25.5	42	$5.0^{+8.0}_{-4.2}$	$2.3^{+3.6}_{-1.9}$	∼ 120	4∼40
	3.5+2.1	31.9	64	$3.8^{+8.2}_{-3.8}$	$2.1^{+4.6}_{-2.1}$	∼ 100	3∼30
B153-1	13.3+6.7	13	90	–	–	∼ 70	∼ 1

Back ground source for abnormal γ ray or delayed π^0 meson may be early decay of strange particles, secondary interactions in the air gap or other interactions in neighbourhood at the target. However, contamination rate may be negrigible because of large air gap and smaller frequency of effective neighbouring interaction.

He got two jet showers accompanied by a γ ray with abnormal direction and two jet showers containing delayed π^0 meson emission in 12 analysed events. To draw a definite production rate of short-lived particles, this raw data should be corrected for bias effects which lead to under or over-estimation. The frequency of the

X particle observation seems, however, to be very high at 100 TeV range.

SUMMARY

In conclusion, the production rate of the X particle is roughly estimated 1∿5% at 10 TeV range and still higher percentage is indicated at 100 TeV range. Uncertainty of bias in detection of jet shower make us somewhat difficult to estimate the cross-section quantitatively, but it may be roughly about one order higher than that of charmed particles at accelerator region. The opening of new channels as well as higher excitation due to higher energy may give us proper understanding of this increase.

Possible new type of cascade decays of short-lived particles are observed rather frequently in the super high energy cosmic ray region. Successive decay of a charmed baryon with higher charm, Ξ_c or Ω_c and decay of new heavy hadron with bottom quark to a charmed hadron followed by a decay of the latter may be considered as the source of those phenomena. Because of large amount of available energy, higher mass states could be copiously produced at the super high energy region.

REFERENCES

1. K.Niu et al., Conf. Paper 12th Int. Conf. on Cosmic Rays, Hobart, 7, 2792 (1971).
   Prog. Theor. Phys. 46, 1644 (1971).
2. T.Hayashi et al., Prog. Theor. Phys. 47, 280 and 1988 (1972).
3. S.Kuramata et al., Conf. Paper of 13th Int. Conf. on Cosmic Rays, Denver, 3, 2239 (1973).
   K.Hoshino et al., DPNU-16, July (1975).
   M.Kaplon et al., Phys. Rev. 85, 900 (1952).
   K.Nishikawa, Jour. Phys. Soc. Japan 14, 880 (1959).
4. H.Sugimoto et al., Prog. Theor. Phys. 53, 1541 (1975).
5. T.Ogata et al., Private communication and paper, this volume.
6. K.Sawayanagi, Doctor Thesis, Waseda University and Phys.Rev.D
7. H.Fuchi et al., this volume.   (to appear).
8. T.K. Gaisser and F. Halzen, Phys.Rev. D14, 3153 (1976).
9. S.Kuramata et al., Conf. Paper of 13th Int. Conf. on Cosmic Rays, Denver, 3, 2239 (1973).
10. M.Kaplon et al., Phys. Rev. 85, 900 (1952).
11. K.Nishikawa, Jour. Phys. Soc. Japan 14, 880 (1959).
12. H.Fuchi et al., Contributed to the XIX Int. Conf. on H.E.Physics, Tokyo, #491 (1978).
13. J.Ellis et al., Nucl. Phys. B131, 285 (1977)
14. N.Arata, Nuovo Cimento 43A, 455 (1978).

DIRECT SEARCH FOR CHARM IN ACCELERATOR PROTON COLLISIONS*

Taiji Yamanouchi
Fermi National Accelerator Laboratory, Batavia, Ill. 60510

ABSTRACT

Experimental evidences concerning the production of charm particles by high energy (200 to 400 GeV) accelerator protons in nuclear emulsions are reviewed. Consistencies among various experimental results are discussed.

CONTENTS

I. INTRODUCTION

II. EXPERIMENTAL RESULTS

    A. Niu group
    B. Kusumoto group
    C. Lebedev group
    D. European collaborations

III. CONCLUSION

I. INTRODUCTION

In the preceding talk Professor Niu has reviewed cosmic ray evidences concerning the production of short-lived particles in nuclear emulsions. I am going to present a similar review for accelerator emulsion experiments.

In the case of cosmic ray experiments discussed by Professor Niu the primary energies are greater than 1 TeV, much higher than the present accelerator beam energy. In such high energy region new particles beyond charm could be produced with non-negligible cross sections. It may complicate the interpretation of the data. In the case of accelerator experiments the most likely candidate for short-lived particles are charm particles. Furthermore, the energy and flux of the incident particle and detection biases are better known for accelerator experiments. For these reasons accelerator experiments are better suited for quantitative study of charm production.

The recent beam dump neutrino experiments at CERN[1] and the calorimeter experiment by Caltech-Stanford group at Fermilab[2] have indicated a source of prompt neutrinos in 400 GeV proton-nucleon collisions which, assuming charm production followed by semi-leptonic

---

* Work supported by the U.S. Department of Energy. Invited talk at Workshop on Charm Production and Lifetimes, Bartol Research Foundation, University of Delaware, October 19, 1978.

decays, requires charm production cross section in the range of 30 to 400 μb. An earlier charm search in emulsion at 300 GeV by a large European collaboration[3] gave an upper limit of 1.5 μb at the 90% confidence level. A similar result has been obtained by another European group.[4] There is an apparent disagreement between the "beam dump" data and the emulsion data. I would like to come back to this question after reviewing more recent emulsion data.

Charm particles have to be produced in pairs in hadronic reactions. In this review I am going to concentrate on those events in which two short-lived tracks are seen. By requiring both decays to be seen one can effectively eliminate backgrounds due to misidentification of secondary interactions or strange particle decays near the primary interaction. If the lifetime of a charm particle is very short ($<10^{-16}$ sec) or very long ($>10^{-11}$ sec) we may miss some of the decays, because they tend to take place too close to or too far from the production vertex. In this case our requirement should reduce detection efficiency significantly. We have to keep in mind such a possibility.

II. EXPERIMENTAL RESULTS

A. Niu group

At the XIX International Conference on High Energy Physics in Tokyo, Niu and his colleagues have reported the first candidate for a pair production of charm particles by accelerator beam.[5] The event was observed in an emulsion chamber[6] exposed to the 400 GeV proton beam at Fermilab. More than 1300 nuclear interactions have been studied before this event. Fig. 1 is a schematic drawing of the event. A vee (V) was observed at 320 μm downstream of an interaction (O). One leg (m) of the vee was identified as an electron, or a positron, because of an electromagnetic shower produced by this track. The other leg (n) was identified as a hadron, because a nuclear interaction of this track was observed. The vee was not coplanar with respect to the line connecting between the interaction vertex and the vertex of the vee. The vee was interpreted as a semileptonic decay of a neutral charm particle. Further study of this event revealed two γ ray showers (d and e) in the downstream of the primary interaction. These

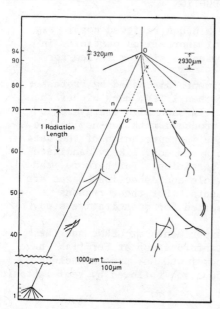

Fig. 1. Schematic drawing of Niu's first event.

γ rays did not aim at the interaction vertex but their lines of flight, defined by the first pair of the shower, did intersect at a point (X) 2930 μm downstream of the interaction. Invariant mass of the γ rays was consistent with that of a neutral pion. No particles were observed in the vicinity of the point X. Additional γ rays were looked for in the area downstream of the point X. No such γ ray was observed. It could be interpreted as a decay of a charm particle into a neutral pion and other neutral particle(s). Table I summarized relevant parameters of these two decay vertices.

Table I Summary of the two vees

	V^0	X^0
Flight length μm	320 ± 20	2930 ± 200
Emission angle rad	$1.89 \cdot 10^{-2}$	$2.83 \cdot 10^{-2}$
Daughters	had + $e^{\pm}$ + ?	π^0 + ?
$\theta_{daughter}$ rad	$\theta_{had} = 1.23 \cdot 10^{-2}$ $\theta_e = 3.90 \cdot 10^{-2}$	$\theta_{\pi^0} = 1.4 \cdot 10^{-2}$
Visible momentum GeV/c	$P_{had} = 52 ^{+8}_{-6}$ $P_e = 6.25 ^{+1.60}_{-0.82}$	$P_{\gamma d} = 11.1 \pm 3$ $P_{\gamma e} = 2.9 \pm 0.9$
Assumed charm meson decay mode (1.863 GeV)	$K^{\pm} e^{\mp} \nu_e$	$\pi^0 K^0$
P (Neutral)	7 GeV/c	1.2 GeV/c
P_t GeV/c	$1.23 ^{+0.21}_{-0.13}$	0.43 ± 0.10
τ sec	$(3.02 ^{+0.39}_{-0.42}) \cdot 10^{-14}$	$(1.18 ^{+0.34}_{-0.20}) \cdot 10^{-12}$
Assumed charm baryon decay mode (2.26 GeV)	$\Sigma^{\pm} e^{\mp} \nu_e$	$\pi^0 \Lambda^0$
P (neutral)	6 GeV/c	4.8 GeV/c
P_t GeV	$1.21 ^{+0.19}_{-0.13}$	0.53 ± 0.12
τ sec	$(3.75 ^{+0.51}_{-1.34}) \cdot 10^{-14}$	$(1.7 ^{+0.34}_{-0.21}) \cdot 10^{-12}$

Recently Niu and his colleagues have observed another candidate for a pair production of charm particles in the same emulsion chamber. (They have studied ~ 300 more nuclear interactions after their first event).[7] Since the new event was discovered very recently and the analysis is still in progress, I simply present a rough sketch (Fig. 2) of the event and some of observed parameters (Table II).

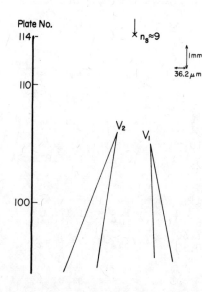

Fig. 2. Niu's second event.

Table II. Summary of the second event (Preliminary)

	V_1	V_2
Flight length μm	4680	4240
Emission angle rad	1.69×10^{-2}	2.93×10^{-2}
θ daughter rad	2.50×10 1.56×10	1.38×10 1.25×10
P daughter (GeV/c)	$9.8 {\,}^{+2.5}_{-1.6}$ $16.0 {\,}^{+6.3}_{-4.0}$	$5.6 {\,}^{+2.4}_{-1.3}$ $4.9 {\,}^{+2.1}_{-1.2}$
Missing P_t (GeV/c)	$0.10 {\,}^{+0.04}_{-0.02}$	$0.09 {\,}^{+.04}_{-.02}$
Associated γ rays	2 or 4	None
Rough life time	8.7×10^{-13}	2.3×10^{-12}

On the basis of the first event (one charm production out of ~ 1300 nuclear interactions) Niu has estimated charm production cross section to be 80 ± 40 μb. In this estimate he has (implicitly) assumed $\sigma_{charm} \sim A^{2/3}$. The cross section will be reduced to ~ 25 μb, if $\sigma_{charm} \sim A^1$ is assumed. If one includes the second event, it will increase to ~ 40 μb.

B. Kusumoto group

Kusumoto and his collaborators have been looking for charm pair production in emulsion stacks exposed at Fermilab.[8] Two stacks of Ilford K5 pellicles were exposed to protons of 200 and 400 GeV. I am going to focus our attention to the 400 GeV exposure, because charm production cross section might be considerably smaller at 200 GeV. They have analyzed ~ 1600 nuclear interactions. Secondary tracks were followed up to 4 mm from the primary vertex. For neutral decays an area which extends 500 μmm along the beam direction, ± 150 μm in projected plane, and ± 200 μm in depth has been scanned with high magnification. When a charm candidate was observed, further scanning was performed to look for associated decays over a distance of 3 mm along the beam line for neutral decays. No associated events were observed. Table III gives a summary of their result.

Table III. Summary of Kusumoto's results

		205	405	total
	momenta of protons GeV/c	205	405	total
	number of proton interactions	2288	1635	3923
charged secondaries	projected track length followed m	43*	49**	92
	number of secondary interactions			
	kinks $\theta \geq 10^{-2}$ rad	6	26	32
	$\theta \geq 5 \times 10^{-2}$ rad	0	19	19[a)]
	others total	120	121	241
	$N_h \geq 1$	105	109	214
	$N_h = 0$	15	12	27
	$n_s = 3$	3	2	5[a)]
	$n_s = 5$	3	0	3[a)]
neutral secondaries	number of proton interactions	2528	1635	4163
	number of secondary interactions	217	99	316
	vees total	217	99	316
	$\theta \geq 10^{-2}$ rad	18	18	36
	$\theta \geq 5 \times 10^{-2}$ rad	9	0	9[b)]
	others $N_h = 0$	0	0	0
	$N_h \geq 1$		not recorded	

Table III continued

* Secondaries emitted within a half angle of 10^{-1} are followed
** Secondaries within a half angle extended to $45°$ are followed.
a) The number of candidates for decays of charged charmed-particles. Total number is 27.
b) The number of candidates for decays of neutral charmed-particles. Total is 9.
θ is the deflecting angle of kink or the opening angle of vee.

For the candidates of single decays, measurements of the grain density and/or multiple scattering has been applied. Some of them have been confirmed as strange particle decays. The effort is still in progress.

On the basis of no associate production candidate out of 1635 nuclear interactions, taking into account of the average detection efficiency, one can get an upper limit of ~ 40 μb (90% confidence level) from this experiment.

C. Lebedev group

Chernyasky et al have also studied charm production in an emulsion stack exposed to the 400 GeV proton beam at Fermilab.[9] Their results, based on the analysis of 1,120 nuclear interactions, have been presented at the Tokyo Conference. Only one possible candidate for the pair production was observed. In addition there were 8 single production candidates. If I take only the pair production candidate, as I have been doing in my talk, the production cross section will be 10 ~ 40 μb. The authors of this work have concluded, on the basis of probability arguments, that all 9 events were true charm events and $\sigma_{charm} \approx 120$ μb.

It is interesting to note that all "decay" vertices were found within 100 μm from primary interactions. Although the scanning was carried out for 1 mm along the beam direction, no candidates were observed beyond 100 μm. From the distribution of the "decay" vertices, they have obtained an estimate of the life time as ~ 2×10^{-14} sec. It is also worthwhile to point out that the frequency of the single production candidates observed in this experiment is about the same as that of other experiments.[3,4,8] Other experimenters have rejected those events because charm particles must be produced in pair and also it is difficult to estimate background contributions for single production candidates. It is possible, as I mentioned earlier, that the life time of the second charm particles is much shorter ($\lesssim 10^{-15}$ cm) than that of the first (observed) one and difficult to separate their decay vertices from the primary interactions. Although such a possibility cannot be ruled out, I would like to stick to my own prejudice, in the remainder of my talk, that only pair productions are the reliable candidates for charm production.

D. European collaborations

More than a year ago, two important results were published by European groups.

Coreman-Bertrand et al, have analyzed ~ 62,000 nuclear interactions in an emulsion stack exposed to the 300 GeV proton beam at Fermilab.[3] No candidate for the pair production of charm particles has been found in this sample. The authors have concluded that the cross section at 300 GeV is less than 1.5 μb at a 90% confidence level.

Bozzoli et al. have exposed two stacks of emulsions to protons of 300 and 400 GeV at Fermilab.[4] They have studied ~ 5,000 nuclear interactions at 300 GeV and ~ 7000 nuclear interactions at 400 GeV. Again no candidate has been observed. They have given an upper limit 7 μb (90% confidence level).

As I mentioned in the beginning there is an apparent disagreement between these emulsion results and the indirect estimate of charm production cross section from the beam dump experiments. Crennel et al have studied the sensitivity of the emulsion experiments to charm lifetime using a Monte Carlo analysis.[10] Fig. 3 shows their results on the upper limit of the cross section as a function of mean lifetime of the charm particles. They have concluded that the emulsion results and the beam dump data do not disagree if the lifetime is either less than 5×10^{-16} sec or greater than $\sim 10^{-12}$ sec.

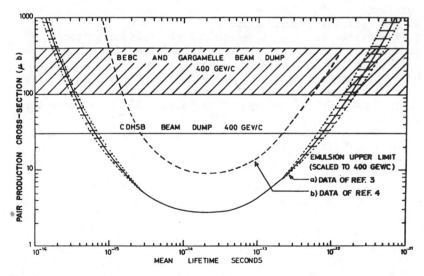

Fig. 3 Monte Carlo calculation of the sensitivity of emulsion experiments to charm particle production as a function of lifetime (Ref. 10).

IV. CONCLUSION

The experimental study of charm particle production in hadronic reaction is still in a preliminary stage. It is difficult to draw a definite conclusion from presently available data. However, there is no serious contradiction among existing data, if

a) $\sigma_{charm} = 10 \sim 30$ μb at 400 GeV, and

b) $\tau \gtrsim 5 \times 10^{-13}$ sec.

These values are also quite consistent with theoretical predictions.

REFERENCES

1. P. Alibran et al., Phys. Lett. <u>74 B</u>, 134 (1978);
   T. Hansl et al., Phys. Lett. <u>74 B</u>, 139 (1978);
   P. C. Bosetti et al., Phys. Lett. <u>74 B</u>, 143 (1978).
2. A. Bodek, This conference.
3. G. Coremans-Bertrand et al., Phys. Lett. <u>65 B</u>, 480 (1976).
4. W. Bozzoli et al., Lett. al Nuovo Cim. <u>19</u>, 32 (1977).
5. N. Ushida et al., "Observation of decays of short-lived neutral particles produced in 400 GeV/c proton interactions", presented at session B7, paper no. 492, XIX International Conference on High Energy Physics, Tokyo, August 23-30, 1978, Proceedings edited by H. Miyazawa, to be published by National Laboratory for High Energy Physics, Tsukuba, Japan; private communication from K. Niu.
6. The construction of the emulsion chamber has been described in detail in K. Niu's talk at this conference.
7. K. Niu, private communication.
8. Y. Yanagisawa et al., "A search for short-lived particles produced by 205 and 405 GeV/c protons in emulsions", October, 1978 (unpublished); private communication from O. Kusumoto.
9. M. M. Chernyavsky et al., "Observation of the new short-lived particles in nuclear emulsion exposed to protons with the momentum 400 GeV/c", presented at Session B7, 1978 Tokyo Conf. op. cit. (Ref. 5), paper no. 308.
10. D. J. Crennell, C. M. Fisher, and R. L. Sekulin, Phys. Lett. <u>78 B</u>, 171 (1978).

OBSERVATION OF ENERGETIC DELAYED HADRONS IN EXTENSIVE AIR SHOWERS--
NEW MASSIVE PARTICLES?

J. A. Goodman[#], R. W. Ellsworth, A. S. Ito[*], J. R. MacFall[**],
F. Siohan[***], R. E. Streitmatter, S. C. Tonwar[+],
P. R. Vishwanath, and G. B. Yodh[++]

Department of Physics and Astronomy, University of Maryland
College Park, Maryland 20742

ABSTRACT

We report the results of a search for massive relatively long lived particles that has been carried out in an experiment performed at the Sacramento Ridge Cosmic Ray Laboratory in Sunspot, New Mexico (2900 meters altitude) during 1975-76. Comparison of the experimental data with detailed Monte Carlo simulations provides evidence for the production of new massive particles at high energies.

INTRODUCTION

Discovery of the Ψ (3.1 Gev/c^2)[1,2] in 1975 and of Υ (9.4 Gev/c^2)[3] recently have added further interest to the searches for massive particles which were initiated earlier by suggestions of Gell-Mann[4] and Zwieg[5] in 1964 about the possible existence of quarks. Since quarks of new flavors are expected to be heavier than about 5 Gev/c^2, their production cross-section at existing machine energies may be negligible as suggested by the experiments of Cutts, et al[6] and Vidal, et al[7]. However it has long been recognized that new massive particles could be produced with significently large cross-sections in ultra-high (10^{14}-10^{15} ev) energy hadron interactions that occur in the cores of extensive air showers initiated by primary cosmic rays in the atmosphere.
A review of earlier experiments has been published recently.[8]
The basic idea[9-13] underlying this search is that massive particles of a given energy travel a little slower through the atmosphere and may arrive at the observational level, say mountain altitude, delayed by tens of nanoseconds relative to lighter particles like electrons, muons, and even nucleons of the same total energy.

This work constituted in part the Ph.D. thesis of J. A. Goodman, University of Maryland, August 1978
*Now at SUNY-Stonybrook, Stonybrook, New York
**Now at Pfizer Corp, Columbia, Maryland
***Now at Nuclear Physics Laboratory, Oxford, England
+Now at Tata Institute of Fundamental Research, Bombay, India
++Now on leave at National Science Foundation, Washington, D. C.

The delay is inversely proportional to γ^2, approximately $1666/\gamma^2$ ns per kilometer of path length, where γ is the Lorentz factor for the particle. The method is most sensitive in the range of Lorentz factors of 5 to 40 and observation of particles with total energies greater than about 40 Gev and arrival delay greater than about 20 ns would have great significance. The search is sensitive to unstable particles also for life times $\gg 10^{-7}$ sec since the decay secondaries would carry the signature of delay which was accumulated by the parent massive particle.

Four experiments[10-13] have attempted to search for energetic delayed hadrons in air showers at mountain altitudes and have reported observing possible candidates for massive particles with varying degree of confidence. Three of these experiments have used the signal from the whole or a part of large area calorimeters to time the arrival of the hadron relative to air shower particles above the calorimeter and to measure the energy of the delayed hadron.

APPARATUS

In our experiment we have attempted to reduce the background effects due to low energy hadrons and side showers. The experimental arrangement is shown in figure 1 and has been discussed in detail elsewhere[14]. Following are the significant features of the experimental setup. The calorimeter of 4 m^2 area has seven layers of iron with two liquid scintillation tanks under each iron layer. The timing detector T3, a plastic scintillation detector of dimensions 0.91m x 0.61m x 1.2 cm viewed edgewise by a 56AVP photomultiplier through an air light guide, has been placed near the center of the calorimeter under 220 g cm^{-2} of iron absorber. The calorimeter has a total absorber thickness of 985 g cm^{-2} including 45 g cm^{-2} of liquid scintillator. In the later stages of the experiment, another scintillation detector T5, very similar to T3, was placed approximately above the detector T3 to provide a check on the signals from T3. Four air shower detectors have been used to measure particle density above and around the calorimeter. Plastic scintillation detectors T1E and T1W, each of dimensions 0.9m x 1.8m x 1.2 cm and viewed by two 56AVP photomultipliers through plexiglass lightpipes, were placed side by side above the top iron layer of the calorimeter. Two other detectors, TSE and TSW, not shown in figure 1, were 0.5m^2, 10 cm deep liquid scintillators, and were located about 3 meters from the center of the calorimeter on opposite corners. A distinguishing feature of this experiment is the presence of four wide-gap spark chambers, one placed above the calorimeter and three located at the depths of 120, 340, and 700 g cm^{-2} inside the calorimeter. The transition radiation detector placed above the top spark chamber was used for a separate experiment discussed elsewhere[15].

All the detectors were calibrated with near vertical muons and their outputs for events collected during the experiment are expressed in number of equivalent muons. Extensive shielding was used for all the detectors and electronics to avoid interference from noise generated by the spark chambers. Detailed studies were made with and without firing the spark chambers to ensure the absence of noise effects in recorded data.

All the experimental data has been collected with basically three requirements: (i) signals from shower detectors T1E, T1W, TSE and TSW corresponding to a certain minimum shower density Δ_m of particles over the calorimeter, (ii) signal from the calorimeter with outputs from all scintillation tanks summed together indicating a deposit of energy $\geq$ 50 Gev, and (iii) a minimum signal from T3 corresponding to N_m muons. The values of Δ_m and N_m were changed during the course of the experiment. Relative arrival times of particles in detectors T1E, T1W, T5, and T3 were measured using an LRS 226A Quad Time to Digital Converter (TDC). The start gate to the TDC was provided by the leading edge of the discriminated signal from T3. Stop pulses for different timing channels were similar signals from T1E, T1W, T5, and T3, the last one providing an on-line check of the system. From the observed time distributions obtained with single unaccompanied charged hadrons, we have determined the error in time measurement to be less than 3.5 ns (one sigma). The arrival time of the shower has been taken as the average of times recorded for T1E and T1W. All events having differences between T1E and T1W of more than 10 ns have been rejected since they are fully understood in terms of chance coincidences.

DATA

Table I summarises the data taken with different selection critera for Δ_m and N_m. In figures 2a, 2b, and 2c are presented the diplots (T3 time delay vs T3 signal in equivalent particles) for the data for the three groups (I, II, and III) defined in Table I. The shower density threshold for events in figures 2a and 2c is 4 m^{-2} and for events in figure 2b is 18 m^{-2}. A glance at these figures reveals striking features. Figure 2a shows two events (labelled A, B) of large delay and large pulse height, which are distinct from all other events. Figure 2c shows one event (C) with similar features. However, figure 2b shows many delayed events, but with small signals from T3. Here also there are three events (D, E, F) which are delayed by more than 40 ns, while the signals from T3 for these events exceeds the 20 particle level. Note that apart from the differences in threshold for T3 signal between group I (and III) and group II, there is also a difference in shower density threshold. This difference makes them sensitive to different regions of core distances, thus rendering comparisons of event rate between different groups hazardous. The three large delay, large pulse height events seen in data groups I and III constitute a fraction 3.3×10^{-4} of the total number in these groups.

Assuming that the hadronic cascades for these delayed events (A, B, and C) are at their maximum at the depth of detector T3, the most probable energy for these hadrons would be greater than 50 Gev. Nucleons of this energy would be delayed relative to shower electrons by about 10 to 20 ns if they were produced in the first few interactions of the cosmic ray primary higher up in the atmosphere. However, they would have to travel nearly 600 g cm^{-2} of air without any significant inelastic interaction and also would have to begin with a small value of transverse momentum, due to the requirement of associated shower density. The expected number of events of this type is less than 0.1. These elementary considerations show that these events are interesting and require detailed calculations taking into account all possible and significant fluctuations for hadronic and electromagnetic shower development. However, it is useful to first examine possible instrumental and local physical processes which could lead to the characteristics observed for these events. Various such processes are listed in Table II, which also shows their estimated contributions to the number of events. It is clear from the numbers shown in Table II that the processes mentioned cannot account for the observations.

Information from spark chamber pictures for these delayed energetic events has not been very useful in giving evidence about cascade structure or the nature of the particles responsible for these events. These pictures do show a dense shower incident on the beam spark chambers, and tracks or a cascade in the spark chambers which are above and below the detector T3, but separated by about 120 g cm^{-2} of absorber. Since the energies of these events are only 40 to 100 Gev and the chambers are separated by 240 g cm^{-2}, the cascades are not expected to show well defined and aligned structure in both chambers.

MONTE CARLO SIMULATION

We have made detailed Monte Carlo simulations of development of air showers in the atmosphere for primary energies in the range of 10 to 100 Tev per nucleon for both proton and iron primaries. Apart from the need to understand the energetic large delay events, these studies were also aimed at understanding the delay distributions of low energy hadrons, such as those seen in figure 2b. Simulations were performed using energy spectra for protons and iron nuclei components, and an attempt was made to match the observed event rates for both delayed and nondelayed events. The hadron interactions in these simulations have been assumed to obey scaling. The interaction parameters used in calculation have been obtained from observations at Fermilab and ISR energies. Details of these calculations and the interesting result that iron group nuclei should begin dominating the primary cosmic rays at energies around 100 Tev are discussed elsewhere[14,16].

It must be noted that apart from simulating the development of air showers, these calculations have also simulated instrumental response, including fluctuations in cascade deveopment in the calorimeter and errors in measuring arrival delay for hadrons.

The simulations reveal that air showers induced by heavier cosmic ray primaries (like iron) are relatively richer in delayed hadrons. The expected delay distribution from iron nuclei generated showers is shown in figure 3, again as a diplot of expected T3 signal delay versus the expected T3 pulse amplitude. This figure shows no event having delay of more than 20 ns and giving a signal in T3 of amplitude greater than 20 particles. Therefore, it seems that the anomalous large delay, large energy events of the data must be interpreted as due to the production of some new massive particle at very high energies.

OTHER EXPERIMENTS

Comparison of our observations with results of similar experiments is difficult due to differences in instrumention and selection systems. Still, it is striking that all four experiments have observed similar phenomena. In 1540 hours of observation, Jones et al have reported observation of six events which deposited energies greater than 30 Gev in their calorimeter and were delayed by more than 30 ns. However, they noted that five of these events deposited their energy in the first two layers of the calorimeter, that is within about 150 g cm^{-2} of iron absorber. This feature led them to suggest these events as due to low energy nucleons or alpha particles. However, Tonwar et al have performed Monte Carlo simulations for their calorimeter and suggest low energy nucleons to be a very unlikely source for these events. Our own calculations of the flux of alpha particles associated with even small size air showers indicate that the events seen by Jones et al could not be caused by "leak-through" of alpha particles. This suggests that the events seen by Jones et al are similar to events seen in the present experiment. Direct comparison of flux is difficult because Jones et al required delay of signal from the whole calorimeter area (1.6 m^2) including the signal from the uppermost detector layer, which was under about 70 g cm^{-2} of absorber. This reduces the experimental sensitivity in near core regions because of the high density of shower particles there.

Tonwar et al have also observed four events in 1280 hours, with similar energy and delay requirements. They have restricted their observations for both energy and delay to the lower half of their calorimeter (area 1.4 m^2), rejecting the top 300 g cm^{-2} in order to eliminate effects due to shower particles and low energy hadrons. They have included in their selection process only those events which are within 20 meters of the axes of air showers with sizes in the range 6.7×10^4 to 1.8×10^6 particles. In their cloud chamber experiment, Tonwar et al observed two events in 2800 hours with an 0.36 m^2 detector placed under 300 g cm^{-2} of iron absorber. These were associated with air showers of size around 10^5 particles.

In all four experiments, there is strong evidence for energetic delayed events which are difficult to understand in terms of known particles and processes. We have attempted to explain these events in terms of the production of massive (5 Gev/c^2) particles in very high energy hadron collisions, and have incorporated this process in our simulations of air showers discussed earlier. Since no firm guidance is available from theory about production or interaction characteristics, our assumptions about these have necessarily been somewhat arbitrary. For the energy dependence of the production cross-section, the functional form suggested by Isgur and Wolfram[17] has been used in the calculations. The massive particles are assumed to have a relatively large interaction (inelastic) mean free path of 300 g cm^{-2}. Their x and p_t distributions in the center of mass system are taken as e^{-10x} and e^{-4p_t} respectively. Air showers have been simulated with both protons and iron nuclei as primaries, with the total flux normalized to the total number of hadrons observed without respect to delay. The experimental shower density requirements and T3 detector response, including fluctuations, have been included in these simulations. In order to reduce computation time, production cross-sections have been taken 1000 times higher than the values suggested by Isgur and Wolfram. Figure 4 shows the expected delay distribution for all particles (both massive and normal hadrons), again as a diplot of delay versus T3 pulse amplitude.

It is seen that the simulations do produce delayed energetic events of a type similar to those observed experimentally. These calculations suggest a normalized value of ~ 50 ub for the production cross-section at an energy of 100 Tev. However, a note of caution is appropriate here. This result should not be treated as evidence for the existance of a particle of mass 5 Gev/c^2, since many arbitrary assumptions have gone into the calculations. These simulations are indicative only, but do suggest that the observations of energetic delayed events in air showers can be understood in terms of the production of massive particles in high energy collisions. It is also guessed that the required production cross-section would be lower if the mass of the particle were higher. Clearly, more simulations are needed to obtain narrower bounds on both the mass and production characteristics of the new particle. In addition, more experimental data is required for these studies. A new cloud chamber experiment of Tonwar (private communication) and a large area experiment being planned by our group should fill this need.

ACKNOWLEDGMENTS

We would like to thank the following individuals who in various ways contributed to the acquisition and analysis of the data: Dr. J.J. Jones, Dr. Thomas G. Morrison, Eldon Vann, Ralph Sutton, Harriet Sutton, Dr. Alan Stottlemyer, Dr. Geeta Tonwar, Calvin Simpson, Sriram Ramaswamy, and Ravi Mani. The research was supported by NSF Grant GP-25303 and by the University of Maryland Computer Science Center.

REFERENCES

1. J. E. Augustin, et al., Phys. Rev. Lett., 33, 1406, (1974)
2. J. J. Aubert, et al., Phys. Rev. Lett., 33, 1404, (1974)
3. S. W. Herb, et al., Phys. Rev. Lett., 39, 252, (1977)
4. M. Gell-Mann, Phys. Lett., 8, 214, (1964)
5. G. Zweig, CERN Report 8182/Th. 401, (1964)
6. D. Cutts, et al., Phys. Rev. Lett., 41, 363, (1978)
7. R. Vidal, et al., Phys. Letters 77B, 344 (1978).

8. L. W. Jones, Rev. Mod. Phys., 49, 717, (1977)
9. G. Damgard et al., Phys. Lett., 17, 152, (1965)
10. B. K. Chatterjee, et al., Proc. 9th Int. Conf. on Cosmic Rays, London, 2, 808, (1965)
11. L. W. Jones, et al., Phys. Rev., 164, 1548, (1967)
12. S. C. Tonwar, et al., J. Phys. A, 5, 569, (1972)
13. S. C. Tonwar, et al., Pramana, (1977)
14. J. A. Goodman, Ph.D. Thesis, University of Maryland (1978)
15. J. R. MacFall, Ph.D. Thesis, University of Maryland (1976)
16. J. A. Goodman, et al., this conference.
17. N. Isgur and S. Wolfram, private communication, (1978)

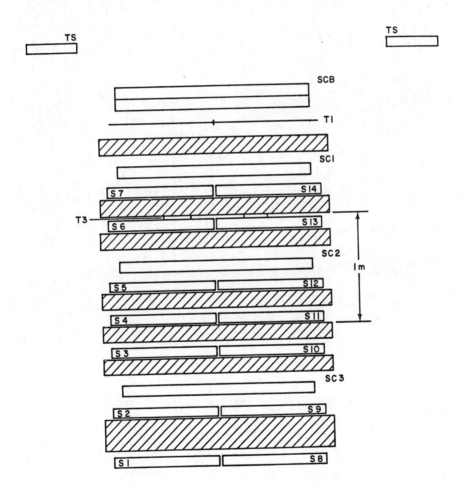

Figure 1. Experimental Apparatus

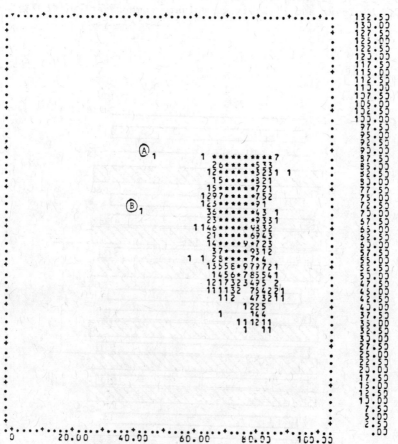

Figure 2a. Time delay vs. T3 particle number for Group I, (4378 events).

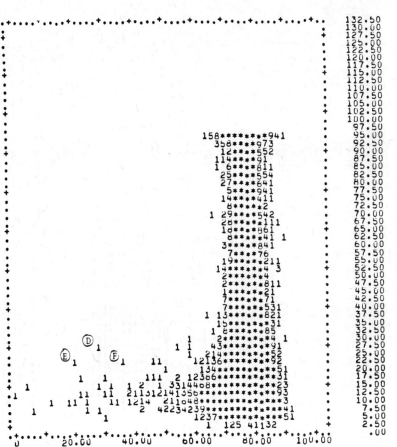

Figure 2b. Time delay vs. T3 particle number for Group II, (21477 events).

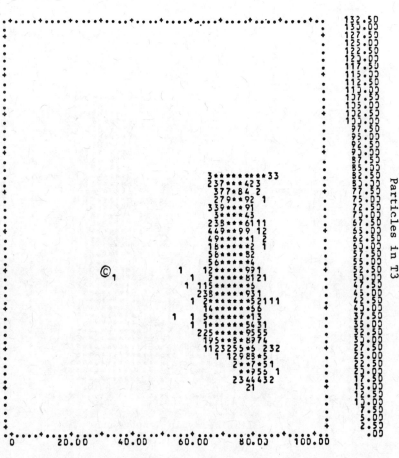

Figure 2c. Time delay vs. T3 particle number for Group III, (4757 events).

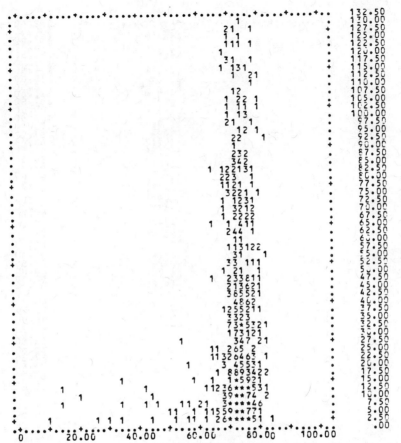

Figure 3. Time delay vs. particle number for simulated iron primaries, (841 events).

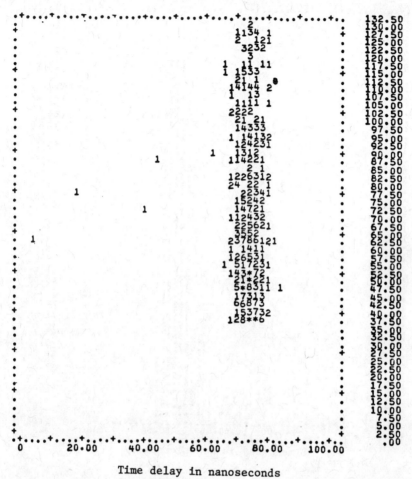

Figure 4 Time delay vs. particle number for simulated proton primaries, massive (5 Gev/c²) particle production included.

TABLE I
Summary of Experimental Data

Group	Dates	Run Time (hours)	T3 Thres. (particles)	Shower Thres. (ptcls./m^{-2})	T1 Thres. (ptcls./m^{-2})	No. Evts.
I	April '75– July '75	1072	32	1.2	4	4378
II	July '75– Jan. '76	3750	3	18	4	21480
III	Jan. '76– May '76	1283	15	1.2	4	4757

TABLE II
Possible Backgrounds

Background	Calculated upper limit to delayed events per trigger, delay 20 ns, T3 ≥ 3 particles
1. Accidental coincidence between T1E, T1W, and T3	3×10^{-6}
2. Other accidentals	3×10^{-7}
3. Backscatter from calorimeter	10^{-8}
4. Neutrons from local interactions	10^{-5}
5. Stopping π^-	10^{-5}
6. Stopping antinucleons from local interactions	10^{-7}
7. Nuclear fragments from heavy primaries	3.7×10^{-8}
8. Production of d, t, He^3 in hadron air collisions	10^{-9}

EVIDENCE FOR PROMPT NEUTRINOS AND INTERPRETATION
OF BEAM DUMP EXPERIMENTS

G. Conforto
Istituto Nazionale di Fisica Nucleare, 00185 Rome, Italy
CERN - European Organization for Nuclear Research,
Geneva, Switzerland

ABSTRACT

The results of the various beam dump experiments performed so far are reviewed. Their implications for the charmed particle lifetime are also presented and discussed.

INTRODUCTION

Beam dump experiments are performed by making the proton beam extracted from an accelerator interact in a very thick and dense target. Particles produced by a primary beam are thus very likely to be absorbed before they can decay. As a consequence, the yield of ordinary neutrinos, coming mostly from π and K decays in normal neutrino beams, is reduced by a typical factor of about 2000. This allows neutrino detectors downstream of the target to be sensitive to possible new sources of "prompt" neutrinos, i.e. neutrinos whose parents have a decay length much smaller than the interaction length in the target material.
A series of beam dump experiments recently performed at CERN [1,2,3,4] at 400 GeV has indeed established the existence of a new prompt source. Two other experiments carried out at Brookhaven and Serpukhov at 28 and 70 GeV respectively have reported at the Tokyo Conference upper limits [5] which indicate the existence of a threshold for prompt neutrino production.
In this talk I will mainly discuss the CERN experiments and the possible interpretation of the positive signal they have observed in terms of charm production.

EXPERIMENTAL FACTS

Figure 1 shows the general layout of the West Experimental Area of the CERN Super Proton Synchrotron. The neutrino beam line is represented by the dashed line at the top of the figure. The three detectors which have contributed to beam dump experiments are labelled as BEBC (Big European Bubble Chamber), WA1 (also referred to as CDHS for CERN, Dortmund, Heidelberg and Saclay collaboration) and GGM (short for Gargamelle).
Figure 2 shows schematically the layout used in the experiments of references [1,2,3]. The dump target, made out of copper, is placed in the position usually occupied by the normal neutrino target in front of the decay tunnel and almost a kilometre upstream of the

detectors. The beam direction coincides with the detectors' axis. This geometry is normally referred to as "0 mr beam dump".

Additional information on the source of prompt neutrinos comes from a Narrow Band Antineutrino Beam experiment [4], in this context renamed as "15 mr Beam Dump". The layout, together with that of the 0 mr beam dump, is shown in figure 3. The proton beam hits a beryllium target at a 15 mr angle. Positive particles produced are deflected away and only negative particles are allowed to enter the decay tunnel to yield the antineutrino beam. Under those circumstances, the neutrino contamination is very small and there is therefore some sensitivity in the experiment to detect a small flux of neutrinos directly produced at the target.

Table 1 contains the most relevant characteristics of the three detectors.

The two bubble chambers are filled with heavy liquids. The muon identification is achieved by means of an External Muon Identifier consisting of an iron absorber preceded and followed by multiwire proportional chambers. The electron identification is instead obtained by visual inspection of the event photograph.

The CDHS apparatus consists of a massive target-calorimeter in which the energy deposited by a neutrino (or antineutrino) is measured by electronic techniques and the momentum of the outgoing muons is determined by magnetic deflection. No direct electron identification is possible and electron events are distinguished from neutral current events on a statistical basis.

The results obtained by the two bubble chambers are shown in figures 4 to 6.

Figure 4 shows the results obtained by BEBC in the 0 mr experiment [3]. The expected number of events, calculated assuming that only pion and kaon decays contribute to the neutrino and antineutrino fluxes, are

$$21 \, \mu^-, \, 1.3 \, e^-, \, 3.4 \, \mu^+, \, 0.3 \, e^+$$

More reliably, the expected number of electron events can be predicted assuming the observed number of muon events to be due to pion and kaon decays only. This calculation depends only on the pion/kaon production ratio, relative Q-values, absorption lengths and branching ratios. Under this hypothesis one obtains:

$$(1.8 \pm 0.4) \, e^- \quad (0.5 \pm 0.25) \, e^+$$

The 15 mr BEBC data [4] are shown in figure 5. The expected number of negative electron events under the two hypotheses above are 1.2 and (1.1 ± 0.4) respectively.

Similarly for the Gargamelle data [1], shown in figure 6, on the basis of the observed 16 muon events, the expected number of electron events is 1.1 to be compared with 9 actually found.

The interpretation of the CDHS data is less straightforward. What the experiment actually measures is the ratio of muon-less to ordinary charge current muon events with the result

$$\left.\frac{\mu- \text{less}}{\geqslant 1\mu}\right|_{E_{h\ vis} > 20\ GeV} = 0.86 \pm 0.08.$$

Subtracting from the observed number of muonless events the contributions due to neutral current events (calculated from the observed number of charged current events) and to standard electron-neutrino sources (Ke_3, Λe_3, etc.), an excess of 236 ± 40 events due to a new source is obtained. Figure 7 shows the shape of the longitudinal shower development for the excess muonless events. This shape is different from that of purely hadronic showers (dashed line) but is consistent (see insert) with that of a superposition of electromagnetic and hadronic showers, thus strongly suggesting that the excess muonless events are indeed electron events.

The energy spectra for single μ^- and μ^+ events in the CDHS experiment are shown in figure 8. The μ^+ expectation from pion and kaon decays only, normalized to the μ^- spectrum, lies below the observed histogram.

In conclusion, it is clear that the experimental results presented above cannot be explained in terms of neutrinos originating from pion, kaon and hyperon decays only. All the available evidence consistently points to the existence of a new, short-lived, neutrino source. All observations are compatible with the hypothesis that this source emits equal fluxes of ν_e, $\bar{\nu}_e$, ν_μ and $\bar{\nu}_\mu$.

The main results obtained in the various experiments are summarized in Table II.

CONSISTENCY OF RESULTS

At this point we are leaving the area of well-established facts to enter that of more personal considerations. The possibility of expressing my own views is of course one of the main attractions of giving a talk like this. I am reminded of a scene in "The importance of being Ernest" where Gwendolen, bickering with Cecily, says: "On an occasion of this kind it becomes more than a moral duty to speak one's mind. It becomes a pleasure"[6].

Before proceeding any further with the interpretation of the results, two comments are in order. The first is on the compatibility of the various 0 mr results. On the basis of fiducial volume tonnages and distances from the target, the ratio between the BEBC and GGM neutrino detection efficiencies is 1.6. This is almost exactly the ratio between the observed events (see table 2). Thus, there is complete agreement between the two bubble chamber results. On the other hand, the ratio between the neutrino detection efficiency of CDHS and that of the two bubble chambers combined is 22, to be compared with an experimental ratio of observed events of 13 ± 4 (see again table 2). There is therefore a discrepancy between the counter and bubble chamber results amounting to about a factor of two. The significance of this discrepancy, however, is a little more than that of a two standard deviation effect.

The second comment refers to the comparison between the 0 mr and the 15 mr data obtained by BEBC. These data are shown in figure 9. What is actually plotted are the ratios of prompt neutrinos to positive pions at various energies. These quantities are derived from the prompt neutrino yields observed by BEBC at 0 mr and 15 mr and the calculated positive pion yields emitted in the dump at the same energy and in the same solid angle of the detected neutrinos. For comparison, the ratio of prompt muons to positive pions as a function of energy is also shown. It can be seen from the figure that the muon to pion ratio lies substantially higher than the neutrino to pion ratio, consistently with the notion[7] that prompt muons are mostly produced in pairs. The striking feature in the figure is that the point obtained in beryllium at 15 mr lies about an order of magnitude higher than the point in copper at 0 mr at the same energy. If pion production scales as a function of the atomic number A as $A^{2/3}$, this result implies that the prompt neutrino source, as observed in the beam dump experiments, scales as a function of A with an exponent smaller than 2/3. I shall come back to this point later on.

INTERPRETATION OF THE RESULTS

The most plausible origin of the observed yields of prompt neutrinos is the production of charmed particles by the proton beam in the target followed by their semileptonic decays. Since charmed particles must be produced in pairs and are expected to have electronic and muonic semi-leptonic decays of similar strength, they generate ν_e, $\bar{\nu}_e$, ν_μ and $\bar{\nu}_\mu$ fluxes of about the same intensity. Furthermore, their lifetimes are expected to be short enough to satisfy the "promptness" condition. Thus the inclusion of charmed particles among the neutrino and antineutrino parents is adequate to qualitatively explain the experimental results.

Charm particle production in strong interactions has been actively searched for in recent times and many upper limits to the production cross section exist in the literature. The question to be answered, then, is whether the observed neutrino yields are compatible with the existing information. To afford this comparison, a determination of the charm production cross section in beam dump experiments is necessary.

Before going any further in discussing cross sections, let me emphasize that, because of the smallness of the solid angles subtended in this kind of experiments, only a very small fraction ($\leqslant 10\%$) of the total cross section in the fragmentation region is detected. Thus, the extrapolation to the total cross section is largely model dependent and is particularly sensitive to the assumed longitudinal form of the invariant cross section.

The procedure used in arriving to a total cross section is, briefly, the following. Charmed particle production is assumed to be dominated by the $D\bar{D}$ channel. D mesons are taken to be produced according to the invariant cross section.

$$E \frac{d^3\sigma}{dp^3} \propto (1 - x_F)^n \cdot f(P_\perp, m)$$

and to decay with a 10% branching ratio into muonic or electronic channels with a center-of-mass momentum $p^* \simeq 500$ MeV/c.

There has been a fair amount of confusion so far on what normalisation procedure to follow in order to arrive to a total cross section on a free nucleon. Giving-in to pressure from my theoretical friends, I have taken the liberty to add to the confusion and I have chosen to normalize all free nucleon cross sections quoted in the following by assuming the total charm production cross section in a nucleus (determined by normalizing the data to the inelastic proton-nucleus cross section) to scale linearly as a function of the atomic number A.

Within the framework of the model outlined above, all physical quantities can be predicted, once the value of n and the transverse form of the invariant cross section $f(P_\perp, m)$ have been chosen. As a typical exemple, figure 10, shows the free nucleon charm production cross section σ and the mean event energy $<E>$ as a function of n calculated for the Gargamelle experiment with $f(p_\perp, m)$ taken according to Bourquin and Gaillard[8]. For $<E>$ = 73 GeV, $4 < \bar{n} < 5$ and $\sigma \simeq 50\mu b$. In spite of its unanimous adoption by all experimenters, the model used to derive the charm production total cross section should actually be taken with a fair degree of caution since it is not really terribly successful in describing even the very meagre data available at present. This point is illustrated in figure 11 which shows the mean event energy calculated for the 0 mr (full lines) and the 15 mr (dashed lines) BEBC experiments as a function of n for various b ($f(p_\perp, m) = e^{-bp_\perp}$) together with the experimental results. The best values for the parameters in the joint analysis of the two results are $3 < n < 5$ and $b \leqslant 2$, but it is clear from the figure that even this choice does not provide a very good description of the experimental observations.

The results on the charm production cross section obtained in the various beam dump experiments are summarized in table 3.

IMPLICATIONS FOR CHARMED PARTICLE LIFETIME

A discussion on whether the charm production cross sections obtained in the beam dump experiments are compatible with the existing limits is in order at this point [9].

I shall limit myself to comparing the beam dump results to cross section upper limits published by emulsion experiments, and this mainly for three reasons. Firstly, the emulsion experiments have claimed the lowest cross section limits so far. Secondly, as it will be clarified shortly, the detection efficiency in emulsion experiments is a function of the charmed particle lifetime. Thus, the comparison between emulsions and beam dump results bears on the question of the charm lifetime. Thirdly, since the mean atomic number of the emulsions is very close to that of copper, any comparison between emulsions and beam dump results is essentially insensitive to the procedure adopted in determining the free nucleon charm production cross section. As far as other experimental results are concerned, let it suffice here that no serious incompatibility exists at present with the cross section values derived from the beam dump data.

The implications for the lifetime of the charmed particles of emulsions searches and beam dump results are illustrated in the charm production cross section versus charm lifetime diagram shown in figure 12.

The requirement of the emulsions experiments [10,11] to observe one of the two pair-produced charmed particles to decay within fixed minimum and maximum distances from the production vertex clearly implies a lifetime dependent detection efficiency. Thus, the negative results obtained reflect themselves in the lifetime dependent upper limits shown in figure 12. For the experiment of reference 11, the curve of figure 12 is trivially derived from the published data. For the experiment of reference 10, the upper limit curve has been determined by means of a Monte-Carlo calculation [12]. In both cases the cross section scale has been normalized following the same procedure used for the beam dump results, taking for emulsions $<A> = 56$. It is perhaps interesting to point out that the two curves of figure 12 should actually coincide on the right-hand side, since for long lifetimes the two experiments have almost exactly the same sensitivity. Thus, the gap between the two curves in this region gives an idea of the theoretical uncertainties involved.

The best upper limit on the charmed particle lifetime ($\tau \lesssim 10^{-12}$ second), obtained from the lack of structure in neutrino induced dimuon events in Gargamelle [13], is represented by the vertical line. The charm production cross section result derived from the Gargamelle beam dump experiment is indicated by the horizontal band.

Taking all experiments at their face values, the over-all picture appears to be rather inconsistent since the beam dump result lies almost everywhere higher than the emulsions upper limits. The situation can be eased somewhat by considering that the emulsion data have been obtained at an energy lower than that used in the beam dump experiments and that the CDHS charm production cross section is lower than those of BEBC and Gargamelle.

One point neglected by all beam dump and emulsion analyses is that charm production actually takes place in a complex nucleus and not on a free proton. Thus, the possibility that charm particles, because of the high density of nuclear matter, may reinteract and be partly "dumped" inside the nucleus in which they have been produced is systematically ignored. Because of charm conservation, secondary interactions do not alter the value of the total cross section. They rather have the effect of reducing the energy and of increasing the transverse momentum of the produced particles with the result that the A dependence of the cross section becomes a function of x_F. In actual fact, this phenomenon is observed to occur in inclusive Λ production at 300 GeV in nuclei [14] where, as shown in figure 13, particle yields decrease with A at large momenta and small angles.

Some evidence that perhaps this mechanism is at play also in charm production is provided by the comparison between the beryllium (15 mr) and the copper (0 mr) beam dump results which, as noted before, suggest a rather weak A dependence of the observed part of the total cross section. Since beam dump experiments, as also pointed out earlier, are sensitive only to a very small fraction of the total cross section in the fragmentation region, there is no inconsistency with the A dependence assumed in deriving the free nucleon cross section.

Coming back now to figure 12, it is clear that the taking into account of secondary interactions produces rather devastating effects as to the consistency of the whole picture. As a result of secondary interactions decreasing the number of produced charmed particles that can decay emitting neutrinos in the region accepted by the detectors, the beam dump band shifts upwards by a factor that, guessing from figure 13, could be about two. Furthermore, owing to the lower velocity of the produced charmed particles, the emulsion upper limits tilt clockwise, thus making the situation for lifetimes in the range 10^{-13}-10^{-12} seconds rather uncomfortable.

Before closing, let me make a last point to illustrate the fragility of any conclusion on charm production cross sections in beam dump experiments. If one assumes that charmed particles are produced with the same dynamics as strange particles, the question of what production mechanism is dominant in the kinematic region accepted by the beam dump experiments can be answered rather easily on the basis of the existing data. Figure 14, taken from reference 15, shows the invariant cross section as a function of x_F for the inclusive production of Λ, K_S^0 and $\bar{\Lambda}$ particles by 300 GeV protons on beryllium. From this figure, it can be seen that strangeness production at large x_F and small angles is dominated by the Λ contribution. Because of the higher masses involved, the forward charmed baryon channel in charm production is certainly less important than the forward Λ channel in strangeness production.

It could nevertheless play a substantial role in the prompt neutrino production observed in the beam dump experiments. Should this be the case, the charm production cross section would turn out to be substantially smaller than that calculated on the basis of the $D\bar{D}$ hypothesis, thus restoring complete compatibility between beam dump and emulsion results even for lifetimes substantially shorter than 10^{-12} seconds.

In conclusion, the prompt neutrino excess observed in beam dump experiments is indeed very likely to be due to the semi-leptonic decays of charmed particles, produced with a cross section of the order of several tens microbarn. No inconsistency can be claimed to exist at present between the beam dump observations and a charm lifetime of the order of 10^{-12} seconds.

ACKNOWLEDGEMENTS

I would like to express my deep gratitude to Professor A. Zichichi, President of INFN, for making the necessary travel support available to me and to Professor T. Gaisser of the Bartol Research Foundation, University of Delaware, for the generous support during my most enjoyable stay in Newark.

I am deeply indebted to Miss M. Mazerand for her careful typing of the manuscript and to Mrs C. Rigoni and Mr. M. Belletieri for their help with the figures.

REFERENCES

1. P. Alibran et al., Phys. Lett, 74B, 134 (1978).
2. T. Hansl et al., Phys. Lett. 74B, 139 (1978).
3. P.C. Bosetti et al., Phys. Lett 74B, 143 (1978).
4. P.C. Bosetti et al., private communication from K.C. Wernhard. See also H. Wachsmuth Proceedings of the Topical Conference on Neutrino Physics, Oxford, U.K. July 1978.
5. K. Kleinknecht, Proceedings of the XIXth International Conference on High Energy Physics, Tokyo, Japan August 1978.
6. O. Wilde, The Importance of Being Ernest, Act 2.
7. H. Kasha et al., Phys. Rev. Lett. 36, 1007 (1976).
8. M. Bourquin and J.M. Gaillard, Phys. Lett., 59B, 191 (1975).
9. Some of the arguments developed in the following can be found, with slightly more details, in G. Conforto, Proceedings of the Topical Conference on Neutrino Physics, Oxford, U.K. July 1978.
10. G. Coremans-Bertrand et al., Phys. Lett. 65B, 480 (1976).
11. W. Bozzoli et al., Nuovo Cimento Lett. 19, 32 (1977),
12. D. Crennell et al., RL-78-051.
13. A. Blondel, Proceedings of the Topical Conference on Neutrino Physics, Oxford, U.K. July 1978.
14. K. Heller et al., Phys. Rev. 16. 2737 (1977)
15. T. Devlin et al., Nucl. Phys. B123, 1(1977).

229

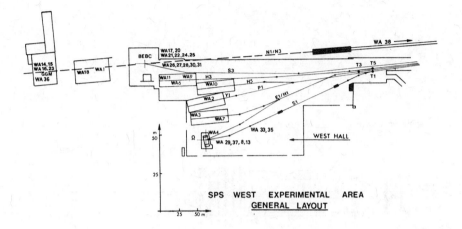

Figure 1

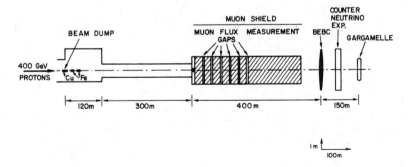

Figure 2: Layout of the 0 mr beam dump experiments

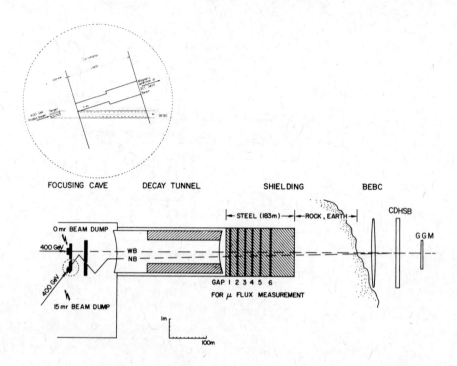

Figure 3: Layout of the 15 mr beam dump experiment compared with that of 0 mr.

TABLE 1: Detectors

	BEBC	GGM	CDHS
MATERIAL	72% Ne-H_2	HEAVY-FREON	Fe-Scint.
FIDUCIAL MASS	13 t	10.5 t	580 t
SOLID ANGLE	10 μsr	1.8 μsr	10.8 μsr
μ IDENTIFICATION	EMI > 98% p_μ > 5 GeV/c	EMI > 98% E_ν > 50 GeV	> 98% p_μ > 5 GeV/c
e IDENTIFICATION	(90±5)%	100%	INDIRECT

231

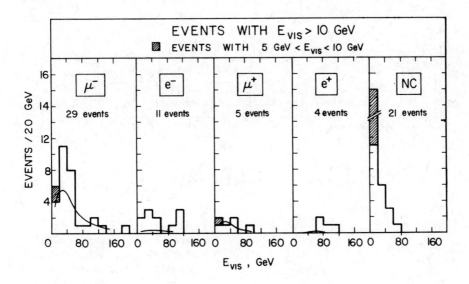

Figure 4: BEBC results at 0 mr

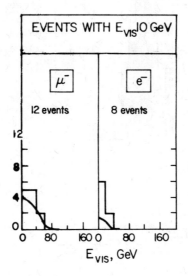

Figure 5: BEBC results at 15 mr

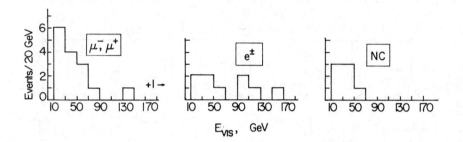

Figure 6: Gargamelle results

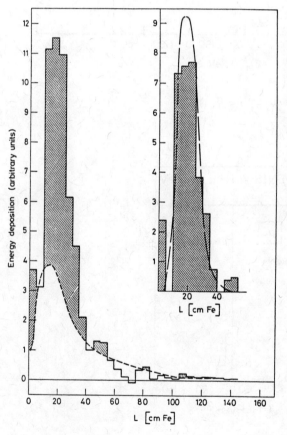

Figure 7: Longitudinal shower development in the iron calorimeter of CDHS for the excess muonless events compared to the shape of hadronic showers (dashed curve) as measured by single muon events. The insert shows the pulse height not due to hadronic showers compared to the shape of purely electromagnetic showers (dashed curve).

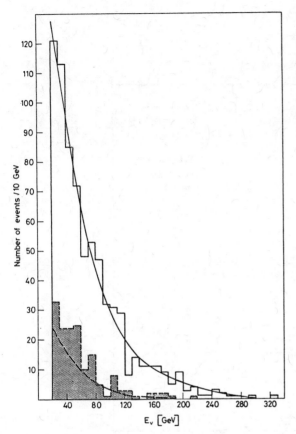

Figure 8: Total visible energy spectra for single μ^- and single μ^+ events (shaded histogram) in the CDHS experiment compared to the expectations from π- and K-decay neutrinos normalized to the total number of negative muons.

TABLE 2:
Results obtained in beam dump experiments at CERN

	BEBC 0 mr	BEBC 15 mr	GGM 0 mr	CDHS 0 mr
Interacting protons $*10^{17}$ at 400 GeV	3.5	3.7	3.5	4.0
e^+, e^- excess events with $E_{vis} > 20$ GeV	11.3	1.7	7.3	236±40
$<E>$ of excess events with $E_{vis} > 20$ GeV (GeV)	71	30	73	85
μ^+, μ^- excess events with $E_{vis} > 20$ GeV				91±40 (μ^+) 210±86 (μ^-)

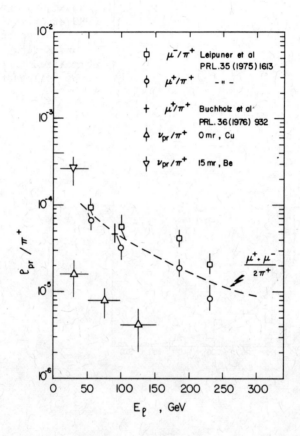

Figure 9: Ratio of prompt neutrinos to positive pions as a function of energy derived from the BEBC data at 0 mr and 15 mr. For comparison, the ratio of prompt muons to positive pions is also shown.

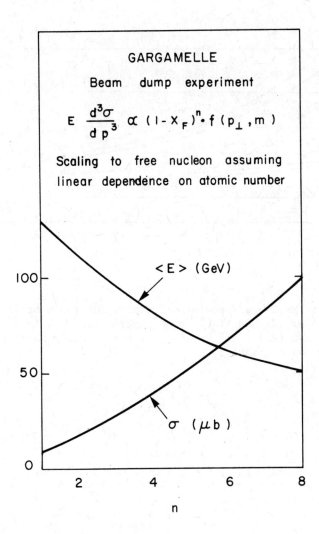

Figure 10: Free nucleon charm production cross section and mean event energy as a function of n in the Gargamelle beam dump experiment.

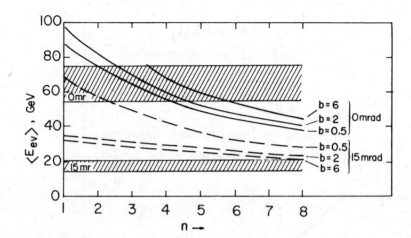

Figure 11: Mean event energy calculated for the 0 mr (full lines) and the 15 mr (dashed lines) BEBC experiments as a function of n for various b (see text). The horizontal shaded bands represent the experimental results.

TABLE 3:

Summary of charm production cross sections obtained in the various beam dump experiments assuming A^1 dependence

Experiment	EPB Energy (GeV)	Particles detected	Cross section (μb)
BEBC	400	ν_e	40-80
CDHS	400	ν_e	∼15
GARGAMELLE	400	ν_e	50±20
IHEP-ITEP	70	ν_e, ν_μ	∼7±6

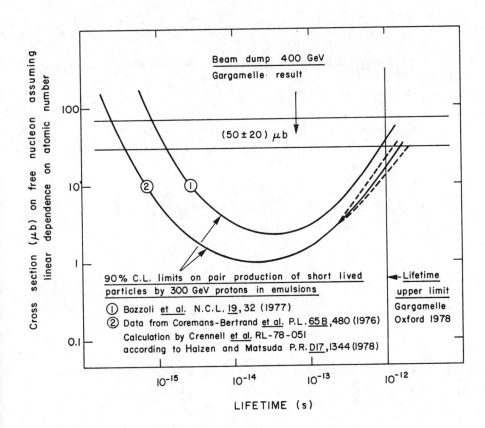

Figure 12: Results from charm searches in emulsions, neutrino induced dilepton events and beam dump experiments displayed in a cross section versus lifetime diagram.

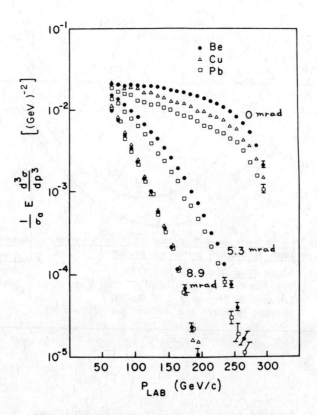

Figure 13: Inclusive cross section for
p + A → Λ⁰ + X at constant angle divided by
the absorption cross section (from ref. 14).

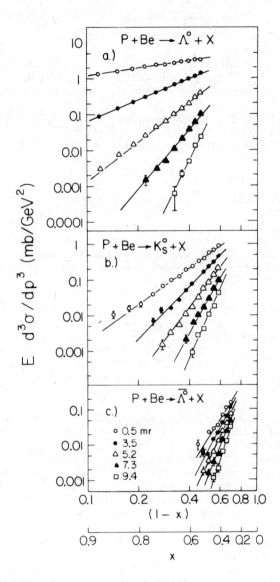

Figure 14: The invariant cross section as a function of x_F for the inclusive production of Λ, K_S^0 and $\bar{\Lambda}$ by 300 GeV on beryllium (from ref. 15).

EVIDENCE FOR PROMPT SINGLE MUON PRODUCTION IN 400 GeV PROTON INTERACTIONS*

B.C. Barish, J.F. Bartlett, A. Bodek**, K.W. Brown, M.H. Shaevitz, E.J. Siskind
California Institute of Technology, Pasadena, CA. 91125

A. Diament-Berger[†], J.P. Dishaw, M. Faessler[††], J.K. Liu[†††], F.S. Merritt, S.G. Wojcicki
Stanford University, Stanford, CA. 94305

ABSTRACT

We have investigated the origin of prompt muon production in the interaction of 400 GeV protons in a steel target. Using a variable density target and a large acceptance muon identifier, we have observed a substantial signal of prompt *single* muon for the kinematic region $X_F \approx 0$ and $0.8 \lesssim p_t \lesssim 1.5$ GeV. Assuming that the major source is charmed $D\bar{D}$ production, a model dependent calculation yields $\sigma(pp \rightarrow charm)$ of about 40 μb.

The discovery[1] in 1974 of large rates for prompt lepton production in hadron-hadron interactions stimulated extensive theoretical and experimental efforts to identify and understand the sources for these leptons. Experiments have indicated that lepton pairs account for most ($\approx 70\%$) of the signal[2], and that the sources of the pairs are electromagnetic decays of vector mesons ($\rho,\omega,\phi,J/\psi,\ldots$) and a large non-resonant continuum[3]. Expected sources for prompt events with only a single charged lepton in the final state are weak semi-leptonic decays of short-lived states, e.g., charmed particles. We present here evidence for such production of single muons in 400 GeV proton interactions.

The experiment was performed in the Fermilab N5 beam[4] using 400 GeV diffracted protons at an intensity of 10^6 protons per accelerator pulse. For each event, the momentum of the incoming proton was tagged using a system of 4 x-y PWC plance and a dipole magnet providing a 16 mr horizontal bend. The apparatus is located in Lab E, where both neutrino[5] and hadron experiments are done. The target is a fine grain calorimeter consisting of 45 counters and steel plates 30" × 30" in transverse dimension for a total thickness of 2 meters of steel. The density of the first meter of steel was varied by changing the plate spacing in order to determine the contribution

---

*Work supported by the U.S. Department of Energy and the National Science Foundation.
**Present address: University of Rochester, Rochester, NY 14627.
[†]Permanent address: Department de Physique des Particules Elementaries, Saclay, France.
[††]Present address: CERN, Geneva, Switzerland.
[†††]Presently with American Asian Bank, San Francisco, CA.

ISSN 0094-243X/79/49240-12 $1.50 Copyright 1979 American Institute of Physic

of meson decays to single muon events. The beam tagging and halo counter system was used to ascertain that only a single 400 GeV proton was incident on the calorimeter, that there was no upstream interaction, and that no additional particles were incident on the calorimeter in the 200 nanosecond period bracketing the event. Other devices monitored the gain of the phototubes and the history of interactions in the calorimeter and enabled us to achieve an intensity independent hadron energy resolution of 3.5% at 400 GeV.

A muon identifier and a toroidal muon spectrometer followed the calorimeter. The muon identifier consisted of two sets of wire spark chambers with ten 48" × 48" × 4" steel plates between them. Each of these steel plates was followed by four 12" × 12" counters. Both spark chamber and counter information were used to identify muons. The efficiency for detecting the second muon from a dimuon pair varied from 87% for masses above 2 GeV to 97% for those below 1 GeV.

The spectrometer (see Fig. 1) which can be moved transversely to the beam axis for use in neutrino or hadron experiments. consisted of 24 11.5' diameter by 8" thick magnetized steel discs interspersed with counters and wire spark chambers. The total $\int B \phi dz$ of the magnet is 3200 kG-inch corresponding to a radial P_t kick of 2.4 GeV and a momentum resolution of 10%. The trigger requirement that a muon traverse the toroid system without crossing the center axis corresponds to a minimum traverse momentum of about 0.8 GeV.

A reconstructed single muon event in the detector is shown in Fig. 1. Such events could originate from three possible sources:
(1) non prompt decay of a long-lived particle (π, K, or hyperon);
(2) a two muon event with a very large-angle or low energy muon;
(3) prompt decay of a short-lived particle (e.g., a charmed particle). The features of the apparatus which allow the separation of these types of events are:

(1) Data were taken in three density configurations to permit a direct measurement of the non-prompt decays from the first meter of steel in the target. The calorimeter plates were expanded about the mean interaction point such that the acceptance for triggering remained the same for each density.

(2) The wide angular coverage of the muon identifier extends to more than 90% of the solid angle in the center of mass for muons of momentum greater than 3.5 GeV/c, so the number of dimuon events with unobserved second muons is small.

(3) A prompt signal from weak decays should have energy carried off by neutrinos. The total hadronic energy and the muon energy are measured for each event in addition to the incident beam energy. Therefore, missing energy associated with the prompt signal can be investigated.

(4) The focussing property of the toroids was used to trigger on positively charged muons with transverse momentum $P_t \gtrsim 0.8$ GeV/c and the Feynman scaling variable $X_F \approx 0$ (i.e., momentum $\gtrsim 15$ GeV/c). This is the region where the contribution of charm decays to the muon/pion ratio is expected[6] to peak. (The investigation of other kinematic regions will be reported elsewhere.)

The procedure for determining the prompt contribution is to plot the observed single muon rates versus inverse density of the target, and extrapolating the rate to infinite density with a linear fit. The slope of the fit measures the non-prompt rate while the intercept is the prompt signal. The data, shown in Fig. 2, display a significant positive intercept indicating a prompt signal. This analysis relies on the assumption that the acceptance is identical for the three density configurations. The geometrical acceptance is expected to be the same because the plates were expanded about the mean interaction point. However, multiple scattering effects can change the acceptance in second order due to the redistribution of the material that the muons multiple scatter through. Monte Carlo studies of this small effect could not explain the large observed non-zero intercept.

The data have been corrected for a small contribution ($\leq 10\%$ of the signal) to the single muon intercept from two muon events in which the μ^- is unobserved. The angular acceptance of the apparatus is so large such that almost all the misidentified events are due to the μ^- ranging out in the calorimeter. The correction is determined by multiplying the number of *observed* dimuon events (see Fig. 2) by the fraction of misidentified events. That fraction was determined using a Monte-Carlo program that generates dimuons according to previously measured distributions[7]. As a check, we have our own data on muon pairs reported elsewhere[8].

Upstream interactions are mostly eliminated in the trigger using the halo and trigger counters information. Other possible background to our prompt signal is the non-prompt decays occurring more than 1 m of steel downstream of the first plate of the calorimeter. Since our target is completely instrumented with scintillation counters dispersed through 3 m of steel, we can establish selection criteria to minimize this background (and also obtain an estimate of its importance); we require that the primary interaction occur near the beginning of the target. We have binned our data according to the location of the incident proton's interaction point in the calorimeter. Figure 3a shows the prompt single muon signal per interacting proton versus the interaction point in the calorimeter. The prompt signal was extracted by doing a density extrapolation for each calorimeter plate. The distribution of proton interactions in the calorimeter plates was obtained from events with no muon (ie., regular proton interactions) taken as a monitor throughout the run. The same ratio is shown for the dimuon events in Fig. 3b. No dependence of the signal on the depth of the interaction in the calorimeter is observed, in contradiction with the large variation expected[9] if the signal came from non-prompt decays in the non-expanded downstream region.

Having established a single muon signal, the hypothesis of the source being a semileptonic decay can be investigated, i.e., is there missing energy indicating a final state neutrino? (Note that there is no missing energy for misidentified dimuon events where the second muon ranged out and deposited its energy in the calorimeter.) Three energies are measured for each event: $E_{proton}(\pm 0.5\%)$, $E_{hadron}(\pm 3.5\%)$ and $E_{muon}(\pm 10\%)$. These resolutions are not sufficient to permit an event-by-event comparison, but at the present level of statistical accuracy in the data, the mean of the total

energy distributions of single muon events are measured to ~0.3 GeV.
The single muon energy distribution exhibits a shift in the mean away
from the beam energy. Because both prompt and non-prompt events contain neutrinos, the size of the shift (the missing energy) is a function of density. An extrapolation to obtain the missing energy of
the prompt signal using an early analysis[10] of 10% of the data yielded $E_{miss} = 8.04 \pm 4.2$ GeV, which is consistent with the energy expected
for a neutrino from the three body decay of a heavy particle. We are
presently performing the missing energy analysis for the full data
sample.

One interpretation of our results is that the single muon signal comes from charm production. The P_t distribution is consistent
with those expected from the three body decay of the $D(1865)$[6].
Assuming that the parent D meson is produced with the same distributions that we have measured[8] for ψ's (i.e. $dN/dx_F dP_t^2 \propto e^{-2P_t}(1-x_F)^5$).
We calculate an acceptance of 3% for observing the semileptonic
decays of D mesons. Using a semileptonic branching ratio for D's of
0.11 and a linear A dependence for the production we obtain a charm
production cross section of 40 µb. Other distributions and assumptions yield cross sections[11] varying from 20 to 80 µb.

We wish to thank Dr. F. Sciulli, who participated in the early
phases of this experiment, and to the members of the Fermilab
neutrino department for their support.

REFERENCES

1. J. P. Boymond et al, Phys. Rev. Lett. <u>33</u>, 112 (1974); J. A. Appel et al, Phys. Rev. Lett. <u>33</u>, 722 (1974).
2. J. G. Branson et al, Phys. Rev. Lett. <u>38</u>, 457 (1977); H. Kasha et al, Phys. Rev. Lett. <u>36</u>, 1007 (1976); L. Kluberg et al, Phys. Rev. Lett. <u>37</u>, 1451 (1976).
3. J. G. Branson et al, Phys. Rev. Lett. <u>38</u>, 1331 (1977).
4. J. Lach and S. Pruss, Fermilab TM-285 and TM-298, 1971 (unpublished).
5. The lab E spectrometer is also used by neutrino experiment E356 (Caltech-Fermilab-Rochester-Rockefeller collaboration).
6. M. Bourquin and J. M. Gailliard, Nucl. Phys. <u>B114</u>, 334 (1976).
7. K. J. Anderson et al, Phys. Rev. Lett. <u>37</u>, 803 (1976).
8. E. J. Siskind et al, CALT 68-665.
9. A crude calculation yields about a factor of e for every interaction length. We are presently performing Monte-Carlo studies to get a more accurate value for this dependence.
10. B. Barish et al, CALT-68-655 (talk presented by M. H. Sheavitz at the Vanderbilt 3rd International Conference, March 1978).
11. The indicated cross sections are in qualitative agreement with the results of beam dump experiments, P. Alibran et al, Phys. Lett. <u>74B</u>, 134 (1978); T. Hansl et al, ibid p.139; P. C. Bosetti et al, ibid p. 143.

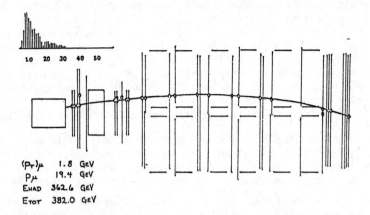

Figure 1 - An example of a single muon event in the detector. The vertical lines above the calorimeter indicate the longitudinal deposition of energy in the calorimeter.

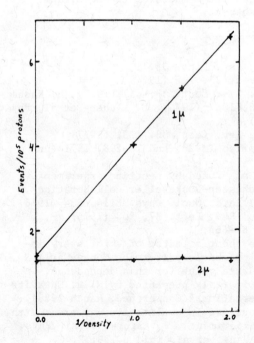

Figure 2 - Extrapolation plots to determine the prompt signal for events with 1 muon or 2 muons observed.

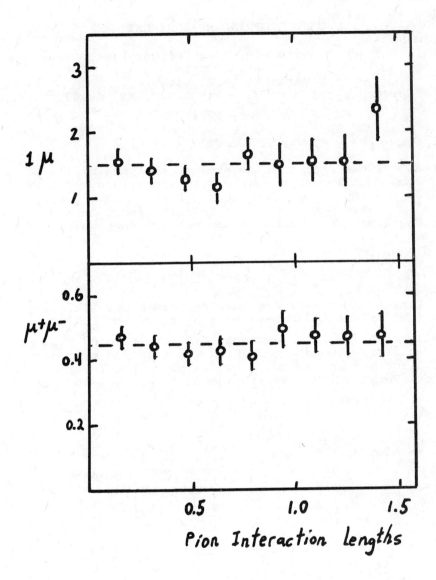

Figure 3 - The prompt signal for single muons and dimuons per 10⁵ interacting protons in each calorimeter plate as a function of the interaction point in the calorimeter.

A PROMPT NEUTRINO MEASUREMENT*

B.P. Roe, H.R. Gustafson, K.J. Heller[†], L.W. Jones
and M.J. Longo
Department of Physics
University of Michigan, Ann Arbor, MI 48109
October, 1978

ABSTRACT

A test has been made to explore the possibility of beam dump neutrino experiments with short target-detector separations and modest detectors. Results have given a positive neutrino signal which is interpreted in the context of various charmed-meson production models. A limit to the lifetime and mass of the axion is also a byproduct of this test.

INTRODUCTION

A test experiment has been performed parasitically in the M2 diffracted proton beam of the Meson Lab at Fermilab. A dimuon experiment, E-439,[1] targeted protons on a thick tungsten target which was followed by a 5.5m solid iron magnet assembly magnetized to 2.1 T (B horizontal). The neutrino detector was a 4-ton iron calorimeter[2] located 22 m from the tungsten target behind an additional 5.4m of steel, as indicated in Figure 1.

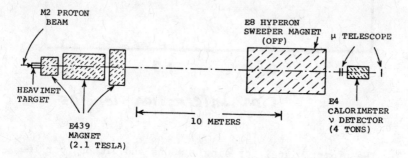

FIGURE 1: Experiment Configuration

*Supported by the U.S. National Science Foundation and the Department of Energy
†Now at the University of Minnesota

Results from beam dump experiments at CERN[3] have indicated a source of prompt neutrinos, and D-pair production has been suggested as the mechanism. This experiment was implemented toward the end of E-439 running so that the running time corresponded to only about 2×10^{15} protons on target, and this was further significantly reduced by deadtime. Nevertheless a positive signal was obtained. If interpreted as D-pair production, this signal corresponds to a D-production cross-section of 60 to 80 μb.

EXPERIMENTAL DETAIL

Even behind the 10.9m of iron the muon flux was very high; a 30×30 cm^2 scintillator telescope straddling the calorimeter on the beam axis recorded 5000 muons per 10^{11} protons on target.[4] As about 10^{-4} of energetic μ's produced an interaction in the calorimeter corresponding to $\gtrsim$ 20 GeV energy release, it was necessary to shield the front face of the calorimeter with anticoincidence counters. Another source of false signal could be cosmic ray events (hadrons or air

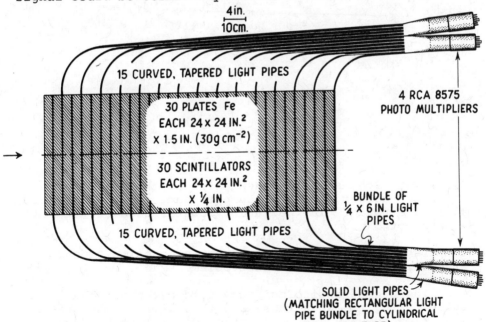

MICHIGAN NEUTRON CALORIMETER

Figure 2

showers) from above. One 60×120cm^2 anticoincidence
scintillator was accordingly put on top of the calorimeter. The experiment was located at the end of the
Meson Detector Building with no overhead shielding.
In order to reduce the calorimeter albedo from desired
events from triggering this top counter, 5 cm of
borated polyethylene were placed between the calorimeter
and this counter.

The calorimeter consisted of 30 steel plates each
61 cm square and 3.9 cm thick with 0.64 cm scintillator
between. The scintillator light was piped to four
phototubes (Figure 2) which permitted left-right and
front-back signal comparisons. Calibration in a test
beam gave a calorimeter resolution $\sigma \simeq (73/\sqrt{E})$% for
hadrons of 20-40 GeV. Cherenkov pulses from the
phototube light pipes were a possible source of concern;
anticoincidence shielding and the requirement of

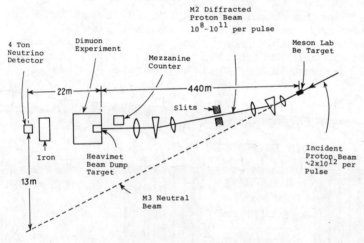

FIGURE 3: Schematic representation of the beam line.
The meson target, the beam dump target, and the beam line
slits are noted. The mezzanine counter is over the beam
line about 3 m.

comparable pulses from phototubes on opposite sides
effectively suppresses this problem.

The summed anticoincidence rate was about 5×10^6 per
pulse during the data runs, and this limited the system
live time to only 22%. The calorimeter threshold was
set to about 10 GeV, or about 3.5 times the most
probable muon pulse height. The neutrino event candidates were separated from spurious triggers by demanding

a ratio of right to left calorimeter phototube pulse heights compatible with the measured resolution and the scatter measured from muon-initiated events. Cosmic ray and other spurious triggers were also reduced by requiring the event time to lie within a 4 ns band relative to the accelerator r.f. signal, again based on the muon calibration. Further, the time difference between the left and right calorimeter signals was required to lie within bounds of 6 ns (low E) and 4 ns (high E). The pulse area from each anticoincidence counter was digitized, in principle providing redundant information as the veto discriminator rejected all events accompanied by pulses in the anticoincidence counters. A cut was made on the veto ADC pulse area corresponding to 1/5 to 1/10 the most probable particle signal; these result in one of the sets of values in Tables I, II, and III. An upper limit on the true neutrino signal is obtained by ignoring the veto ADC signals and assuming that the anticoincidence electronics functioned ideally. These values are also noted in the tables.

The fiducial volume of the calorimeter is not certain. It is probable that vertices within 5 cm of the side edges of the calorimeter are recorded at significantly lower efficiencies; hence we take the area to be 2500 cm^2 (50×50 cm). The depth will be less that the 900 g/cm^2; again because cascades close to the back of the calorimeter will be detected inefficiently. This effect was studied with muon-induced electromagnetic cascades from which it was deduced that a depth of 700 $g\ cm^{-2}$ was the effective fiducial volume. Together with the edge cut, the effective mass of the detector was 1.75 metric tons.

Because of the anticoincidence counter which lay over the calorimeter to veto cosmic ray air showers, there was some probability that a neutrino event in the calorimeter would produce a scattered particle into this counter and veto itself. This was evaluated by looking both at the fraction of muon initiated events in which this counter fired, and by operating the system in a hadron beam and determining the faction of events in which this counter fired. A value of 30% was obtained from the muon-initiated events. The hadron data for this fraction fit an empirical function

$$f_i = 0.32 + 0.006\ (E_i - 20),$$

so that a corrected number of events can be obtained by scaling with a factor

$$K = \frac{1}{N} \sum_{i=1}^{N} \frac{1}{1 - f_i}.$$

RESULTS

Data were taken under four conditions: (1) high intensity on E-439 (data run) with about 1.5×10^{11} protons per pulse (runs 54 and 58); (2) low intensity on E-439 target at about 10% the data run intensity, or about 1.3×10^{10} protons per pulse (runs 60 and 71); (3) very low intensity, less than 10^9 protons per pulse (runs 74 through 79); and (4) cosmic rays (accelerator off; run 104). During (1), (2), and (3) the beam on the Meson Area target was similar; about $2 \; 10^{12}$ per pulse.

The primary data from the high intensity run (1) contained 8 events of over 20 GeV if all cuts are applied, or 14 events if the digitized veto counter levels are ignored. The energies of these events are tallied in Table I.

TABLE I

Energies of prompt neutrino candidates form high intensity beam dump

E(GeV)	20	21	24	26*	29	34*	46
	20*	21*	26	27	33	39*	98*

*Accompanied by small veto signals.

The cosmic ray rate provides a reasonably certain (and statistically sound) background which may be substracted from each of the data sets. It corresponds to about 10% of the high-intensity event rate. The two lower-intensity runs provide somewhat contradictory data on backgrounds although statistics are sufficiently modest to render an apparent contradication rather insignificant. The background may be due either to the protons on the meson area target or to the scraping and to the collimators in the M2 beam line. The latter was monitored by a scintillation counter on the mezzanine of the Meson Detector Builidng and by the muon flux. The muon flux correlated reasonably with the mezzanine counter and these two were adjudged to be a more reliable monitor of background neutrino events than protons per pulse on the meson area target. The measured muon flux per pulse was actually about 1 1/2 times greater during the low intensity runs 69 and 71 than during the data runs 54 and 58. On the other hand, there were over twice as many neutrino events per muon in the data runs as in the low intensity runs. Thus we assume that most of the observed muons and a proportionate fraction (~40%) of the neutrino events may be from upstream beam scraping. The proton beam direction at the Meson Area target makes an angle of 27 mr with respect to a line from that target to our detector. However the first bends in the M2 beam line would effectively channel some π^+ and K^+ along trajectories directed more nearly toward our detector.

TABLE II

Runs	Protons per pulse	Total targetted protons[a]	Pulses[a]	Mezzanine[a] counts	Muon telescope counts[a]	Events[c]
54,58	1.4×10^{11}	4.28×10^{14}	2,929	16×10^6	19.3×10^6	8(14)
69,71	1.3×10^{10}	2.73×10^{13}	2,091	19.4×10^6	29.3×10^6	6(9)
74,79	2×10^8	2.5×10^{12}	12,052	14.5×10^6	---	6(7)
104	0 (cosmic rays)		239,000[b]	---	---	74(103)

a. Corrected for dead time
b. Equivalent pulses
c. Total events without anticoincidence ADC cut in parentheses.

The effective number of beam dump events, with and without the veto ADC data, are corrected for cosmic ray background in Table IIIa, and then for background from neutrinos produced upstream of the beam dump in Table IIIb. In that table it was assumed that (1) the mezzanine counter rate, or (2) the number of pulses (proportional to protons on the meson area target) is proportional to the true background. On grounds of both plausibility and self consistency, it was subjectively decided to weight the corrected number of events 3:1 in favor of the mezzanine-corrected results. When averaged over both sets of lower-intensity runs and corrected for the self-veto effect, a pair of best-estimated net numbers of beam dump neutrino events are obtained: 6.2±4.8 (including veto ADC's) and 13.8±6.2 (ignoring veto ADC's). With the obvious uncertainties reflected by the diverse entries in Table III, we will take 10±5.5 as the best estimate of true neutrino events.

Table IIIa Events in each set of runs corrected for cosmic ray background

RUN	54,58		69,71		74-79		69-79	
	(a)	(b)	(a)	(b)	(a)	(b)	(a)	(b)
a) INCLUDE VETO ADC								
b) IGNORE VETO ADS								
Net events corrected for cosmic ray rate	7.1	12.8	5.4	8.2	2.3	2.3	7.7	10.5

Table IIIb Beam dump events corrected for upstream neutrino sources based on the indicated low intensity runs (columns) and the background assumptions (rows).

RUN	69,71		74-79		69-79	
Constant background per pulse	-.5	1.4	6.5	12.2	5.5	10.6
Background correlated with mezzanine flux	2.8	6.1	4.7	10.2	3.4	7.9
Weighted 3/4 mezzanine background, 1/4 pulse background	2.0	5.3	5.2	10.7	3.9	8.6
Overall net beam dump events weighted by self-veto correction factor K=1.6	3.2	8.5	8.3	17	6.2	13.8

ERRORS

The data in Table III indicate the uncertainties in background and true beam-dump neutrino event rate and emphasize both the need for careful beam preparation (to avoid upstream sources of π- and K-decay neutrinos) and of careful measurements to appraise it. The best we can say from Table III is that our true signal appears to be 10±5.5 events. Various sources of normalization error besides the background subtraction and veto ADC uncertainty remain.

The fiducial mass of the calorimeter is uncertain to ±100 g cm^{-2}, or ±14% in depth, and ±2.5 cm in radius, or ±22% in area (although this is less significant in rates due to the radial fall-off in neutrino flux). Overall, the effective, radially-weighted fiducial mass is uncertain by ±30%. The absolute calibration of the calorimeter is uncertain by ±15%. In view of the observed neutrino event energy spectrum this reflects as a ±30% uncertainty in rate. The various timing cuts and the cut on the ratio of pulse heights from the two halves of the calorimeter cuase little uncertainty. Nevertheless, there is perhaps a ±15% uncertainty due to the cumulative uncertainty of these criteria. The self-veto effect from the cosmic ray anticoincidence counter may be uncertain by ±20%.

All of these effects taken together add up to a 50% uncertainty in the results of Table III. They do not, however, modify the 2σ signifigance of the evidence for a positive beam-dump signal. An estimate of neutrinos from π and K decay within the beam dump based on the CERN BEBC data indicates that $\lesssim 7\%$ of our net beam dump signal may be from this source.

INTERPRETATION

Essentially all of the incident protons interact in the tungsten (Heavimet) target, so that the number of neutrinos produced is

$$N_\nu = N_p \left\{ \frac{\sigma_\nu(pW)}{\sigma_I(pW)} \right\} F(pW) \qquad (1)$$

where σ_ν and σ_I are neutrino production and inelastic cross sections, respectively, for protons on tungsten and F(pW) is an enhancement factor to account for neutrino production by degraded nucleons which continue beyond a first target interaction. The number of detected events is

$$N_{ev} = N_\nu \, \rho \ell \sigma(\nu Fe) \, G(E) \, \Delta\Omega \qquad (2)$$

where $\sigma(\nu Fe)$ is the neutrino interaction cross section on iron, and $G(E)\Delta\Omega$ is the fraction of the produced neutrinos which pass through the fiducial volume of the calorimeter and which, upon interaction, are detected in the calorimeter with a signal corresponding to >20 GeV

In order to interpret direct neutrino production in terms of a specific model, it was assumed that all neutrinos come from D-decay, and that the branching ratio for semi-leptonic D-decay is 20%,[5] so that

$$\sigma_\nu(pp) = 0.4\sigma_D(pp)$$

where $\sigma_D(pp)$ is the production cross section for D pairs in nucleon-nucleon collisions.

It is necessary to interpret production processes in tungsten in terms of elemental pp processes. As it appears that production of ψ's, direct μ's, and large $p_\perp$ mesons is proportional to $\sim A^{1.0}$, it is reasonable to make the same assumption for direct neutrino production. Then

$$\sigma_\nu(pW) = A_W \sigma_\nu(pp)$$

so that Eq. (1) may be rewritten as

$$N_\nu = N_p \left[\frac{A_W \sigma_I(pp)}{\sigma_I(pW)} \right] \frac{\sigma_\nu(pp)}{\sigma_I(pp)} \cdot F(pW). \qquad (3)$$

The σ_I's are total interaction or inelastic cross sections and the factor in brackets in Eq. (3) has a value of 3.6 for tungsten. The intra-target cascading factor, $F(pW)$ is estimated to be 1.12 for Drell-Yan processes for $m_{\mu\mu} > 7$ GeV, and does not include effects due to secondary pions. To the extent that the more copious lower energy pions are important in D production, $F(pW) = 1.12$ is an underestimate, and our deduced cross sections are correspondingly overestimates.

The value for $\rho\ell$ for the 700 g cm^{-2} (fiducial length) calorimeter is 4.2×10^{26} cm^{-2}. The neutrino interaction cross section is taken as $\sigma(\nu Fe) = A_{Fe}\,\sigma(\nu N)$. The values of cross sections are taken as

$$\sigma(\nu N) = 0.6\ E(GeV) \times 10^{-38}\ cm^2,$$
$$\sigma(\bar\nu N) = 0.25\ E(GeV) \times 10^{-38}\ cm^2. \qquad (5)$$

Equal numbers of neutrinos and antineutrinos are assumed. The interaction cross section was further scaled by 1.32 to include neutral current events, so that the effective $\sigma(\nu N)$, averaged over ν and $\bar\nu$, is $0.55 E \times 10^{-38}$ cm^2 (E in GeV). The function G(E) is derived by folding the calorimeter resolution function with the calculated hadronic plus electro-magnetic products of the neutrino interactions (assuming equal numbers of ν and ν̄). This distribution is sketched in Figure 4 for 40 GeV neutrinos.

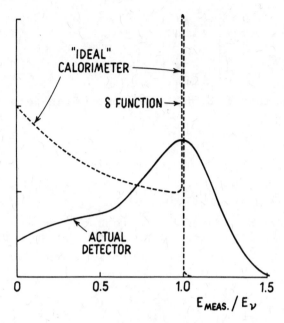

FIGURE 4: The expected response of an ideal detector (dashed line) and of a calorimeter with the experimental resolution to the assumed equal mix of ν_e and ν_μ. For neutral current events and ν_μ charged current events only the hadron cascade is detected with $0 \leq E_h \leq E_\nu$; for ν_e charged current events (∿ 1/3 of the total) the entire neutrino energy appears in the cascade.

The fraction of neutrinos within the calorimeter solid angle fraction $\Delta\Omega$ was calculated assuming that all ν came from D decays, D→K+ℓ+ν or K*+ℓ+ν using the measured ℓ spectrum. A sample of 30,000 events was run through a Monte Carlo program for each of several assumed D production models. The results of these calculations are tabulated in Table IV. From the observed neutrino events, the resulting calculated D-production cross sections are also tabulated for the different production models. We have also compared our results with the CERN BEBC 0 mr observations, considering only the electron neutrino events. The CERN beam dump target was copper, for which the factor $A\sigma_{pp}/\sigma_{pA} = 2.54$. A factor F(pCu) of 1.12 is applied to the CERN data to determine the cross section values of Table IV. The assumptions of the various D-production models are spelled out below.

Model I

$$\frac{d^2\sigma}{dydp_\perp^2} \propto e^{-6m_\perp} \Big]_{-y_{lim}}^{+y_{lim}} \quad ; \quad m_\perp^2 = p_\perp^2 + m_D^2; \tag{5}$$

where the y distribution was assumed flat between the cm y limits of ±0.5, ±1.5, or ±2.5.

Model IIa

$$\frac{d^2\sigma}{dxdp_\perp} \propto e^{-(1.75p_\perp + 10|x|)} \quad ; \quad x = (p_\parallel/p_\parallel(max)_{cm}). \tag{6}$$

This model has been used by Lauterbach[6] as a fit to ψ production. Using this form and examining μ polarization data[7], he sets a limit of 1 μb to D production by 400 GeV protons.

Model IIb

$$\frac{d^2\sigma}{dxdp_\perp^2} \propto e^{-(1.75p_\perp + 10|x|)}, \tag{7}$$

in accord with a more commonly accepted $p_\perp$ distribution.

Model IIIa

$$E\frac{d^3\sigma}{dp^3} \propto (1 - |x|)^4 e^{-1.6p_\perp}, \tag{8}$$

as a fit to J/ψ production[7,8].

Model IIIb

$$E\frac{d^3\sigma}{dp^3} \propto (1 - |x|)^4 e^{-2.2p_\perp}. \tag{9}$$

Model IV

$$\frac{d^2\sigma}{dxdp_\perp^2} e^{-(2.2p_\perp + 9.7x')} \tag{10}$$

where $x' = p_{lab}/p_{beam}$. This model is used to fit data from another J/ψ production experiment.[9] The results are shown in Table IV.

Table IV

Cross Sections for Production of D-Pairs[1]

Model[2]	y_{limit}	This Experiment Probability $G(E)\Delta\Omega$(M.C.) for $E>20$ GeV released in cal.	$\bar{E}_\nu$	$\sigma(\mu b)$[4]	CERN BEBC $\sigma(\mu b)$[5] 0 mr
Ia	0.5	0.008	26	700	500
Ib	1.5	0.052	51	60	50
Ic	2.5	0.131	87	17	5
IIa		.020	52	202	128
IIb		.05	54	76	46
IIIa		.102	55	72	50
IIIb		.115	54	64	44
IV		0.042	49	75	45

1. Semileptonic decays of D's of 20% assumed source of ν; equal numbers of $\nu_e, \bar{\nu}_e, \nu_\mu, \bar{\nu}_\mu$
2. See text for details of models
3. Based on Monte Carlo calculation of 30,000 events, $G(E)\Delta\Omega$(M.C.) defined in text.
4. Based on a signal of 10 events. See text for systematic errors.
5. Based on 15 e^+e^- events. (see Ref. 3)

The most sensitive published counter search for D's from hadronic interactions by Ditzler et al.[10] determined 95% c.l. upper limit cross sections for $K^-\pi^+$ ($K^+\pi^-$) production at the D^0 mass. With $d\sigma/dp_\perp^2 \propto e^{-1.6p_\perp}$, they determined $B\, d\sigma/dy < 360$ nb (290 nb) at $y = -0.4$ for $D^0 \to K^-\pi^+$ ($\bar{D}^0 \to K^+\pi^-$). If D production is flat in dσ/dy over $-1.5 < y < +1.5$ (equivalent to a hybrid of our models Ib and IIIa), these results are equivalent to an upper limit for Bσ of approximately 1 μb. If a branching ratio for $D° \to K^-\pi^+$ of 1.8±0.5% is included, we have $\sigma(D°) < 48\text{-}60$ μb per nucleon. If $\sigma(D°) = \sigma(\bar{D}°) = \sigma(D^+) = \sigma(D^-)$, and all contribute equally to the neutrinos observed in this and the BEBC experiment, the limits correspond to an upper limit for D-pair production of about 100 μb, comfortably compatible with most of the values of Table IV.

As can be seen in Table IV, our limits vary enormously depending on the model. For all but Model Ic, our results are consistant with the CERN 0 mr BEBC data. Both our result and the BEBC result are consistant with reasonable models which assume that prompt neutrinos are from D-decays.

AXIONS

The results of this experiment may also be interpreted to set limits on axion lifetimes and hence mass. The observed number of axions would be given by relations analogous to Eqs. (1) and (2). Again, since axion production is a semi-weak process, we may expect the production of axions in tungsten, $\sigma_a(pW)$ to be given by

$$\sigma_a(pW) = A_W \, \sigma_a(pp)$$

so that

$$N_a = N_p \left(A_W \frac{\sigma_I(pp)}{\sigma_I(pW)} \right) \frac{\sigma_a(pp)}{\sigma_I(pp)} F'(pW), \qquad (11)$$

analogous to Eq. (3). The intra-target cascading factor $F'(pW)$ may be considerably larger than for neutrino production via D-pairs as the axion threshold is presumably quite low. We will take $F'(pW) = 3.0$ assuming that first-generation pions and nucleons are effective in producing axions of over 20 GeV. There is also a factor for the decay of the axion, $\exp(-7.3\times10^{-8}/\gamma\tau)$ over our 22m target-detector separation. For $E_{axion} = 24$ GeV, the exponent is unity for $\tau/m = 1.8\times10^{-12}$ sec MeV^{-1}. The observed number of axion interactions would then be (neglecting decay)

$$N'_{ev} = N_p \frac{\sigma_a(pp)}{\sigma_I(pp)} F'(pW) \left[\frac{A_W \sigma_I(pp)}{\sigma_I(pW)} \right] \left\{ \rho \ell \sigma_I(ap) \Delta\Omega \right\} \qquad (12)$$

where $\sigma_I(ap)$ is the axion interaction cross section per nucleon in the calorimeter. The solid angle fraction, $\Delta\Omega$ was determined from $d^2\sigma/dydp_\perp^2 \propto \exp(-6m_\perp)$, with $m = 0.1\ m_\pi$ and a uniform y distribution over $-2.5 < y < +2.5$. This gave $\Delta\Omega = 0.09$. The resulting number of detected axions is then

$$N_{ev} = 5 \times 10^{66} \sigma_a(pp)\ \sigma_I(ap)\ e^{-7.3 \times 10^{-8}/\gamma\tau}.$$

If our 10 events are all axions, our results would yield

$$\sigma_a(pp)\ \sigma_I(ap) = 2 \times 10^{-66}\ e^{+7.3 \times 10^{-8}/\gamma\tau}\ cm^4.$$

This may be compared with the prediction by Ellis and Gaillard[11] of

$$\sigma_a(pp)\ \sigma_I(ap) \geq 9 \times 10^{-66}\ cm^4.$$

Even considering our large uncertainties our results appear to be incompatible with axions of very low mass. As the axion lifetime is given[12] as

$$\tau > 10^{-10}\ (sec),$$

our data may be interpreted as setting a lower limit to the axion mass of ~ 25 MeV.

CONCLUSION

A two σ positive signal for direct neutrino production is observed, although the background is $\sim 1/3$ of the signal, and normalization uncertainties are considerable. If the data are interpreted in terms of D production and semi-leptonic decay, they agree with the CERN BEBC beam dump results when reasonable D production kinematics are assumed. If on the other hand the neutrinos had an angular distribution characteristic of π and K decays, they would be produced in too small a solid angle to account for the signal we observe and the BEBC results. Our results are also in agreement with a preliminary report on a Fermilab measurement of direct muon production[13] and with upper limits to D production set by Fermilab counter experiments. We are in disagreement with the negative results on charm production from some hadron emulsion experiments.

Our new results may be interpreted in terms of axion production only if the axion mass is greater than ~ 25 MeV.

REFERENCES

1. W.P. Oliver et al., TUFTS PUB 78-1601, to be published in the proceedings of the 3rd Int'l Conf. at Vanderbilt Univ. on New Results in H.E. Physics (1978), and S. Childress et al., "A High Statistics Study of Dimuon Production by 400 GeV/c Protons" Sub. to XIX Int'l. Conf. on H.E. Physics. Tokyo (1978)(unpublished).
2. L.W. Jones et al., Nucl. Instr. and Methods, 118, 431 (1974).
3. P. Alibran et al., Phys. Lett. 74B, 134 (1978); T. Hansl et al., Phys. Lett. 74B, 139 (1978). P.C. Bosetti et al., Phys. Lett. 74B, 143 (1978). The data referenced in Table IV are from Bosetti et al.
4. L.W. Jones et al., UM HE 78-34, "The Muon Flux from E439 Beam Dump Targeting" (1978)(unpublished) and L.W. Jones UM HE 78-42 "Production Processes from Protons Incident on Thick Metal Targets"(1978) (unpublished).
5. Review of Particle Properties LBL 100 and Phys. Lett. 75B, 1, (1978).
6. M.J. Lauterbach, Phys. Rev. D17, 2507 (1978).
7. H.D. Snyder et al., Phys. Rev. Lett. 36, 1415 (1976).
8. M. Binkley et al., Phys. Rev. Lett. 37, 574 (1976).
9. G.J. Blanar et al., Phys. Rev. Lett. 35, 346 (1976).
10. W.R. Ditzler et al., Phys. Lett. 71B, 451 (1977).
11. J. Ellis and M.K. Gaillard, Phys. Lett. 74B, 374 (1978).
12. J. Ellis et al., Nucl. Phys. B106, 292 (1976).
13. K. Brown et al., to be published in the Proceedings of the Third International Conference on New Results in High Energy Physics, Vanderbuilt University (1978). See also preceding paper in this volume.

HADRONIC PRODUCTION OF CHARMED AND OTHER FAVORITE PARTICLES*

Francis Halzen
University of Wisconsin, Madison, WI 53706

ABSTRACT

We review well known facts about charmed particles (i) that in the G.I.M. model the expected lifetime of a charmed particle is in the range ($4 \times 10^{-13} - 2 \times 10^{-12}$) seconds (ii) that charm production cross-sections in the ($5 \sim 20$) μb range for 400 GeV proton-proton interactions are expected from quantum chromodynamics. We show that accumulating (and seemingly contradictory!) information regarding charmed particle lifetimes and cross-sections from counter, beam-dump and accelerator or cosmic-ray emulsion experiments converges on values consistent with these theoretical prejudices. The confirmation of perturbative Q.C.D. for calculating hadronic cross-sections of heavy particles is good news for the future hadron colliders discussed at this conference: they might not only discover the weak intermediate bosons but also the t-quark and possibly the next flavor doublet.

Why go through the considerable effort of observing charmed particles in sparse numbers in hadron collisions when e^+e^- machines have blessed us with a charm factory at the $\psi(3.77)$? There are two obvious answers to this question, they will be our guideline in developing this talk.

(1) As emulsion (exposed to hadron beams or cosmic rays) detection techniques are used in several experiments, we can get a direct handle on the lifetime of the charmed particles. A measurement of the lifetime τ would directly probe the hitherto untested normalization of the G.I.M.current. The fact that the weak charmed current shares the Fermi coupling G as an overall normalization with the weak current responsible for β-decay leads to the prediction that $\tau=(4 \times 10^{-13}-2 \times 10^{-12})$ seconds depending on the value of the charmed quark mass($m_c=1.1 \sim 1.87$ GeV corresponds to the range quoted). These estimates will be discussed in detail by Rosner.[1]

(2) Measurements of the charmed particle production cross section and its energy dependence check the validity of perturbative chromodynamics (Q.C.D.) calculations.[2] As a test of Q.C.D. they probe a new kinematic range in hadron interactions and are therefore complementary to lepton pair production and high-p_T phenomenology.

To establish the validity of Q.C.D. calculations of the hadroproduction of heavy particles is crucial as it is really the only quantitative guide we have for modelling the signature and backgrounds of weak boson or heavy flavor production in the design of detectors for the hadron colliders presently under construction.[3]

A. Theoretical Expectations

As Rosner[1] will discuss the theoretical justifications for the lifetime range quoted for charmed particles, we will concentrate our efforts on reviewing the present technology[4-18] for computing their production cross sections. The hope is that due to the presence of a large mass ($2m_c$) in the final state the "short distance" or "impulse approximation" in the high energy collision is guaranteed. In such interactions the constituent quarks and gluons behave as nearly free particles and their interactions can be studied in perturbation theory. Indeed the quark-gluon effective coupling constant α_s becomes small in the impulse approximation.[2] Typical diagrams for producing a $c\bar{c}$ quark pair are sketched in Fig. 1. They fall into two classes labelled as single (Fig. 1a) and double (Fig. 1b) gluon contributions. Set (a) dominates the cross section near threshold (typically $2m_c/\sqrt{s} > 0.15$) and can be directly computed by the observation[4-6] that each diagram has an electromagnetic equivalent. The gluon (materializing in a quark pair) can be substituted by a photon (converting to a lepton pair) and the diagrams of Fig. 1a can therefore be simply computed from the lepton pair data by the rescaling of the coupling constant $\alpha \rightarrow \alpha_s$. The argument is pictorially represented in Fig. 2 and the resulting cross sections for c and b quarks is shown in Fig. 3. Further above threshold gluons (which carry on the average a small fraction of the momentum of the incident hadrons) themselves become energetic enough to make heavy quark pairs and the diagrams of Fig. 1b take over (see Fig. 3).[7-18] Standard structure functions describing the momentum distributions of quarks and gluons inside the incident hadrons were used in computing the results shown in Fig. 3. They are remarkably insensitive to this aspect of the calculation.

The warning is now timely that these calculations could <u>underestimate</u> the real cross section in several ways:

(1) The results are very sensitive to the threshold mass. It is important to realize that even if the combined mass of the $c\bar{c}$ pair is below threshold (i.e., $2m_c \leq m(c\bar{c}) \leq 2m_D$) a $D\bar{D}$ final state is possible through a boost in energy from the residual quarks from the initial hadrons.[17-18] The effective threshold can be lowered further by the Fermi-motion of the target nucleons in the nuclear targets used in most experiments.[19] Both effects can be summarized by saying that the threshold mass usually taken to be $m_c = m_D = 1.87$ GeV could be less. This will increase the cross-section: by a factor of 4 for $m_c = 1.5$ GeV and more than an order of magnitude for $m_c = 1.1$ GeV in the case of $p_{lab} = 400$ GeV collisions. Fig. 4 illustrates the m_c-dependence of the cross section at different energies.

(a)
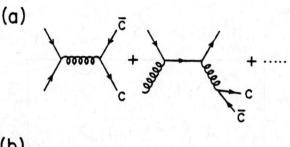

(b)
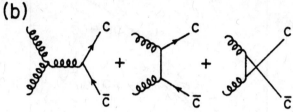

Fig. 1. Typical Q.C.D. diagrams contributing to the associated production of heavy quarks in hadron collisions. Solid and curly lines represent quarks and gluons, respectively.

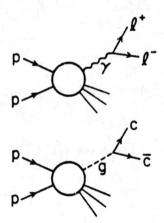

Fig. 2. The quark-lepton connection: a way to evaluate the diagrams of Fig. 1a directly from lepton pair data (Refs. 4-6).

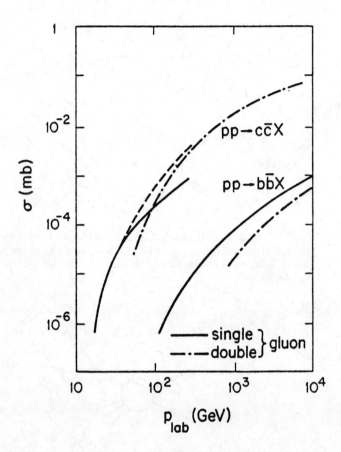

Fig. 3. Production cross sections of charmed particles obtained from the diagrams of Fig. 1a (full line) and Fig. 1b (dash-dotted line). Also shown is a computation of the associated production of the b-quark bound in the $\gamma(9.5)$

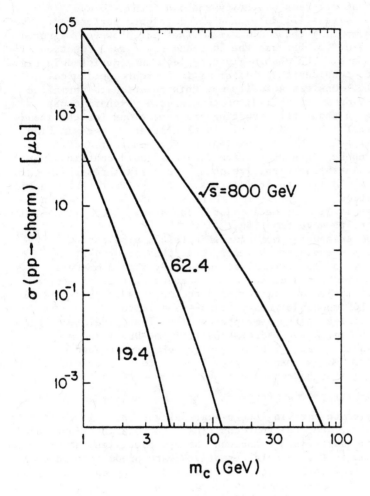

Fig. 4. Dependence of the production cross-sections of heavy quarks on the quark mass m_c is shown at some representative energies.[18]

(2) It has been pointed[20] out in the context of photoproduction calculations that diagrams of the type shown in Fig. 5 might contribute. If we were still doing the old fashioned parton model, we would all be computing those diagrams. They are, however, ignored in perturbative Q.C.D. because the $\bar{c}u$ bound state wave function explicitly appears. The photon-gluon replacement (depicted in Fig. 5) forces us to consider these diagrams in hadroproduction calculations as well.[21] One can actually use this relation to normalize their contribution in pp collisions to measured $\gamma p \to$ charm cross sections. Fig. 6 shows the resulting cross sections for the assumptions $\sigma(\gamma p \to charm) = 5\mu b$ and $0.5\mu b$. As the latter is more in line with present experimental information on charm photoproduction we conclude that these non-perturbative diagrams could contribute at the same level as the perturbative ones. As a final remark in this connection we would like to draw attention to the fact that Combridge's "flavor excitation" diagrams[18] are similar (and so are his conclusions) to those in Fig. 5; he is just using a one gluon exchange model for the bound state wave functions.

(3) Parton transverse momentum effects can also increase the cross section. We estimate, however, that they are small.

We conclude the theoretical discussion by some comments:

- For the cosmic ray audience at this conference we present a total flux calculation of stable (or detectable as delayed particles, typically $\tau > 10^{-8}$ seconds) heavy hadrons and leptons generated by cosmic ray particles in the atmosphere. The flux calculated[22] from the diagrams of Fig. 1 for hadrons and from the Drell-Yan model for leptons, is shown in Fig. 7 as a function of the produced mass. Notice, however, that we expect heavy leptons to be predominantly produced in decays of hadrons. For the τ lepton we calculate that

$$\sigma(pp \to B \to \tau) \simeq 40 \times \sigma(pp \to \gamma^* \to \tau^+\tau^-) \qquad (1)$$

where B is the state carrying the b-quark flavor bound in the $\gamma(9.4)$

- For sceptics, who feel that the final state interactions turning colored charmed quarks into colorless charmed particles cannot be ignored, we show in Fig. 8 an alternative charm production calculation where the fragmentation $c \to D$ (as measured in lepton induced interactions) has been explicitly included in the calculation.[16]

- Finally some comments regarding ψ production and heavy quark-antiquark bound states in general. Fig. 9 is a reminder that, although facing the color recombination problem of the final $c\bar{c}$ pair is now imperative, Q.C.D. estimates can be made. The claim that the $c\bar{c}$ pair produced in a 2-gluon amalgamation (see Fig. 1b) loses its color by forming a χ state and then adjusts its charge conjugation to form a ψ via a radiative decay $\chi \to \psi\gamma$ (see Fig. 10) was taken seriously by few.[7-10] Q.C.D. predicts that some of the time ψ's have to be produced in conjunction with photons with the ψ-γ invariant mass $m(\psi\gamma) = m_\chi$. The data obliged (Fig. 10). Also a brief reminder of the useful relation that the hadroproduction cross-section for a flavor bound state V of mass m at energy $\sqrt{s} = (2m_N P_{lab})^{1/2}$ is given by[23]

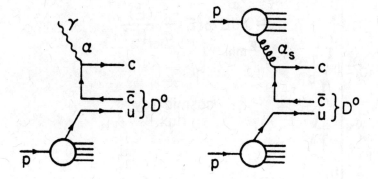

Fig. 5. Possible non-perturbative contributions to the photo- and hadroproduction of heavy quarks.

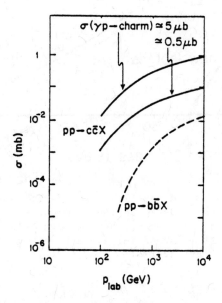

Fig. 6. Production cross sections of charmed and B ($\bar{b}d$) particles in pp collisions obtained from the diagrams of Fig. 5 assuming $\sigma(\gamma p \to \text{charm}) = 5$ μb. Also shown is an evaluation of the charmed particle cross section assuming $\sigma(\gamma p \to \text{charm}) = 0.5$ μb.

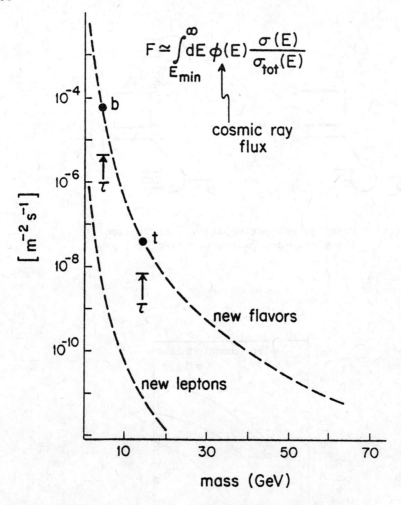

Fig. 7. The total cosmic ray flux of heavy hadrons and leptons produced in the atmosphere is shown as a function of their mass.[22] Also shown is the flux of τ-leptons from the decay of heavy quarks.

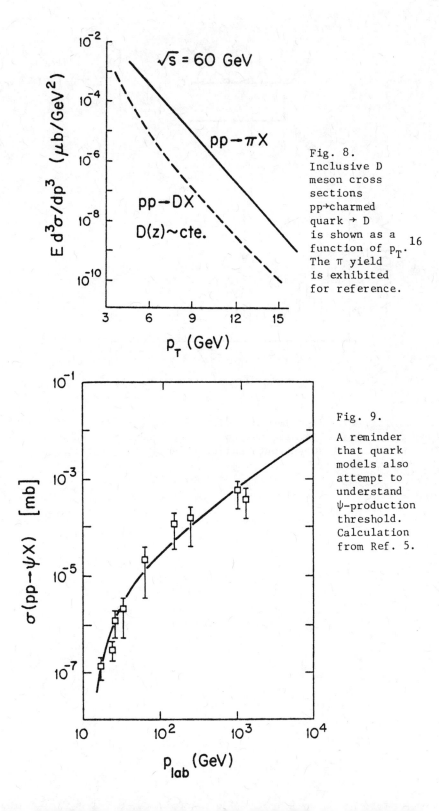

Fig. 8. Inclusive D meson cross sections pp→charmed quark → D is shown as a function of p_T.[16] The π yield is exhibited for reference.

Fig. 9. A reminder that quark models also attempt to understand ψ-production threshold. Calculation from Ref. 5.

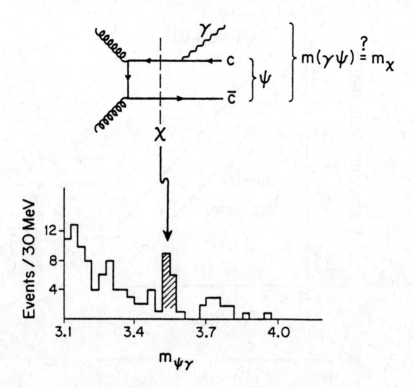

Fig. 10. Experimental evidence (see Ref. 25) for ψ's originating via the production and subsequent radiative decay of χ states.

$$\sigma_v = \frac{\Gamma_v}{m^3} F(\frac{m}{\sqrt{s}}) \qquad (2)$$

Here Γ_v is the direct hadronic width of V and F is a universal function for all V. Although there is some theoretical support[6] for Eq. (2), the empirical support is impressive (see Fig. 11). We illustrate the experimental verification of Eq. (2) by showing the universal shape of the quantity $\frac{m^3}{\Gamma_v} \sigma_v(\sqrt{s})$ for $\phi, \psi, \psi', \gamma$. The data points in Fig. 11 would scatter over more than 7 orders of magnitude without rescaling!

B. A (Biased) Experimental Review

At p_{lab}= 400 GeV/c we therefore expect to find charmed particles with $\tau = (4 \times 10^{-13} - 2 \times 10^{-12})$ seconds and cross sections in the ($5 \sim 25$) μb range with possible apologies for somewhat higher values. At $E \simeq 10-20$ TeV the cross sections have increased to "a fraction of one mb" and even B states are produced at the 1 μb level (see Fig. 3). Let us immediately remark that old[24] and new (see Table I) cosmic ray experiments in the TeV range are in agreement with expectations in the TeV energy range.

The experiments can be classified in 3 categories, we list them quoting some recent results.[25]

TABLE I

Emulsion experiments (exposed to accelerators and cosmic rays)

First author	beam type	p_{lab}	σ (μb/nucleon)	τ (units 10^{-13} sec)
Bannik	p, π⁻	70/60	5 ± 3	0.6
Coremans	π⁻, p	300	< 1.5	
Bozzoli	p	300/400	< 7	
Ushida	p	400	30	1
Chernyavsky	p	400	100	0.2
Fuchi	?	$(1-2) 10^4$	500	4/12
Diabrim-Palazzi	γ	20-80	0.5	.02∿.05
Burhop	ν	400	-	6
Europe collaboration	ν	350	-	-

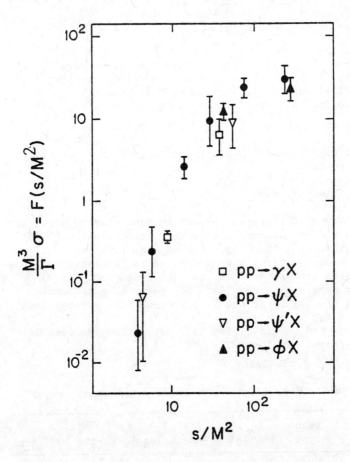

Fig. 11. The figure demonstrates that the quantity $m^3/\Gamma\,\sigma$ is a universal function of s/m^2 for different flavor bound states. m, Γ and σ are respectively the mass, direct hadronic width and production cross section of the flavor bound state.

TABLE II

Non emulsion searches

First author	signature		$\sqrt{s}$	σ (μb per nucleon)
Antipov	$K\pi$	π^-Be	10	< 9
Albrecht	$K^0\pi^\pm$, $K\pi\pi$	nC	10	< 25
Jonckheere	$\mu, \mu V^0$	π^-C	20	< 22
Abdins	$K\pi$	nBe	22	< 15
Spelbring	$K\pi, \mu$	nBe	24	< 100
Lipton	e, μ	nBe	24	< 30
Lauterbach	μ(polariz.)	pCu	27	< 1
Ditzler	$K\pi$	pBe	27	< 30
Baum	e, μ	pp	55	< 76
Clark	e, μ	pp	55	$\begin{cases} 76\pm36 \\ 25\pm10 \end{cases}$
Brown	single μ	pFe	27	30-80

BR(D→Kπ) = 2% and BR(D→e,μ) = 10% assumed

TABLE III

Beam dump experiments - prompt neutrinos

		$\sqrt{s}$	σ (μb per nucleon)
Serpukhov	pFe	12	19±15
BNL	pCu	7	< 1
CDHS	pCu	27	40
GGM	pCu	27	200-400
BEBC	pCu	27	100-200
FNAL	pCu	27	30

A cursory glance at the tables reveals cross sections ranging from 1 μb upper limits to 400 μb signal, and lifetimes ranging from the expected to the totally unacceptable[1] (10^{-14} - 10^{-15} seconds). We will show that there are actually no striking disagreements between these experiments and that, in the end, they bracket values for cross sections and lifetimes consistent with our expectations. The entries in the table have to be interpreted with care, and the following guidelines should be kept in mind:

- emulsions
  - (i) scanning biases towards high multiplicity events (σ can be overestimated)
  - (ii) backgrounds from hyperon decay and secondary interactions (diffraction dissociation, elastic scattering)
  - (iii) have to make acceptance corrections if the lifetime is as long as 10^{-13} seconds or larger (e.g. Coremans' 1.5 μb can be multiplied by a factor 4∼5 for $\tau \simeq 10^{-12}$ seconds)

- counter experiments
  - (iv) have to make assumptions regarding the momentum distributions of the produced charmed particles.
  - (v) or assumptions regarding polarization of colliding quarks and gluons have to be made (e.g. Lauterbach's 1 μb limit can be relaxed to approximately 10 μb)[25]

- beam dumps
  - (vi) same as (iv).
  - (vii) might have to consider[26] in their data analysis the presence of a small (but possibly not negligible) component of diffraction dissociation of charm to which they would be particularly sensitive (large Feynman x_F).
  - (viii) should make their nuclear reduction to pN interactions using A^1 not $A^{2/3}$.

After applying rules (i)-(viii) and checking individual publications we claim that data and expectations agree, with possibly two exceptions which we will discuss next. Let us also refer the reader to the two beautiful events recently found by Niu in an emulsion exposed at FNAL.[27] As both of the associatively produced particles are observed, these events cannot be explained by any backgrounds. The tracks confirm the expected lifetime range and their rate of observation is consistent with our expectation for the cross section.

The large cross section values from the CERN bubble chamber for the observation of prompt neutrinos assumed to come from the decay of charmed particles have to be divided by a factor 3.5 because of (viii). We feel that $(1-x)^4$ and e^{-p_T} are the preferred choices for the Feynman x and p_T dependence of the D secondaries. Indeed, we expect from Q.C.D. an x-dependence ranging from $(1-x)^2$ for $\sqrt{s}=20$ GeV to $(1-x)^5$ for $\sqrt{s}=63$ GeV. A peak develops at x=0 from the increasing contribution of the diagrams of Fig. 1b and is responsible for this energy dependence. Previous arguments[28] that no enhancement from diffraction dissociation exists when x→1 involve ad hoc assumptions and one should watch out (see (vii))! Gaisser has been studying[29] Das-Hwa diagrams[30] where a fast valence quark combines with a c quark from the sea of one of the colliding particles, subsequently forming a forward D. Their presence cannot be eliminated (they are in fact similar to the diagrams of Fig. 5). They would display a "softened" $(1-x)^3$ distribution ($(1-x)^3$ characterizes the valence quark). More importantly this mechanism would strongly favor forward $q\bar{c}$ (rather than $\bar{q}c$) which would lead primarily to antineutrinos and hence to positive leptons in the detector.

The Lebedev emulsion exposure at Fermilab (Chernyavsky et al.) poses a double threat to the theory with $\sigma=100$ μb, $\tau=2.10^{-14}$ seconds. If their signature is real it cannot come from charmed particles. We would like to point out, however, that the 9 events observed are all single tracks, the associated partner is not observed. There are backgrounds faking such events: hyperon decay, diffraction dissociation or elastic scattering of a secondary in the emulsions are the most prominent ones. These backgrounds are fortunately calculable[24] from accelerator data. E.g., what is the probability that a high mass diffraction dissociation of a secondary proton into a proton and π^0 pair within the emulsion, fakes a charmed particle?

$$P = <n> \left(\frac{\Delta}{\lambda_{int}}\right) \frac{\sigma_{dissociation}}{\sigma_{inelastic}} N \qquad (3)$$

$<n>$: probability of producing a secondary proton

$\frac{\Delta}{\lambda_{int}}$: probability that the produced secondary interacts in an emulsion of thickness Δ (λ_{int} is the interaction length)

$\frac{\sigma_{diss}}{\sigma_{inel}}$: probability that this interaction is an appropriate diffraction dissociation faking a charmed particle

N : number of events

Such backgrounds range[24] from a few to 50%. The Lebedev exposure flirted with these backgrounds by scanning for secondary stars and by scanning at small angles.

The innocent entry in Table III from the BNL beam dump experiment is actually interesting. It establishes a threshold in the prompt neutrino signal ruling out light sources (e.g. the axion).

C. A Few Afterthoughts

The confirmation of Q.C.D. estimates of charm production threshold is good news, we can now hope that Q.C.D. can qualitatively anticipate features of the physics at the next generation of hadron colliders discussed at this meeting by Cline, Rubbia and White.[3] As examples we discuss the production of t-quarks and weak intermediate bosons.

W-estimates deserve renewed attention because of progress in perturbative Q.C.D. For the first time we can compute[31] their cross section, signature and competing backgrounds in the same framework. A cartoon of the calculation and its implications is shown in Fig. 12.

(a) The total production rate is determined by the usual quark-antiquark annihilation diagram.

(b) The transverse momentum signature of a lepton from the decay $Z \to e^+e^-$ or $W^\pm \to e^\pm \nu$ is determined by the fact that Z, W recoil against quarks and gluons in ways computable in perturbation theory.

(c) The dominant background competing with the lepton signature comes from internal conversion of high p_T real photons.

The favorable signal/background ratio is a very encouraging result. Remember that the equivalent Q.E.D. calculation, relevant for W, Z masses of a few GeV, predicted the weak boson to be unobservable in a direct lepton search.[32] Therefore, even though it means that the W is one accelerator away, $M_W > \sqrt{\frac{\alpha}{G}}$ is a definite blessing. The discovery of the W might be something of an anticlimax, but it will definitely be an ideal vehicle to test Q.C.D. ideas as illustrated by Fig. 12.

The t-quark provides us with another illustration of the potential of future hadron colliders. One might expect $m_t \simeq 15$ GeV from

$$\frac{m_c}{m_s} \simeq \frac{m_t}{m_b} \simeq 3, \qquad (4)$$

although Bjorken[33] reminds us that masses in gauge theories follow logarithmic rules and he expects $m_t \simeq 30$ GeV. This might put t-flavor out of reach of e^+e^- machines and whether the t-quark is accessible with hadron colliders becomes an important question. The answer is positive (even for machines of moderate luminosity); we deduce from standard Q.C.D. prejudices the results shown in Table IV [we use the notation $(t\bar{t})$ for the t-bound state or topsilon and T for the states carrying t-flavor, units are KeV and cm^2].

Yodh[34] observed delayed particles in cosmic ray interactions at the 1 μb level that could be B or T states. B is unlikely, however, because of the recent FNAL that $\tau > 10^{-8}$ sec for B states.[1]

As a final remark we would like to remind you that threshold or even pre-threshold production of heavy particles using Fermilab's energy doubler on a heavy nuclear target is another game in town. The situation regarding definite estimates is fluid and we will only draw your attention to the fact that some exploratory calculations have been made.[35] It has recently been pointed out that this idea should definitely be implemented by the use of secondary $\bar{p}$ or π^+ beams, no matter how pessimistic one is about the loss one takes in intensity.[36]

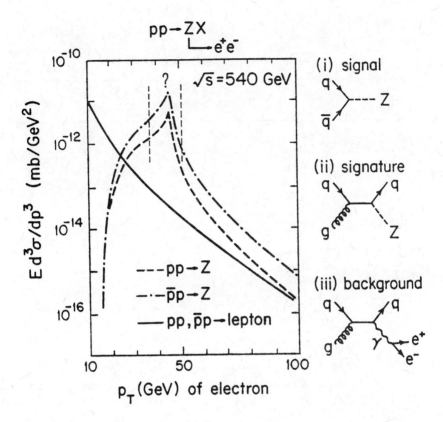

Fig. 12. Transverse momentum signature of the leptonic decay of the weak intermediate boson in hadron collisions with $\sqrt{s}$ = 540 GeV. Also shown is the background from directly produced leptons. The question mark reminds us that because of infrared divergences Q.C.D. makes no clear prediction of the shape of the peak at $p_T = \frac{M_W}{2}$.

TABLE IV

t-physics at hadron colliders

	m_t = 15 GeV	m_t = 30 GeV
Leptonic width	5.6	5.9
Direct hadronic width	18	14
Decay via virtual photon	25	26
Total hadronic width	43	40
Leptonic branching ratio B	0.09	0.1
$\sigma(\bar{p}p \to (t\bar{t})X)$ at y=0, $\sqrt{s}$ = 540 GeV	2×10^{-33}	2×10^{-34}
$\sigma(\bar{p}p \to TX)$ at $\sqrt{s}$ = 540 GeV	3×10^{-32}	10^{-33}
$\sqrt{s}$ = 2 TeV	4×10^{-31}	2×10^{-32}

*Notice that $B^2 \sigma(\bar{p}p \to TX)$ is not necessarily smaller than $B\sigma(\bar{p}p \to (t\bar{t})X)$ [units KeV, cm^2].

ACKNOWLEDGEMENTS

C. Carlsson, D. Cline, G. Conforto, T. Gaisser, L. Jones, J. Leveille, K. Niu, S. Pakvasa and D. Scott contributed information and comments to this review. Thanks!

Work supported in part by the University of Wisconsin Research Committee with funds granted by the Wisconsin Alumni Research Foundation, and in part by the Department of Energy under contract E(11-1)-881, COO-881- 71.

By acceptance of this article, the publisher and/or recipient acknowledges the U.S. Government's right to retain a nonexclusive, royalty-free licence in and to any copyright covering this paper.

REFERENCES

1. J. Rosner, these proceedings.
2. For recent reviews of phenomenological application of Q.C.D. see, e.g., R. Field, F. Halzen and L. Lederman in Proceedings of the XIX International Conference on High Energy Physics, Tokyo, Japan (1978).
3. D. Cline, C. Rubbia and H. White, these proceedings.
4. H. Fritzsch, Phys. Lett. $\underline{67B}$, 217 (1977).
5. F. Halzen, Phys. Lett. $\underline{69B}$, 105 (1977).
6. F. Halzen and S. Matsuda, Phys. Rev. $\underline{D17}$, 1344 (1978).
7. M. K. Gaillard, B. W. Lee and J. Rosner, Rev. Mod. Phys. $\underline{47}$, 277 (1975).
8. F. Halzen in Proceedings of the International Conference on Production of Particles with New Quantum Numbers, ed. by D. B. Cline and J. J. Kolonko (University of Wisconsin, Madison, 1976).
9. C. E. Carlson and R. Suaya, Phys. Rev. $\underline{D14}$, 3115 (1976).
10. S. D. Ellis, M. B. Einhorn and C. Quigg, Phys. Rev. Lett. $\underline{36}$, 1263 (1976).
11. L. Jones and H. Wyld, University of Illinois Report ILL-(TH)-77-32.
12. J. Babcock, D. Sivers and S. Wolfram, Argonne preprint report ANL-HEP-PR-77-68.
13. M. Gluck, J. F. Owens and E. Reya, Florida State University preprint report FSU-HER-770810.
14. M. Georgi, S. L. Glashow, M. E. Machacek, and D.V. Nanopoulos, Harvard preprint HUTP-78/A008 (1978).
15. M. J. Teper, talk at the 12th Rencontre de Moriond, Flaine, 1977 [LPTPE 77/15].
16. K. Hagiwara and T. Yoshino, TMUP-HEL-803 (1978).
17. C. Carlson and R. Suaya, SLAC-PUB-2212 (1978).
18. R. Combridge, CERN TH.2574 (1978).
19. T. K. Gaisser and F. Halzen, Phys. Rev. $\underline{D11}$, 3157 (1975).
20. F. Halzen and D. M. Scott, Phys. Lett. $\underline{72B}$, 404 (1978).
21. H. Fritzsch and K. H. Streng, CERN TH 2520 (1978).
22. N. Isgur and S. Wolfram, Oxford preprint (1977).
23. T. K. Gaisser, F. Halzen and E. A. Paschos, Phys. Rev. $\underline{D15}$, 2572 (1977).
24. T. K. Gaisser and F. Halzen, Phys. Rev. $\underline{D14}$, 3153 (1976).
25. Compilation taken from R. Diebold, Proceedings of the XIX International Conference on High Energy Physics, Tokyo, Japan (1978).
26. G. Conforto, these proceedings.
27. K. Niu, these proceedings.
28. T. K. Gaisser, F. Halzen and K. Kajantie, Phys. Rev. $\underline{D12}$, 1968 (1975); V. Barger and R.J.N. Phillips, Phys. Rev. $\underline{D12}$, 2623 (1975); R. Field and C. Quigg (unpublished).
29. T. K. Gaisser, private communication.
30. K. P. Das and R.C. Hwa, Phys. Lett. $\underline{B68}$, 459 (1977).

31. F. Halzen and D. M. Scott, Phys. Lett. in print (1978).
32. Y. Yamaguchi, Nuovo Cimento $\underline{43}$, 199 (1966).
33. J. Bjorken, SLAC-PUB 2195 (1978); S. Pakvasa and H. Sugawara, University of Wisconsin, COO-881-66 (1978).
34. G. Yodh, these proceedings.
35. K. E. Lassila and J. D. Vary, Iowa State, IS-M-131 (1978); F. Halzen, University of Wisconsin, COO-881-6 (1977); Y. Afek et al., Technion-PH-51 (1977).
36. F. Halzen and P. McIntyre, FNAL preprint (1978).

CHARM PRODUCTION BY NEUTRINOS AND THE CHARM PARTICLE LIFETIME

David Cline
Fermilab
Batavia, Illinois
and
University of Wisconsin
Madison, Wisconsin

ABSTRACT

The production of charm particles by neutrino beams is well established. We briefly review the present evidence. The production of charm particles by neutrinos is a useful way to study the properties of charmed particles. The lifetime of some charmed particles has been estimated from two events observed in nuclear emulsions to be $3 - 6 \times 10^{-13}$ sec.

The experimental subject of charm particle production by neutrinos started with the report of two neutrino induced $\mu^-\mu^+$ events in 1974 at the Penn and London Conferences.[1] These events were particularly convincing from an experimental view point and provoked Fubini to state in his London Conference summary talk "Let me finally report an experimental finding which may play an important role in future conferences. The production of two $\mu^+\mu^-$ pairs in high energy neutrino production has been reported. If other events of this kind will be found, this might be the beginning of some new phenomenon whose interpretation will keep many of us busy".[2]

Since the observation of the first few dimuon events there has been convincing evidence that these events represented the weak production and weak decay of some new massive particles.[1,3,4] Thus the best guess for the kind of particle has always been charm; only recently has there been entirely convincing evidence that the bulk of the opposite sign dileptons actually come from charm mesons.[4]

From an experimental point of view there has been remarkable progress in this field: from the observation of a couple of $\mu^-\mu^+$ events in 1974 to greater than 10,000 in 1978-79 and from a few 3μ events in 1977 to 123 in 1978 as well as a few 4 lepton events. Nevertheless it is important to recall that these processes represent relative rates of $10^{-3} - 10^{-6}$ compared to normal neutrino interactions and they are exceedingly rare events.

Most of the qualitative features of $\mu^-\mu^+$ and μ^-e^+ production is now known, and this is in good agreement with the expectation that diagrams in Fig. 1 dominate charm production. The amount of experimental data on $\mu\mu$ production is increasing rapidly and is summarized in Table 2.

The μ^-e^+ events are accompanied by a substantial strange particle component as expected in the charm model (Fig. 1). There has been a lively controversy about the actual number of strange particles per μ^-e^+ event (summarized in Table 2a). Table 2b summarizes the latest results presented at the Oxford Conference.[5] It appears that the world average favors a ratio of ~ 1.5 strange particle/μ^-e^+ event at present. This is an excellent agreement with the charm model (Fig. 1).

The e^+ and Vee mass spectrum is a further test of the origin of the μ^-e^+ events. Figure 2 shows the $\binom{e^+}{\mu^+}K^\circ$ and $\binom{e^+}{\mu^+}\Lambda$ mass spectrum from the data reported at this meeting. The $e^+\Lambda$ data are likely to arise from charm baryons. The e^+K° spectrum is in remarkable agreement with the expectations of $D^+ \to e^+ + \nu_e + K^\circ$ or $K^*(890)$. The data actually favor the K^* decay slightly.

New $\mu^-\mu^+$ data from the FHOPRW group has been analyzed in the context of the charm model in order to extract the $s-\bar{s}$ components of the ocean.[6] Table 3 lists the results of this experiment. The results are in agreement with a similar analysis by the CDHS group and the μ^-e^+ data from the bubble chamber.[3,4,5] The x and y distribution for this data, compared to the charm model are shown in Fig. 3.[6] Again the agreement is very good.

Charm baryon production is expected to lead to the states $\nu\mu + N \to \mu^- + e^+ + \Lambda + X$ with a rate larger than expected for charm meson production and associated strange particle production. (Note that charm meson production on the ocean leaves behind an $\bar{s}$ with strangeness + 1 and thus can't contribute to single Λ production). There is now convincing evidence for $\mu^-e^+\Lambda$ production in the Columbia - BNL experiment at Fermilab. Several events have been reported in other experiments, for example, by the GGM group. Fig. 2b shows a plot of the $e^+\Lambda$ mass for the data reported at the Oxford Conference.[5] Since there are events with $Me_\Lambda > 2.4$ GeV we conclude that these events do not come from the lowest mass charm baryons.

Another channel to search for charm baryons is through the production of single Λ's by neutrinos. Such an event was presented a long time ago by the BNL group.[7]

The third method used to search for charm baryons is through effective mass plots. This is far less sensitive than the previous techniques because of the many possible combinations, but has the virtue of directly obtaining the charm baryon mass. Unfortunately, the data reported to the Oxford Conference have so far shown no evidence for charm baryon production. A mass spectrum from the CB experiment is reproduced in Fig. 4.[5] There is no evidence for any important structure in the vicinity of the expected Λ_c. However direct evidence for D meson production in a neutrino experiment was reported by C. Baltay et al.[8]

A very nice analysis of Λ production in $\bar{\nu}$ reactions was presented by F. Messing at the Oxford Conference.[9] A careful estimate of the expected associated production cross section was made using electro production results. The observed yield of Λ when plotted vs. (x,y) is shown in Fig. 5 and compared with the expectations of the charm model. There is excellent agreement.

In summary all evidence for the production of charm particles by neutrinos is in excellent agreement with the simple GIM model.

There is no example of a convincing explicit charm particle signal in a hadron production experiment to date of this meeting. The most sensitive direct searches have been made using nuclear emulsion experiments. Direct searches for $k\pi$, $k\pi\pi$ final states have not been conclusive to date. There are, however, two sources of indirect indication of charm particle production through the observation of prompt neutrinos at CERN and through the observation of single prompt muons at Fermilab. When these experimental results are interpreted as due to charm production the resulting cross section is estimated to be in the vicinity of 40 - 200μb. G. Conforto and others presented a summary of the CERN results at this meeting. On the other hand the emulsion experiments have presumably set limits on charm particle production at a considerably lower cross section. Is there an inconsistency between these two results?

A Monte Carlo calculation of the effective cross section limits as a function of the assumed charm particle mass has been carried out by Crennell et al. The results, when compared with the estimated charm particle cross section from the beam dump experiments limit the possible charm particle lifetime to $\tau > 10^{-12}$ sec or $\sim 10^{-15}$ sec as shown in Fig. 6.

Two experiments reported a search for long lived charm particles in the BEBC and GGM bubble chamber. The vertex of several muon events were investigated in these experiments. The resulting lifetime limits were quoted as

$$\tau < 3 \times 10^{-12} \text{ sec at 90\% C.L.}$$
(BEBC experiment)

$$\tau < 0.9 \times 10^{-12} \text{ sec (GGM experiment)}$$

Assuming all the experimental results shown in Fig. 5 are correct, these results limit charm lifetime very near 10^{-12} sec or in the vicinity of $\sim 10^{-15}$ sec. Both values of the lifetime are somewhat unexpected from a theoretical standpoint of the GIM model.

The observation of a likely example of the decay of such a short-lived charged particle produced in a high-energy neutrino interaction in nuclear emulsion has been reported by a large group of European and American physicists. The experiment was performed in the wide-band neutrino beam at Fermilab. A neutrino interaction occurring in the emulsion can be located by predicting for the tracks of secondary particles observed in the spark chambers their point of origin in the emulsion. Stacks containing 19 liters of Ilford K5 emulsion, comprised of pellicles of dimensions 20 cm x 8 cm x 0.6 mm were placed in association with a double wide-gap spark chamber followed by a detector of electromagnetic showers and a rudimentary muon identifier. A veto counter upstream discriminated against interactions in the emulsion produced by charged particles. The experimental setup is shown schematically in Fig. 7.

The stacks were exposed to neutrinos produced by a total of about 7×10^{17} protons of energy 400 GeV on target. From the known neutrino flux it is estimated that some 150 neutrino interactions should have been produced in the emulsion. Approximately 250 candidates for these interactions have been seen in the spark chamber pictures and a search has been made for about one third of them to date. The search involves scanning a volume of the emulsion averaging about 0.7 cm^3, around the position of the vertex predicted from the measurements of the spark chamber film. So far, 31 interactions have been located in the emulsion and analyzed. Among these has been found an event in which one of the secondary particles presents the features expected of the decay of a short-lived particle. Figure 8 shows a photomicrograph of the event as seen in the emulsion. Fig. 8 also shows a schematic drawing based on the spark chamber photographs associated with the event. At the neutrino interaction, vertex A in Figure 8, are seven tracks due to low-energy protons or nuclear fragments. There are in addition five tracks with a specific ionization, as measured in the nuclear emulsion, compatible with minimum ionization. One of these, track 4, is found to give rise to three further tracks of minimum ionization at vertex B after a distance of 182 μm. As there is no sign of a nuclear recoil track, the event has the characteristics expected of the decay of an unstable particle. The direction of the neutral V particle derived from the spark chamber pictures makes it a good candidate for the missing neutral particle from the vertex B, removing at least part of the imbalanced transverse momentum. The muon identifier recorded the passage of at least one particle, correlated in time with the spark chamber pictures, through the 1.35-m lead screen. The total pulse height recorded by the electron multipliers of the shower detector indicates the absence of any high-energy electron cascade.

The conclusion is that the event most probably represents a charmed particle decaying after an observed flight time of $\sim 3 \times 10^{-13}$ sec with characteristics compatible with those expected for the decay of a charmed particle.

A similar experiment has been performed at CERN. In this case the emulsion stack was placed in front of the large BEBC bubble chamber. A schematic of the detector is shown in Fig. 9. At the Oxford Conference this group presented the data on 37 neutrino interactions but no charmed particle event was observed at that time.[13] More recently at least one event has been observed. We don't have the details of this event yet but it apparently is very clean and convincing.[13] The estimated lifetime is reportedly $\sim 3 \times 10^{-13}$ sec.

With the observation of a second event in nuclear emulsions and the extremely small backgrounds for neutrino produced charm as well as the beautiful agreement with the GIM model it seems very likely that at least some charm particles have a lifetime in the few x 10^{-13} sec vicinity. If we compare these results with the expectations shown in Fig. 6 it seems likely that the hadronic production of charm may be underestimated. This is confirmed by the results of Niu and collaborators where two charmed pair events have been observed at 400 GeV and the estimated production cross section is $80 \pm 40\mu b$.[14]

REFERENCES

1. A. Benvenuti et al., Phys. Rev. Lett. $\underline{34}$, 419 (1975);
   A. Benvenuti et al., Phys. Rev. Lett. $\underline{35}$, 1199 (1975); ibid $\underline{35}$, 1203 (1975).

2. S. Fubini, summary talk at the London Conference, 1974.

3. B. C. Barish et al., Phys. Rev. Lett. $\underline{36}$, 939 (1976); M. Holder et al., Phys. Lett. $\underline{69B}$, 377 (1977).

4. J. Blietschau et al., Phys. Lett. $\underline{60B}$, 207 (1976); J. Von Krogh et al., Phys. Rev. Lett. $\underline{36}$, 710 (1976); P. Bosetti et al., Phys. Rev. Lett. $\underline{38}$, 1248 (1977); C. Baltay et al., Phys. Rev. Lett. $\underline{73B}$, 380 (1978).

5. See the reports of D. Cline and C. Baltay in the Proceedings of the Oxford Neutrino Conference, July 1978.

6. A. Benvenuti et al., submitted to Phys. Rev. Lett. (1978).

7. E. Cazzoli et al., Phys. Rev. Lett. $\underline{34}$, 1125 (1975).

8. C. Baltay et al., Proceedings of the Oxford Neutrino Conference.

9. F. Messing, Report on the Proceedings of the Oxford Neutrino Conference.

10. See the talks of G. Conforto and L. Jones at this meeting.

11. B. J. Crennell, Rutherford Preprint RL-78-051, 1978.

12. E.H.S. Burhop et al., Phys. Lett. $\underline{65B}$, 299 (1976) and A.L. Read et al., Fermilab Pub. 78/56 (1978).

13. See the report of the Ankaro, Bruxellas, CERN, Dublin - U. C. London - Open. U., Piza - Roma - Torino Group to the Oxford Neutrino Conference.

14. K. Niu et al., report to this meeting. See also the paper of T. Yamanouchi, this volume.

TABLE 1

Summary of Number
$\mu^-\mu^+/\mu^-\mu^-$ Candidates at Oxford Conference

Exp.	$\mu^-\mu^+$ (now)	$\mu^-\mu^+$ (final sample)	$\mu^-\mu^+$ (final sample)
	(All E_{vis})		$E_{vis} > 200$ GeV
FHOPRW	308	∼ 3,000	∼ 400
CDHS	716	∼ 10,000	∼ 100
15'	40 (26)		
GGM	46 (30 ± 8)		
BEBC	22		

Exp.	$\mu^-\mu^-$ (now)	$\mu^-\mu^-$ (final)	$\mu^-\mu^-$
PHOPRW	67		$E_{vis} > 200$ GeV
		∼ 300	∼ 40
CDHS	289	∼ 1,000	∼ 4

TABLE 2a

Dilepton Production by Neutrinos in the 15 Foot B.C. $\nu\mu + Ne \to \mu^- + \mu^+ + e^+ \ldots$

Experiment	Liquid	Pictures	μ^-e^+ Events	Pet Cut MeV/c	Rate $\frac{\mu^-e^+}{\mu^-}$ %	Observed vees	Neutral S.P./event	Total S.P./event
Berkeley CERN Hawaii	21% Ne	90,000	17	800	.8±.3	11	1.84±.6/.5	3.7
Wisc. Columbia BNL	64% Ne	50,000[a]	81 83 ——— 164	300	.5±.15[b]	15 18 ——— 33	0.6±0.2	1.2

a) 50,000 analyzed out of a total of 150,000 total pictures
b) With a Pe+ ≥ 800 MeV/c cut, Columbia-BNL ser 61 μ^-e^+ events with 11 vees, for a rate $\mu^-e^+/\mu^- = 0.4 \pm 0.15$

TABLE 2b

Recent Results on Strange Particle Production in Dileptons

Experiment	Beam	Average E_ν	Total Dileptons Events	Total Vees
Fermilab 15'	Wideband Fermilab	25 BeV	6 μ^-e^+	1
E172 E180	Fermilab	25 BeV	6 μ^-e^+	1
Fermilab 15' E460	Wideband Fermilab	25 BeV	9 $\mu^-\mu^+$	1
CERN BEBC	Wideband SPS	25 BeV	21 μ^-e^+ 40 μ^-e^+	6
CERN BEBC	Narrowband SPS	75 BeV	11 $\mu^-\mu^+$ 5 μ^-e^+	6 2
Gargamelle	Wideband PS	3 BeV	14 μ^-e^+	3

TABLE 3

Values of the fractional momentum carried by the strange quarks and by the ordinary antiquarks obtained from the data above 80 GeV in the FHOPRW experiment.

    Ratio   Value

Ratio	Value
$\bar{S}/U$	0.067 ± 0.024
S/D	0.080 ± 0.027
$\bar{S}/D$	0.066 ± 0.03
$\bar{D}/D = \bar{U}/U$	0.14 ± 0.03

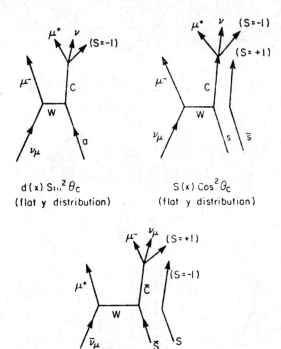

Figure 1

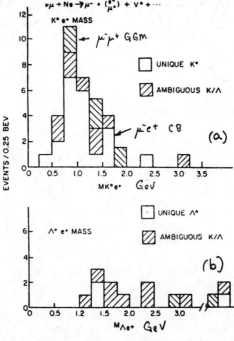

Figure 2

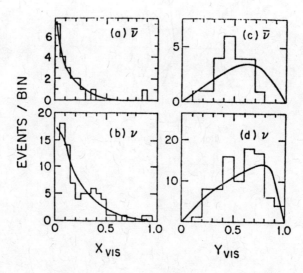

Figure 3

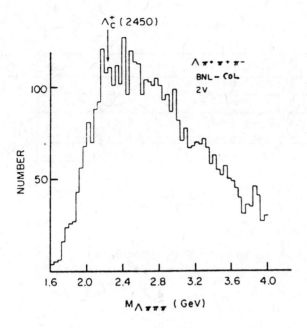

Figure 4

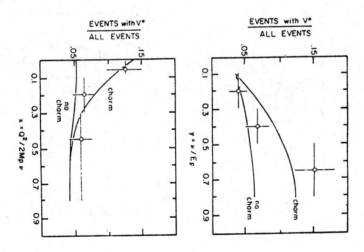

Figure 5

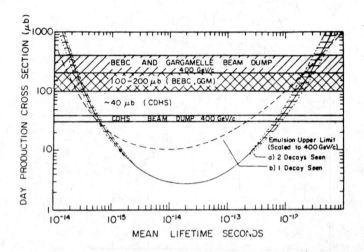

Figure 6

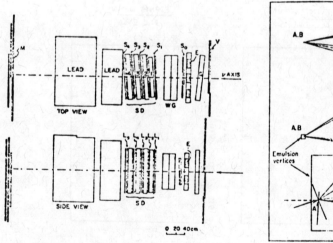

Figure 7

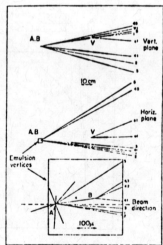

Figure 8

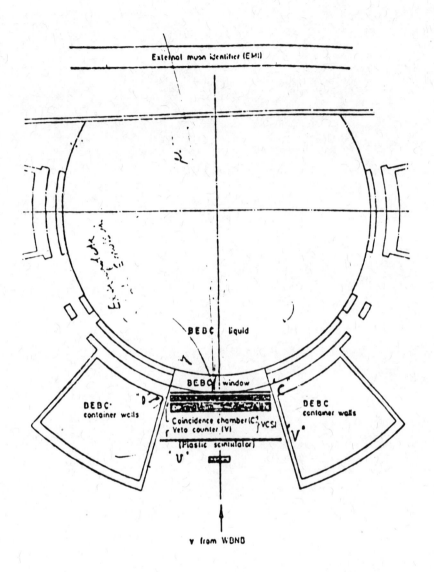

Figure 9

CHARMED PARTICLE LIFETIMES[*]

Jonathan L. Rosner

School of Physics and Astronomy
University of Minnesota, Minneapolis, Minnesota 55455

ABSTRACT

Conventional estimates are reviewed for charmed particle lifetimes. Free-quark models give values of (a few) $\times 10^{-13}$ sec to (a few) $\times 10^{-12}$ sec. The shorter of these values also follows from an extrapolation based on $D \to Ke\nu$. Possible differences among the lifetimes and production rates of D^0, D^+, F^+, C_0^+, the heavy lepton τ, and the fifth quark b are discussed. Extreme values of mixing angles in a six-quark model could extend charmed particle lifetimes by a factor of at most three from the above estimates, while shorter lifetimes than those predicted could occur for some species like D^0 or F^+ if their nonleptonic decays were enhanced. The predictions are discussed in the light of some current experimental results, and it is estimated that $\sigma(pp \to \text{charm}) \simeq 10\,\mu\text{b}$ at 400 GeV/c.

CONTENTS

I. INTRODUCTION

II. THEORETICAL PREDICTIONS
   A. Lifetime estimates.
      1. Free quarks.
      2. Extrapolation from $D \to Ke\nu$.
   B. Possible differences among species.
      1. D^+ vs. D^0 (charmed non-strange mesons).
      2. F^+ (charmed-strange meson).
      3. C_0^+ (charmed baryon).
      4. τ (heavy lepton).
      5. b (fifth quark, $e_b = -1/3$, $m_b \simeq 5$ GeV/c^2).
   C. Production ratios.
      1. D^+ vs. D^0.
      2. F^+ vs. D.
      3. C_0^+ vs. charmed mesons.
      4. τ.
      5. b.

---

[*]Work supported in part by the U. S. Department of Energy under Contract No. E-(11-1)-1764. Invited talk at Workshop on Charm Production and Lifetimes, Bartol Research Foundation, University of Delaware, October 19, 1978.

III. SOURCES OF DEVIATIONS
  A. If $\tau_{observed} \gtrsim$ few x 10^{-12} sec.
    1. Hadronic effects.
    2. Suppressed c-s interaction (mixing).
    3. Non-charmed particles.
  B. If $\tau_{observed} \lesssim$ few x 10^{-13} sec.
    1. Form factor enhancements.
    2. Nonleptonic enhancements.
    3. New decay mechanisms.
    4. Non-charmed particles.

IV. COMPARISON WITH EXPERIMENTS.
  A. Lifetimes.
  B. Cross sections.

V. SUMMARY.

I. INTRODUCTION.

The experimental study of charmed particle lifetimes is still in its infancy. The present article is intended as a guide to theoretical predictions of these lifetimes, drawing primarily on the data that have accumulated over the past few years but also on some of the detailed calculations that have been performed since the initial estimates.[1,2]

The particles which will be discussed here are those whose decays proceed via the weak interactions:[3] the charmed mesons $D^0(1863) = c\bar{u}$, $D^+(1867) = c\bar{d}$, $F^+(2030) = c\bar{s}$, the charmed baryon $C_0^{\pm}(2260) = c[ud]_{I=0}$, the heavy lepton $\tau(1782)$, and the fifth quark b.[4] "Standard" theoretical estimates, based on the four-quark model of Ref. 1, are given in Sec. II. Possible sources of deviation from these estimates are mentioned in Sec. III. Mean lifetimes of the order of 10^{-12} sec. (give or take a factor of 5) and cross sections $\sigma(pp \to charm) \simeq 10\mu b$ (give or take a factor of 2) at 400 GeV/c are compatible with most experiments (Sec. IV). A summary is contained in Sec. V.

II. THEORETICAL PREDICTIONS.

In this section we shall assume the weak charged currents couple with the same V-A form to each of the doublets $(e^- \bar{\nu}_e)$, $(\mu^- \bar{\nu}_\mu)$, $(\tau^- \bar{\nu}_\tau)$, $(d'\bar{u})$, and $(s'\bar{c})$. Here $d' \equiv d\cos\theta + s\sin\theta$, $s' \equiv s\cos\theta - d\sin\theta$, and $\sin\theta = 0.228 \pm 0.003$.[5]

A. Lifetime estimates.
  1. Free quarks. A free charmed quark would decay to $se^+\nu_e$ or $de^+\nu_e$ with the rates

$$\Gamma\left(c \to \begin{Bmatrix} s \\ d \end{Bmatrix} e^+ \nu_e \right) = \left(\frac{m_c}{m_\mu}\right)^5 \Gamma_\mu \, f\left(\frac{m_{s,d}^2}{m_c^2}\right) \begin{Bmatrix} \cos^2\theta \\ \sin^2\theta \end{Bmatrix} \quad (1)$$

where

$$\Gamma_\mu = 4.55 \times 10^5 \text{ sec}^{-1} \quad (2)$$

is the muon decay rate and

$$f(x) \equiv 1 - 8x + 8x^3 - x^4 - 12 x^2 \ln x \ . \tag{3}$$

If nonleptonic decays are not enhanced, one would estimate

$$\Gamma(\text{charm}) \simeq 5 \ [\ \Gamma(c \to se^+\nu_e) + \Gamma(\ c \to de^+\nu_e)], \tag{4}$$

where the factor of 5 refers to the channels $e^+\nu$, $\mu^+\nu_\mu$ and three colors of $u\bar{d}$. On this basis one would estimate $\Gamma(c \to e^+ + \text{all})/\Gamma(c \to \text{all}) \simeq 20\%$. In fact, it appears that[6]

$$B_e \equiv \frac{\Gamma(c \to e^+ + \text{all})}{\Gamma(c \to \text{all})} = (9.3 \pm 1.7)\ \% \ . \tag{5}$$

On the basis of the free quark model, one can then calculate the total rate for charm decays if one knows the masses of the quarks.

"Current" quark masses (at typical momentum transfers of several GeV) are negligible on the present scale of interest for u and d, and of the order of 0.1 GeV for s.[7] The "constituent" masses of s, u, and d (relevant for low momentum-transfer processes) can be estimated from magnetic moments of hadrons,[8] or by averaging over spin-spin splittings between pseudoscalar and vector mesons and neglecting binding effects. Thus, we obtain

$$m_u \simeq m_d = \frac{m_p}{\mu_p} = 0.34 \text{ GeV}$$

$$\simeq \frac{1}{2} [\ \frac{3m_\rho + m_\pi}{4}\ ] = 0.31 \text{ GeV} \tag{6}$$

$$m_s = m_d\ (\frac{\mu_n}{2\mu_\Lambda}) = 0.52 \text{ GeV}$$

$$\simeq [\ \frac{3m_{K^*} + m_K}{4}\ ] - m_u \simeq 0.48 \text{ GeV} \ . \tag{7}$$

As for the charmed quark, "current" quark estimates would rely on a chiral SU(4) x SU(4) type of theory[9] which is probably too badly broken to be reliable. Reasonable estimates of the "constituent" charmed quark mass range from

$$m_c \simeq 1.1 \text{ GeV} \ , \tag{8}$$

obtained (for example) via inverse-scattering methods,[10] to

$$m_c \simeq \frac{3m_{D^*} + m_D}{4} - m_u \simeq 1.66 \text{ GeV} \ , \tag{9}$$

as obtained (for example) in Ref. 11. The discrepancy between (8) and (9) is a major source of uncertainty in the naive lifetimes based on Eqs. (1) - (3) and (5). These are plotted as a function of m_c in Fig. 1, for representative "current" and "constituent" masses of d and s. Judging that the use of Eq. (9) probably requires one to take constituent quark masses for s and d as well, we would estimate that if m_c lies between (8) and (9),

$$4 \times 10^{-13} \text{ sec} < \tau_{charm} < 7 \times 10^{-12} \text{ sec}. \tag{10}$$

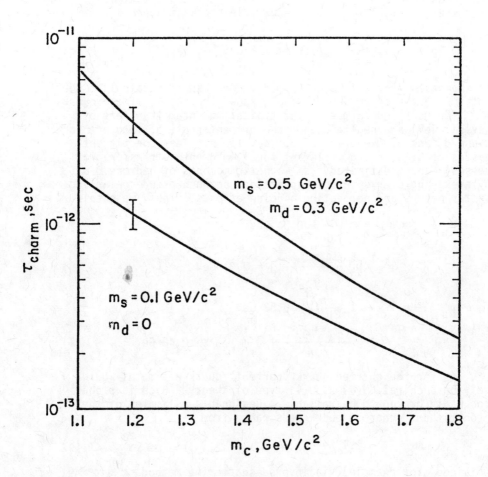

Figure 1. Average charmed particle lifetimes (from a free-quark model) as function of charmed quark mass m_c. Based on Eqs. (1) - (3) and (5) of text.

A by-product of this calculation is the ratio $\Gamma(c \to se^+\nu_e)/\Gamma(c \to de^+\nu_e)$, plotted in Fig. 2 as a function of m_c. This ratio should be $\cot^2\theta = 18.2 \pm 0.5$ if masses were unimportant. Its measurement could indicate what effective quark masses should be important in such calculations.

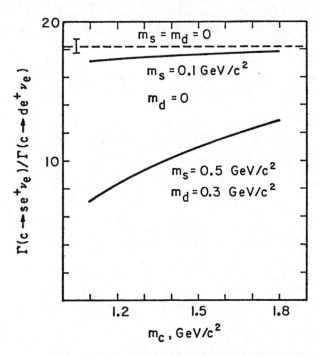

Figure 2. Quark mass dependence of ratio of $\Delta S \neq 0$ to $\Delta S = 0$ semi-leptonic charmed quark decays. Based on Eqs. (1) - (3) and (5) of text.

2. **Extrapolation from $D \to Ke\nu$.** Another means of estimating charmed particle lifetimes is by comparison of the known decay rates[12]

$$\Gamma(K^+ \to \pi^0 e^+ \nu_e) = (3.90 \pm 0.04) \times 10^6 \text{ sec}^{-1} \quad (11)$$

or

$$\Gamma(K_L^0 \to \pi^- e^+ \nu_e) = (3.74 \pm 0.05) \times 10^6 \text{ sec}^{-1} \quad (12)$$

with $\Gamma(D^0 \to K^- e^+ \nu_e)$. One expects, for example,

$$\frac{\Gamma(D^0 \to K^- e^+ \nu_e)}{\Gamma(K^+ \to \pi^0 e^+ \nu_e)} = 2 \cot^2\theta \left(\frac{m_D}{m_K}\right)^5 \frac{f(m_K^2/m_D^2)}{f(m_\pi^2/m_K^2)}, \quad (13)$$

where $f(x)$ is defined in Eq. (3) and the form factors f_+ in the decays are taken equal and constant. Then one estimates

$$\Gamma(D^0 \to K^- e^+ \nu_e) \simeq 1.1 \times 10^{11} \text{ sec}^{-1} , \qquad (14)$$

with the major uncertainty coming from the approximation $f_+^D = f_+^K = \text{const.}$

The rate (14) can be converted to an upper bound on $\tau(D^0)$ with the help of information on tagged D^0 decays.[13] It is found that

$$\tau(D^0) \Gamma(D^0 \to K^- e^+ \nu_e)$$
$$\equiv B(D^0 \to K^- e^+ \nu_e) \leq 3.6\% \text{ (1 } \sigma \text{ limit)}, \qquad (15)$$

which, when combined with Eq. (14), leads to

$$\tau(D^0) \leq 3.3 \times 10^{-13} \text{ sec (1 } \sigma \text{ limit)} . \qquad (16)$$

A model-dependent fit of electron energy spectra[14] gives

$$\frac{B(D^0 \to K^- e^+ \nu_e)}{B(D^0 \to e^+ + \ldots)} = (35 \pm 30)\% . \qquad (17)$$

If $B(D^0 \to e^+ + \ldots) = B(D^+ \to e^+ + \ldots) = (9.3 \pm 1.7)\%$ (note that this is a stronger statement than Eq. (5)), one can then use Eqs. (14) and (17) to estimate

$$\tau(D^+) = (3.0 \pm 2.6) \times 10^{-13} \text{ sec} . \qquad (18)$$

B. Possible differences among species.

1. D^+ vs. D^0. The $\Delta I = 0$ property of the $\Delta S = \Delta C$ decay $c \to s \ell^+ \nu_e$ means that there is a pairwise equality[15,16]

$$\Gamma(D^0 \to e^+ \nu_e X^-) = \Gamma(D^+ \to e^+ \nu_e X^0) \qquad (19)$$

for any pair of $S = -1$ final states X^- and X^0, e.g., K^- and $\bar{K}^0$ or $(\bar{K}\pi)^-$ and $(\bar{K}\pi)^0$. Then, simply from the definition of a branching ratio,[16]

$$\frac{\tau(D^+)}{\tau(D^0)} = \frac{B(D^+ \to e^+ \nu_e X^0)}{B(D^0 \to e^+ \nu_e X^-)} \qquad (20)$$

To measure the right-hand side of (20) one has to search for situations in which the D^0 and D^+ production ratios are expected to differ and see if the inclusive e^+ signals differ. At $E_{c.m.} = 3.772$ GeV, a phase-space estimate predicts[17]

$$\psi" \longrightarrow \begin{matrix} D^0\bar{D}^0: & (56 \pm 3)\% \\ D^+D^-: & (44 \pm 3)\% \end{matrix} \qquad (21)$$

and one observes

$$B_{e^+} = \begin{cases} (7.2 \pm 2.8)\% & \text{(Ref. 18)} \quad , \quad (22) \\ (11 \pm 2)\% & \text{(Ref. 19)} \quad . \quad (23) \end{cases}$$

At high $E_{c.m.}$ (4 GeV and above), a simple model with D^* and D produced in their ratio of statistical weights (3:1) (Sec. II.C.1) would predict that $\sigma(D^0): \sigma(D^+) = 3:1$, compatible with the ratio extracted from the observed D^* and D production ratios,[20]

$$\text{charm} \longrightarrow \begin{matrix} D^0: & (67\pm8)\% \quad , \\ D^+: & (33\pm8)\% \quad . \end{matrix} \qquad (24)$$

At high energies in e^+e^- annihilations one observes

$$B_{e^+} = \begin{cases} (8.2 \pm 1.9)\% & \text{(Ref. 21)} \quad , \quad (25) \\ (11 \pm 3)\% & \text{(Ref. 14)} \quad . \quad (26) \end{cases}$$

The constraints due to Eqs. (21) - (26) are plotted in Fig.3. These are the most stringent ones possible since, for illustration, we have assumed $\sigma(D^0) = 3\sigma(D^+)$ at high energies. If $\sigma(D^0)/\sigma(D^+)$ is more like 2,[20] the bands will overlap even more. Figure 3 permits considerable latitude in $B(D^+ \to e^+)/B(D^0 \to e^+) = \tau(D^+)/\tau(D^0)$ at present. A measurement of $B(D^+ \to e^+ + ...)$, for example by using tagged D^+'s at $\psi"(3.772)$, could be very helpful in pinning down $\tau(D^+)/\tau(D^0)$.

The hadronic final state in $\Delta C = \Delta S$ nonleptonic D^+ decays must be exotic: it is a positively charged state with $S = -1$ (e.g.,$[\bar{K}+n\pi]^+$) and must have $I = 3/2$. It has been suggested that this exoticity suppresses D^+ $\Delta C = \Delta S$ nonleptonic decays, and hence lengthens the D^+ lifetime.[2,22,23] Charged "X-particles" (produced in pairs by 10-20 TeV cosmic rays) may indeed live longer than neutral ones:[24,25]

$$\tau_{X^\pm} = (10-15) \times 10^{-13} \text{ sec} , \qquad (27)$$

$$\tau_{X^0} = (3-5) \times 10^{-13} \text{ sec} . \qquad (28)$$

Equation (28) could represent an estimate of the D^0 lifetime, while (27) might represent a weighted average of mostly $D^\pm$ and some $F^\pm$, C_0^+ or $\bar{C}_0^-$, and $\tau^\pm$. (We shall discuss production ratios in Subsection C).

Another possible indication that D^+ $\Delta S = \Delta C$ decays may be suppressed with respect to those of D^0 comes from comparison of the inclusive $D^+ \to \bar{K}$ branching ratios,[26]

$$B(D^+ \to K^- + ...) = (10 \pm 7)\% , \qquad (29)$$

$$B(D^+ \to \bar{K}^0 + ...) = (39 \pm 29)\% , \qquad (30)$$

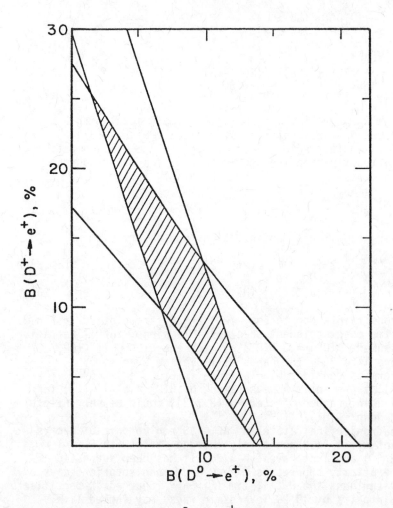

Figure 3. Constraints on D^0 and D^+ semi-electronic branching ratios due to e^+e^- data at ψ'' (3.772) (Eqs. (21) - (23)) and at higher energies (Eqs. (25) - (26)), where a $D^0 : D^+$ ratio of 3 has been assumed. Shaded area shows the allowed range of values.

with those of D^0:

$$B(D^0 \to K^- + ..) = (35 \pm 10)\% \quad , \tag{31}$$
$$B(D^0 \to \overline{K}^0 + ..) = (57 \pm 26)\% \quad . \tag{32}$$

The small value of (29) is surprising in comparison with

expectations, based on statistical models, of at least 30% both for semileptonic[27] and for nonleptonic[27,28] decays. At the same time, the observed ratio $B(D^+ \to K^- + \cdots)/B(D^+ \to \bar{K}^0 + \cdots)$ is consistent with statistical expectations.[26-28]

2. F^+. The dominant $\Delta S = \Delta C$ part of H_{NL} may be a V-spin singlet.[23, 9-31] If so, one would expect $F^+ = c\bar{s}$ and $D^0 = c\bar{u}$ to have similar $\Delta S = \Delta C$ nonleptonic decay rates. Moreover, the semileptonic decay rates of F^+ and D^0 probably are not too dissimilar, so that we might expect

$$\tau(F^+) \simeq \tau(D^0) . \tag{33}$$

Various effects which could spoil equality of partial or total F^+ and D^0 rates are discussed, for example, in Refs. 31 and 32.

3. C_0^+. The semileptonic decay rate for a charmed isoscalar baryon of mass $\simeq 2.25$ GeV/c^2,[33,34] has been estimated:[35]

$$\Gamma_{SL}(C_0^+) \simeq 10^{12} \text{ sec}^{-1} . \tag{34}$$

This could be 20% to 40% of all C_0^+ decays, the lower figure a guess based on doubling Eq. (5) ($SL \equiv e + \mu$) and the upper one based on the naive estimate (4). Thus we would expect

$$\tau(C_0^+) = (2-4) \times 10^{-13} \text{ sec.} \tag{35}$$

This is compatible with a neutrino-induced emulsion event,[36] which can be interpreted[27] as a charmed baryon C_0^+ living between 2 and 5×10^{-13} sec. There are also two particles produced in a very high-energy interaction,[37] which, if interpreted as $C_0(\bar{C}_0)$ candidates,[38,39] live about 7×10^{-13} sec and 5×10^{-12} sec.[39]

4. τ. Since $\Gamma(\tau \to e \nu\nu) = (m_\tau/m_\mu)^5 \Gamma_\mu$ for $m(\nu_\tau)=0$, and since $B(\tau \to e \nu\bar{\nu}) \simeq 18\%$,[40] we expect (for $m_\tau = 1.782$ GeV/c^2 and a full-strength $\tau - \nu_\tau$ coupling):

$$\tau(\tau) = 2.9 \times 10^{-13} \text{ sec.} \tag{36}$$

If the lifetime of τ turns out to be longer than this, important mixing effects will have been demonstrated for the leptons for the first time. $\tau(\tau)$ is unlikely to be shorter than Eq. (36).

5. b. Hadrons containing a b quark should be produced in pairs at least as copiously as T states in high-energy hadronic interactions.[41] Two experiments[42,43] show that any charged particles with lifetimes longer than 5×10^{-8} sec are produced in pN-interactions at 400 GeV/c with a cross section less than 1/10 that for T production. Thus, if the lowest-lying meson containing b is $D_b^- \equiv (b\bar{u})^-$,[44] it must live less than 5×10^{-8} sec. If the lowest-lying meson containing b should happen to be $D_b^0 \equiv (b\bar{d})^0$, the $b\bar{u}$ state can beta-decay to it: $D_b^- \to D_b^0 e^- \nu_e$. The lifetime[45] for this process is longer than 5×10^{-8} sec unless $m(D_b^-) - m(D_b^0) > \mathcal{O}(m_\pi)$, an unreasonably large splitting. Hence the limits in Refs. 42 and 43 can be construed as rough limits on the lifetime of the b quark itself:

$$\tau(b) \lesssim 5 \times 10^{-8} \text{ sec} \tag{37}$$

The b quark has been suggested to belong to a weak isodoublet along with a quark t of charge 2/3.[46] The absence of prominent signals aside from the Υ family in $\mu^+\mu^-$ mass distributions indicates that $m_t > 8$ GeV.[47] Hence the b quark cannot decay to t (with which it is likely to couple most strongly), and must resort to the transitions b → u or b → c, which are likely to be suppressed by angles analogous to the Cabibbo angle. Opinions differ on the possible range of these angles;[5,48] as an absolute minimum on $\tau(b)$ we note only that a b → u coupling of full strength or less would give[49]

$$\tau(b) > 1.3 \times 10^{-15} \text{ sec} . \tag{38}$$

This is a conservative bound; further refinements by factors of 2-3 are probably possible.[5,48]

C. Production Ratios.

1. D^+ vs. D^0. Suppose one imagines the production of a "bare" charmed quark c which then must "dress" itself with $\bar{q}$ or qq to become a hadron. Taking the probabilities for "dressing" with u and d equal, and counting the possible spin degrees of freedom, one expects the S-wave $c\bar{q}$ states produced in this "dressing" process to be in the ratio

$$D^{*0} : D^{*+} : D^0 : D^+$$
$$= 3 : 3 : 1 : 1 . \tag{39}$$

The D^*'s then decay: D^{*0} always gives D^0, while[13,20]

$$B(D^{*+} \to D^0) \simeq 2/3$$
$$B(D^{*+} \to D^+) \simeq 1/3 . \tag{40}$$

One then finds $\sigma(D^0) = 3 \sigma(D^+)$, as mentioned. This model is certainly an oversimplification of the details of resonance cascade decays to $D^{0,+}$ but both it and Eq. (24) suggest that one should be prepared for a predominance of D^0 over D^+ production in high energy e^+e^-, hadron-hadron, and neutrino interactions.

2. F^+ vs. D. One would expect $F^{*+} : D^{*0} : F^+ : D^0 = 3 : 3 : 1 : 1$, if the above statistical weight arguments can be extended in an SU(3)-invariant way to the "dressing" of c by $\bar{s}$. Using Eqs. (39) and (40) and noting that F^{*+} always should decay to F^+, we would then find

$$\sigma(F^+) = \frac{2}{3} \sigma(D^0) \quad [\text{SU(3) limit}] . \tag{41}$$

How badly could SU(3) be broken? At worst one might expect the suppression factor to be of order $\sigma(K^{*+})/\sigma(\rho^0)$, which gives an indication of the relative difficulty of "dressing" via strange and non-strange q production. At the highest energies studied,[50] this ratio is about 0.3. Thus we should expect

$$0.2 < \frac{\sigma(F^+)}{\sigma(D^0)} < \frac{2}{3} \tag{42}$$

3. C_0^+ vs. charmed mesons. The relative amount of charmed meson and baryon production may depend on the incident beam, the kinematic region, the energy, and other variables much more than the ratios of charmed meson cross sections discussed above. We shall take only one illustrative example. In e^+e^- annihilations the cross-section for inclusive baryon and antibaryon production appears to rise by $\Delta R = 0.3 - 0.4$ in the range $4.4 \leq E_{c.m.} \leq 5$ GeV.[51] It is tempting to ascribe this rise to the onset of charmed baryon production. The total amount of charm production at 5 GeV is probably a bit higher than the free-quark value associated with $\Delta R = 4/3$. I would guess that at 5 GeV the data indicate

$$\frac{1}{5} \lesssim \frac{\sigma(\text{charmed baryons})}{\sigma(\text{charm})} \lesssim \frac{1}{3} . \qquad (43)$$

The nonstrange charmed baryons are expected to end up as C_0^+ as a result of strong or electromagnetic decays.[52] The strange charmed baryons are probably less copiously produced than the nonstrange ones; their properties are described in Ref. 52.

4. τ. In general this heavy lepton has to be produced electromagnetically (in pairs) and thus is not likely to account for a large fraction of short-track events in most interactions. One possible exception may arise if F^+ has an appreciable branching ratio to $\tau \nu_\tau$:[53] then, one may expect to see τ wherever F is produced, and (moreover) neutrino beams produced in beam dump experiments may have a ν_τ or $\bar{\nu}_\tau$ contamination from processes $F^+ \to \tau^+ \nu_\tau$, $\tau^+ \to \pi^+ \bar{\nu}_\tau$, etc.,[54] giving rise in turn to τ production.

A straightforward calculation[53,55] gives

$$\Gamma(F \to \tau \nu) \simeq 6 \times 10^{10} \text{ sec}^{-1} \qquad (44)$$

for $m_\tau = 1.782$ GeV/c^2, $m_F = 2.03$ GeV/c^2, with $f_F = f_K$. If $\tau(F^0) \simeq \tau(D^0)$ as given in Eq. (18), we find

$$B(F \to \tau \nu) = (1.9 \pm 1.6)\% . \qquad (45)$$

This is not a particularly large number; nonetheless $F \to \tau \nu$ may be the best source of τ or ν_τ under some circumstances.

5. b quarks. The production of $b\bar{b}$ in hadronic interactions is discussed more extensively in Refs. 41 (Halzen) and 56. It is expected to be far less frequent than hadronic $c\bar{c}$ production, and thus unlikely to be responsible for many of the short-track events already seen in high-energy cosmic ray interactions.[57]

III. SOURCES OF DEVIATIONS

A. If $\tau_{\text{observed}} > $ few $\times 10^{-12}$ sec.

1. Hadronic effects. Suzuki has estimated that the rate for $c \to se^+\nu_e$ could be reduced by as much as 35% with respect to its free-quark value by perturbative QCD effects.[58] Moreover, it has been suggested[59] that non-perturbative effects could be

even more important. I suspect that many of these effects in fact are crudely included when using constituent quark masses in the free-quark estimates of Sec. II.A.1.

It could happen that, in contrast to the assumption in Sec. II.A.2, f_+^D for $D \to K\ell\nu$ is less than f_+^K for $K \to \pi\ell\nu$, as a result of a failure of the D and K wave functions to overlap. One would then expect the weak coupling strength lost in $D_{\ell 3}$ decays to appear somewhere else, as in (e.g.) $D \to K\pi\ell\nu$ or $K2\pi\ell\nu$ and the free quark calculations of Sec. II.A.1 should still apply with a slightly elevated effective value of m_s.

2. Suppressed c-s' interaction. This would imply, within the context of models of universal weak couplings, that the charmed quark has significant coupling to a new $Q = -1/3$ quark (not d or s), such as the b quark mentioned earlier.[4,46,48] One cannot turn down the c-s' coupling strength to an arbitrarily small level,[48] since one then loses the cancellation between charmed quark and u quark contributions to quantitites like the K_L-K_S mass difference.[60] The role of the charmed quark then must be taken up to some extent by a heavier $Q = 2/3$ quark t, whose mass must be at least 8 GeV (see Sec.II.B.5).[47] I find that by manipulating mixing angles within the rather permissive bounds of Ref. 5 one can lengthen the average charmed particle lifetime by at most a factor of 3. However, such drastic mixing will entail appreciable $b \to u$, $b \to c$ couplings, will lower $\Gamma(c \to s)/\Gamma(c \to d)$ significantly from the predictions of Fig. 2,[61] and may already be ruled out by the appreciable dilepton signal in the reaction[62]

$$\nu N \to \mu^+\mu^- + \ldots \qquad (46)$$

3. Non-charmed particles. They could be new heavy leptons or new quark flavors with greatly inhibited couplings to lighter quarks. (For b, note the restriction (37).) A problem would arise in producing heavy leptons or new quark flavors (even b) copiously enough to confuse them with charmed particles.

B. If $\tau_{observed}$ < few x 10^{-13} sec.

1. Form factor enhancements of charmed particle decays. The c-s transition may be dominated by form factors; the F^* and corresponding axial vector poles may play more significant roles than their counterparts in K decays. These effects could enhance the predicted $D_{\ell 3}/K_{\ell 3}$ ratio by some 20-30%.[2,63]

2. Enhancement of nonleptonic decays for some species. Eq. (5) and Fig. 3 place constraints upon this possibility; both D^0 and D^+ nonleptonic decays cannot be enhanced arbitrarily.

3. New decay mechanisms. In an SU(2) x U(1) model with more than one Higgs doublet, charmed quarks could, in principle, decay semiweakly to charged Higgs bosons of lower mass.[64] Couplings then would seem to require considerable arbitrary tinkering to avoid destroying the original motivation for charm.[1] Hence particles living substantially less than the estimates in Sec. II (say, with τ < few x 10^{-14} sec) are, almost by

definition, not charmed.

4. Non-charmed particles. (See also subsection A.3). The b could be one of these: cf. Eq. (38). The t, if it exists, will probably decay very rapidly to b, with a lifetime [$< \mathcal{O}(10^{-15}$ sec) for $m_t > 8$ GeV/c^2] which is easily estimated once m_t is known.

IV. COMPARISON WITH EXPERIMENTS.

A. Lifetimes.

A complete summary of emulsion experiments has been given in Refs. 57 and 65. We have already mentioned the results of Refs. 24 [eqs. (27), (28)], 36, and 37, as being compatible with the predictions of Sec. II. A,B: observed charmed particle lifetimes of the order of 10^{-12} sec, give or take a factor of 5. The original $X^{\pm}$ event found by Niu et al.[66] corresponds to $t = (2-4) \times 10^{-14}$ sec; this is shorter than present theoretical estimates but of course does not constitute a mean lifetime! Other more recent emulsion experiments include:

(1) One event, 400 GeV/c protons, associated production: $t_1 \simeq (3-4) \times 10^{-14}$ sec, $t_2 \simeq 10^{-12}$ sec;[65,67]

(2) Another (preliminary) event in the same exposure, associated production, with path lengths several mm; this event and the previous one lead to an estimate $\sigma(pp \to \text{charm}) \simeq 25$ μb.[65,68]

(3) Nine events, 400 GeV/c protons, non-associated production, $\tau \simeq 2 \times 10^{-14}$ sec, $\sigma \simeq 100$ μb/nucleon.[69]

(4) Two possible events, 20-80 GeV photons, non-associated production, $t = 2, 5 \times 10^{-15}$ sec, $\sigma \simeq 1/2$ μb/nucleon.[70]

In view of the lifetime estimates of Sec. II, only the first two of these can be regarded as candidates for charm. The mean lifetime in the third is uncomfortable short, and the cross section (see below) uncomfortably big for charmed particles. The failure to observe any events with $t > 10^{-14}$ sec in the fourth means that the experiment almost certainly is not detecting charm. Note that backgrounds tend to be easier to estimate in associated production events.[39]

B. Cross sections.

Figure 4 shows some cross sections (and limits) for charm production as a function of $\sqrt{s}$.[25,68,71-81] The crosses are an attempt to interpolate an estimate of charm production by assuming it to be a constant multiple of J/ψ production at 3/4 $\sqrt{s}$. (The $c\bar{c}$ effective masses most important for J/ψ production probably are about 3/4 those for charm production: 3.1 - 3.6 GeV for J/ψ and 4-5 GeV for charm. We are implicitly using a form of scaling in terms of $\sqrt{s}/M$, as proposed in Ref. 82.) Cross sections for J/ψ production are taken from Bamberger et al., Ref. 68.

From Figure 4 and from what we have heard today, a value

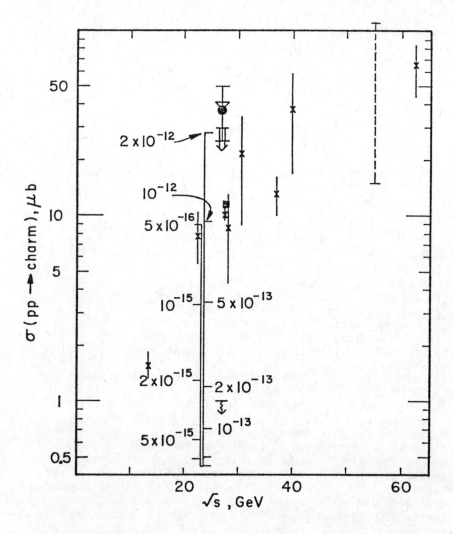

Figure 4. Cross sections and limits for charm production as a function of $\sqrt{s}$. (See Ref. 68). ⊥ : upper bound in Ref. 71; ●: Refs. 72, 78, 80; ∇: Ref. 73; I: Ref. 74 (as quoted in Ref. 25); numbered scale: upper bounds in Ref. 75 (more restrictive than those in Ref. 76, not shown), as calculated in Ref. 77 for various mean lifetimes given in seconds; ▩: Ref. 79; ⇩: upper bound in Ref. 81; ✳: estimates of charm production based on J/ψ data (see text).

$$\sigma(pp \to charm) \simeq 10 \ \mu b$$
$$(p_{Lab} = 400 \ GeV/c) \tag{47}$$

seems to be a likely compromise among various experiments. Slightly smaller values are favored by some[39,56,83] (but by no means all[84]) theorists; slightly larger values are favored by some[66,78,80] (but by no means all[71]) experiments. As long as $\tau_{charm} > 10^{-12}$ sec, present emulsion exposures to hadron beams[75,77] do not contradict Eq. (47),[77] since their sensitivity is greatly impaired for long lifetimes. Indeed, it is probably unwise to take the conclusions of a single model calculation[77] too literally : the absence of signals in Refs. 75 and 76 might well be compatible with lifetimes somewhat shorter than 10^{-12} sec for certain admixtures of charged and neutral charmed particles. (The efficiencies for detecting these might well differ.)

V. SUMMARY.

In this review, the lifetimes of various charmed particles and the heavy lepton τ have been calculated using standard assumptions. Bounds on the lifetime of the b quark also have been presented. The results are summarized in Table I.

Table I. Summary of theoretical results on particle lifetimes.

Particle	Lifetime, sec	Method	See Eq. no.:
charm (average)	4×10^{-13} to 7×10^{-12}	Free quarks (m_c=1.1-1.7 GeV)	(10)
D^0	$\leq 3.3 \times 10^{-13}$ (1 σ limit)	Extrapolation from $D^0 \to K^- e^+ \nu$	(16)
$D^{+,0}$	$(3.0 \pm 2.6) \times 10^{-13}$	$D^0 \to K^- e^+ \nu$; if $B(D^+ \to e^+)$ = $B(D^0 \to e^+)$	(18)
F^+	$\simeq \tau(D^0)$	Assumption that H_W has V = 0	(33)
C_0^+	$(2-4) \times 10^{-13}$	Ref. 35; B_{SL} = 20 - 40%	(35)
τ	2.9×10^{-13}	Universality	(36)
b	(1.3×10^{-15}) -(5×10^{-8})	Universality; Refs. 42,43	(38) (37)

The "average" charmed particle lives about 10^{-12} sec if we believe free quark estimates. Specific calculations for most individual particles give slightly shorter lifetimes: (a few) $\times 10^{-13}$ sec. However, we have presented arguments that the D^+ could well be longer-lived than the "average" charmed particle.
    It is interesting to compare present expectations for

charmed particle lifetimes with those of several years ago,[1,2] which were about a factor of 10 shorter. Two effects have acted to lengthen the predicted values. First, the charmed quark mass appears to be quite low, much in accord with early theoretical estimates[60] but to the surprise of some of us who were prepared to be more conservative in estimating the range of possible charmed particle masses. Second, the enhancement of nonleptonic decays of charmed particles is not as substantial as for strange particles, as shown in Eq. (5) and as careful theoretical reflection showed fairly early.[31,85] Nonetheless there is still the possibility that the nonleptonic decays of one or more charmed particles (like D^0, or D^0 and F^+) could be enhanced more than the average in Eq. (5) would indicate. Questions like these are the reason the experimental study of charmed particle lifetimes is still of considerable interest, even though the expected range has narrowed considerably.

There is one very bright prospect for detection of particles with lifetimes $\simeq (10^{-12}$ sec$)$: a high pressure streamer chamber[86] which already has operated at Fermilab. There are also several emulsion experiments in neutrino beams at Fermilab and CERN either with preliminary results or in the planning stage.[87]

Since the arguments we have presented lead one to expect[88]

$$\tau_{charm,\tau} \simeq (2 - 10) \times 10^{-13} \text{ sec} \qquad (48)$$

and

$$\sigma(pp \rightarrow charm) \simeq 5 - 10 \text{ μb}$$
$$\text{at Fermilab and SPS energies,} \qquad (49)$$

the following effects, discussed at this conference, are ones for which we have no explanation, at least in terms of charm:

Beam-dump or calorimetric experiments which, if interpreted in terms of charm production, would imply $\sigma(pp \rightarrow charm) > 20$μb at $p_{Lab} = 400$ GeV/c.[72,73]

Suppressed production and/or polarization of prompt single muons,[71] implying cross sections for charm production of 1 μb or less at 400 GeV/c.

Mean lifetimes (averaged over all species) of less than 10^{-13} sec.[69] (It is still possible that some charmed hadrons have $\tau \simeq 10^{-14} - 10^{-13}$ sec but this degree of nonleptonic enhancement is now unlikely.)

There is a narrow window of expected charmed particle lifetimes, but for anything else (b and t quarks, leptons heavier than τ, etc.) the field is wide open for studies of short-lived particles. Many of the techniques developed for charm undoubtedly will be useful for these studies as well.

ACKNOWLEDGMENTS

I wish to thank Tom Gaisser for the opportunity to review this subject and for fruitful discussions over the course of several years. Conversations with G. Feldman, F. Halzen, C. Quigg,

J. Sandweiss, T. Yamanouchi, and W. Wilson also are gratefully acknowledged. Part of this work was performed during a visit to Fermi National Accelerator Laboratory.

REFERENCES

1. S. L. Glashow, J. Iliopoulos, and L. Maiani, Phys. Rev. $\underline{D2}$, 1285 (1970).
2. M. K. Gaillard, B. W. Lee, and J. L. Rosner, Rev. Mod. Phys. $\underline{47}$, 277 (1975).
3. Recent summaries of properties of these particles have been given by Gary J. Feldman, SLAC Summer Institute Lectures, 1977, SLAC-PUB-2000, proceedings edited by Martha C. Zipf, SLAC 1977; in Lectures at Banff Summer Institute on Particles and Fields, Banff, Alberta, Canada, 1977, SLAC-PUB-2068, proceedings edited by D. H. Boal, to be published; in rapporteur's talk, XIX International Conference on High Energy Physics, Tokyo, August 23-30, 1978, proceedings edited by H. Miyazawa, to be published by National Laboratory for High Energy Physics, Tsukuba, Japan, 1978; and by G. Flügge, rapporteur's talk, 1978 Tokyo Conference, op. cit.
4. The evidence for the fifth quark b comes from properties of the T and T', which are assumed to be the $3S_1$ $b\bar{b}$ ground state and first excited state, respectively. See J. D. Jackson, C. Quigg, and J. L. Rosner, "New particles-theoretical", presented at 1978 Tokyo Conf., Session B8, op. cit. (Ref. 3), available as Lawrence Berkeley Laboratory report LBL-7977.
5. Robert E. Shrock and Ling-Lie Wang, "Bounds on certain mixing angles in the Weinberg-Salam model with arbitrary numbers of quarks", August, 1978 (unpublished).
6. G. J. Feldman, rapporteur's talk, 1978 Tokyo Conf., op cit. (Ref. 3).
7. See, e.g., P. Langacker and H. Pagels, DESY report 78/33, subm. to Phys. Rev. D, for a recent discussion of current quark masses.
8. Harry J. Lipkin, Phys. Rev. Letters, to be published.
9. Cf. M. Gell-Mann, R. Oakes, and B. Renner, Phys. Rev. $\underline{175}$, 2195 (1968) for the chiral SU(3) x SU(3) version.
10. H. Thacker, C. Quigg, and J. Rosner, Phys. Rev. $\underline{D18}$, 274, 287 (1978).
11. A. De Rujula, H. Georgi, and S. L. Glashow, Phys. Rev. $\underline{D12}$, 147 (1975). There are even models which permit m_c to be as large as 2 GeV. See, for example, W. Celmaster, Howard Georgi, and M. Machacek, Phys. Rev. $\underline{D17}$, 879, 886 (1978).
12. C. Bricman et al., Phys. Letters $\underline{75B}$, 1 (1978).
13. G. Feldman, Banff lectures, Ref. $\underline{3}$.
14. R. Brandelik et al., Phys. Lett. $\underline{70B}$, 387 (1977).
15. M. Peshkin and J. L. Rosner, Nucl. Phys. $\underline{B122}$, 144 (1977).
16. A. Pais and S. B. Treiman, Phys. Rev. $\underline{D15}$, 2529 (1977).
17. I. Peruzzi et al., Phys. Rev. Lett. $\underline{39}$, 1301 (1977).
18. J. M. Feller et al., Phys. Rev. Lett. $\underline{40}$, 274 (1978).

19. W. Bacino et al., Phys. Rev. Lett. $\underline{40}$, 671 (1978).
20. G. Feldman, SLAC-PUB-2000, Ref. 3.
21. J. M. Feller et al., Phys. Rev. Lett. $\underline{40}$, 1677 (1978).
22. Jackson, Quigg and Rosner, Ref. 4.
23. M. B. Einhorn and C. Quigg, Phys. Rev. $\underline{D12}$, 2015 (1975); Phys. Rev. Lett. $\underline{35}$, 1114 (C) (1975).
24. H. Fuchi et al., "X particle search in super high energy interactions observed by the emulsion chambers", submitted to 1978 Tokyo Conf., op. cit. (Ref. 3), paper no. 491.
25. R. Diebold, rapporteur's talk, 1978 Tokyo Conf., op cit. (Ref. 3).
26. V. Vuillemin et al., Phys. Rev. Letters $\underline{41}$, 1149 (1978).
27. J. L. Rosner, in Deeper Pathways in High-Energy Physics, edited by B. Kursunoglu, A. Perlmutter, and L. F. Scott, New York, Plenum Press, 1977, p. 489.
28. C. Quigg and J. L. Rosner, Phys. Rev. $\underline{D17}$, 239 (1978).
29. G. Altarelli, N. Cabibbo, and L. Maiani, Nucl. Phys. $\underline{B88}$, 285 (1975); Phys. Letters $\underline{57B}$, 277 (1975).
30. R. L. Kingsley, S. B. Treiman, F. Wilczek, and A. Zee, Phys. Rev. $\underline{D11}$, 1919 (1975).
31. J. Ellis, M. K. Gaillard, and D. V. Nanopoulos, Nucl. Phys. $\underline{B100}$, 313 (1975).
32. T. F. Walsh, in Proceedings of the International Symposium on Lepton and Photon Interactions at High Energies, Aug. 25-31, Hamburg, edited by F. Gutbrod, Hamburg (DESY), 1978, p. 711.
33. E. G. Cazzoli et al., Phys. Rev. Lett. $\underline{34}$, 1125 (1975).
34. B. Knapp et al., Phys. Rev. Lett. $\underline{37}$, 882 (1976); for more recent results see W. Y. Lee, presented at 1978 Tokyo Conf. op cit, ref. 3.
35. A. Buras, Nucl. Phys. B109, 373 (1976).
36. E. H. S. Burhop et al., Phys. Letters $\underline{65B}$, 299 (1976); A. L. Read et al., Fermilab-Pub-78/56-$\overline{EXP}$, June, 1978, submitted to Phys. Rev.
37. H. Sugimoto, Y. Sato, and T. Saito, Prog. Theor. Phys. $\underline{53}$, 1541 (L) (1975), and in Proceedings of the 14th International Cosmic Ray Conference, Munich, Aug. 15-29, 1975, Max-Planck-Institut, 1975, paper no. HE 5-6, p. 2427.
38. B. W. Lee, C. Quigg and J. Rosner, Phys. Rev. $\underline{D15}$, 157 (1977).
39. T. K. Gaisser and F. Halzen, Phys. Rev. $\underline{D14}$, 3153 (1976).
40. G. Feldman, Tokyo Conf. rapporteur's talk, Ref. 3.
41. See, for example, F. Halzen, this conference, and footnote 14 of C. Quigg and J. L. Rosner, "Multilepton final states and the weak interactions of the fifth quark", Lawrence Berkeley Laboratory report LBL-7961, July, 1978, to be published in Phys. Rev. D.
42. D. Cutts et al., Phys. Rev. Lett. $\underline{41}$, 363 (1978).
43. R. Vidal et al., Phys. Lett. $\underline{77B}$, 344 (1978).
44. This would be our expectation; other hadron masses indicate that $m_u < m_d < m_s$.

45. See, e.g., Robert N. Cahn, Phys. Rev. Lett. $\underline{40}$, 80 (1978). Further discussions of decays of particles containing b quarks may be found in Quigg and Rosner, Ref. 41, and R. N. Cahn and S. D. Ellis, Phys. Rev. $\underline{D16}$, 1484 (1977).
46. M. Kobayashi and K. Maskawa, Prog. Theor. Phys. $\underline{49}$, 652 (1973). For a comprehensive review, see H. Harari, Phys. Reports $\underline{42C}$, 238 (1978).
47. Leon M. Lederman, rapporteur's talk, 1978 Tokyo Conf., op. cit. (Ref. 3).
48. J. Ellis, M. K. Gaillard, D. V. Nanopoulos, and S. Rudaz, Nucl. Phys. $\underline{B131}$, 285 (1977).
49. Quigg and Rosner, Ref. 41.
50. H. Kichimi et al., 1978, Tokyo Conf., op.cit. (Ref. 3), paper no. 545, quoted by R. Diebold, Ref. 25.
51. M. Piccolo et al., Phys. Rev. Lett. $\underline{39}$, 1503 (1977).
52. B. W. Lee et al., Ref. 38; A. De Rujula et al., Ref. 11.
53. I. Karliner, Phys. Rev. $\underline{36}$, 759 (C) (1976).
54. I am grateful to T. Yamanouchi and C. Quigg for raising this possibility.
55. J. L. Rosner, Nucl. Phys. $\underline{B126}$, 124 (1977).
56. H. M. Georgi, S. L. Glashow, M. E. Machacek, and D. V. Nanopoulos, Ann. Phys. $\underline{114}$, 273 (1978).
57. K. Niu , this conference.
58. M. Suzuki, Lawrence Berkeley Laboratory report LBL-7948, July 1978, submitted to Nucl. Phys. B.
59. N. Cabibbo and L. Maiani, Phys. Letters $\underline{79B}$, 109 (1978).
60. B. W. Lee and M. K. Gaillard, Phys. Rev. $\underline{D10}$, 897 (1974); I. Vainshtein and I. B. Khriplovich, Zh. Eksp. Teor. Fiz. Pis'ma Red. $\underline{18}$, 141 (1973). [Sov. Phys. JETP Letters $\underline{18}$,83 (1973)]; E. Ma, Phys. Rev. $\underline{D9}$, 3103 (1974); A. Gavrielides, Phys. Rev. $\underline{D11}$, 1884 (1974).
61. While the $c \to s$ coupling strength can be lowered appreciably, the extent to which the $c \to d$ coupling can be reduced by such mixing effects turns out to be negligible.
62. The production of dileptons (i.e., of charmed particles decaying semileptonically) by antineutrinos probably proceeds mainly via the s "sea": $\bar{\nu}s \to \bar{c}\mu^+$. The s-c weak coupling then must be of a certain minimum strength to reproduce the observed rate without an excessive s "sea". For analyses see C. Baltay, rapporteur's talk, 1978 Tokyo Conf., op. cit., Ref. 3; A. Benvenuti et al., Phys. Rev. Lett. $\underline{41}$, 1204 (1978); and D. Cline, this conference. I thank D. Cline for making this point during the discussion session.
63. W. Wilson, private communication.
64. E. Golowich and T. C. Yang, to be published in Phys. Letters B.
65. T. Yamanouchi, this conference.
66. K. Niu , E. Mikumo, and Y. Maeda, Prog. Theor. Phys. $\underline{46}$, 1644 (1971).

67. N. Ushida et al., "Observation of decays of short-lived neutral particles produced in 400 GeV/c proton interactions", presented at session B7, 1978 Tokyo Conf. op. cit. (Ref. 3), paper no. 492.
68. Cross sections have been extracted from those on nuclei by assuming that $\sigma(\rho A \to charm) \simeq A\, \sigma(pp \to charm)$. This scaling law characterizes J/ψ production: See M. Binkley et al., Phys. Rev. Letters $\underline{37}$, 571 (1976); J. G. Branson et al., Phys. Rev. Lett. $\underline{38}$, 1331, 1334 (1977); A. Bamberger et al., Nucl. Phys. $\underline{B134}$, 1 (1978).
69. M. M. Chernyavsky et al., "A search for new particles with lifetime $10^{-12} - 10^{-14}$ sec in interactions of protons with nucleons and nuclei in emulsion at 400 GeV/c", presented at Session B7, 1978 Tokyo Conf., op. cit. (Ref. 3), paper no. 308.
70. J. M. Bolta et al., "Preliminary results of a search for charmed particles photoproduced in nuclear emulsions coupled with the omega system of CERN", presented at session B7, 1978 Tokyo Conf., op. cit. (Ref. 3), paper no. 974.
71. M. Lauterbach, Phys. Rev. $\underline{D17}$, 2507 (1978). H. Kasha has questioned, in remarks at this conference, the stringent cross section limits in this work.
72. G. Conforto, this conference.
73. A. Bodek, this conference.
74. A. G. Clark et al., Phys. Lett. $\underline{77B}$, 339 (1978).
75. G. Coremans-Bertrand et al., Phys. Lett. $\underline{65B}$, 480 (1976).
76. W. Bozzoli et al., Lett. al Nuovo Cim. $\underline{19}$, 32 (1977).
77. D. J. Crennell, C. M. Fisher, and R. L. Sekulin, Phys. Lett. $\underline{78B}$, 171 (1978).
78. P. Alibran et al., Phys. Lett. $\underline{74B}$, 134 (1978).
79. T. Hansl et al., Phys. Lett. $\underline{74B}$, 139 (1978).
80. P. C. Bosetti et al., Phys. Lett. $\underline{74B}$, 143 (1978).
81. W. Ditzler et al., Phys. Lett. $\underline{71B}$, 451 (1977): limits for $D^0 + \bar{D}^0$ only.
82. F. Halzen and S. Matsuda, Phys. Rev. $\underline{D17}$, 1344 (1978).
83. F. Halzen, this conference.
84. D. Sivers, private communication.
85. A. I. Vainshtein, V. I. Zakharov, and M. A. Shifman, Pis'ma v Red. Zh. Eksp. i Teor. Fiz. $\underline{22}$, 123 (1975) [Sov. Phys. - JETP Letters $\underline{22}$, 55 (1975)].
86. J. Sandweiss, Physics Today $\underline{31}$, 40 (1978).
87. D. Cline, this conference.
88. The present subject also is treated in "Weak decays of heavy quarks," Fermilab-Conf-78/64-THY, August, 1978, by Mary K. Gaillard. See in particular, the estimates $\tau_D = (1-4) \times 10^{-13}$, $\tau_B \gtrsim 10^{-13}$ sec, and predictions for specific charmed and b-particle final states given there.

Chap. 6. Interactions Around 1000 TeV and Above

A NEW TYPE OF NUCLEAR INTERACTIONS IN THE $\Sigma E_\gamma > 10^{14}$ eV REGION

Brasil-Japan Emulsion Chamber Collaboration

J.Bellandi Filho, J.L.Cardoso Jr., J.A.Chinellato, C.Dobrigkeit, C.M.G.Lattes, M.Menon, C.E.Navia O., A.M.Oliveira, W.A.Rodrigues Jr. M.B.C.Santos, E.Silva, E.H.Shibuya, K.Tanaka, A.Turtelli Jr.
Instituto de Fisica Gleb Wataghin-Universidade Estadual de Campinas. Campinas, São Paulo, Brasil

N.M.Amato, F.M.Oliveira Castro
Centro Brasileiro de Pesquisas Fisicas, Rio de Janeiro, RJ, Brasil

H.Aoki, Y.Fujimoto, S.Hasegawa, H.Kumano, K.Sawayanagi, H.Semba, T.Tabuki, M.Tamada, S.Yamashita
Science and Engineering Research Laboratory, Waseda University, Shinjuku, Tokyo, Japan

N.Arata, T.Shibata, K.Yokoi
Department of Physics,Aoyama Gakuin University,Setagaya, Tokyo, Japan

A.Ohsawa
Institute for Cosmic Ray Research, Tokyo University, Tanashi, Tokyo, Japan

ABSTRACT

Centauro-type and Mini-Centauro-type events are reported and discussed. They are events in which nuclear-active particles (i.e. hadrons which do not undergo rapid photon decays) are produced without any sign of accompanying neutral pion emission. Estimated hadron multiplicity is ~100 for Centauro-type and ~15 for Mini-Centauro-type.

Up to the present moment, 5 examples of Centauro and 13 of Mini-Centauro have been observed in the two-storey emulsion chambers exposed on Mt.Chacaltaya,Bolivia(5220 m above sea level).

The experimental results on energy and angular distributions are in agreement with the hypothesis that this type of events occur through production and decay of a fire-ball with a rest energy of (200~300) GeV/c^2 for Centauro and ~30 GeV/c^2 for Mini-Centauro events.

INTRODUCTION

In 1972, Chacaltaya emulsion chamber recorded a strange cosmic-ray event which shows arrival of bundle of more than fifty hadrons accompanied by very few electrons or γ-rays[1]. The analysis showed that the event is a nuclear interaction of extreme energy, about 10^{15} eV, in the atmosphere near the chamber, in which hadrons of multiplicity about one hundred are produced without associated π°-meson emission. Such large imbalance in charge states of the produced particles suggests that the event does not contain production of π-meson in an example of a new type of nuclear interaction, such as Multiple Production of Baryons and Anti-Baryons.

We put the name "Centauro" for the event because of its exotic appearance, and the Brasil-Japan Collaboration experiment group have been since then engaged in the systematic survey for further examples of the same characteristics as Centauro. Through three successive exposures of emulsion chambers at Chacaltaya Observatory, we were able to observe a number of events of similar type. All of them are a bundle of high energy cosmic-ray particles, rich in hadronic component and poor in the electro-magnetic component. Their abnormality in the composition is far from the expected fluctuation in the usual Multiple Meson Production processes. The study shows that the association of The few electrons and γ-rays in the bundle comes from secondary nuclear interactions in the air, and the parent multiple production of hadrons is without emission of π°-mesons.

Within the collected samples of this particular kind, we are able to recognize existence of two types. One is called "Centauro" type, which is similar to the first observed event. It is characterized by the large multiplicity of hadrons, about one hundred[2]. The other is named "Mini-Centauro" type, which has smaller multiplicity of about fifteen[3].

EXPERIMENTAL METHOD

A) EMULSION CHAMBER

Our detector is two storey emulsion chamber exposed at Mt.Chacaltaya, Bolivia, 5220 m above sea level.

The upper chamber, located at the top of our apparatus, detects atmospheric electrons and γ-rays. A fraction of arriving atmospheric hadrons will interact inside the apparatus (Pb-jet upper) and can be detected in the upper chamber. These showers can be identified either by their starting point (below 10 c.u.) or by the double maxima in their transition curves showing successive nuclear interactions. Then showers in the upper chamber are mixture of cascade showers from atmospheric electrons and/or γ-rays and those from Pb-jets upper generated by hadrons.

Below the upper chamber, there are the target layer of pitch, the air gap of 1.5 m and the lower chamber. Since atmospheric electrons and γ-rays are absorbed by Pb plates of the upper chamber, showers observed in the lower chamber come from nuclear interactions in the target layer (C-jet) or in the chamber itself (continuation of Pb-jet upper or Pb-jet lower). Microscopic observations on shower tracks in the nuclear emulsion plates allow one to distinguish between these types.

B) ENERGY MEASUREMENT

Energy measurement of showers is made either counting their electron tracks in emulsion plates at each depth in the Pb-sandwich, or by microscopic photodensitometry of their dark spots in X-ray films. For Pb-jets, the measurement gives the sum of energies liberated into gamma rays, ΣE_γ, which we call observable energy of interaction. The incident hadron energy E_N is given as $E_N = \Sigma E_\gamma / k_\gamma$ with γ-ray inelasticity k_γ, which is estimated as $k_\gamma \simeq 0.2$. For C-jets, the energy measurement can be made on individual secondary γ- rays, the sum of which gives the observable energy of interaction ΣE_γ.

CENTAURO-TYPE EVENTS

For demonstration of characteristics of Centauro-type events, Fig.1 presents illustration of the events Centauro I and IV penetrating through the whole chamber.

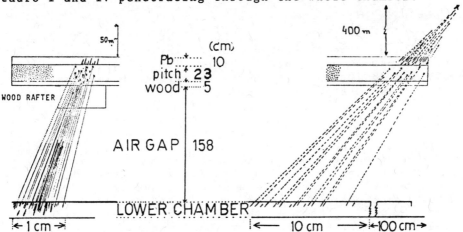

Fig.1 Schematic illustration of Centauro I and CentauroIV. Centauro I passed through the supporting beam of wood which worked as additional target layer. Shower direction converges to a point 50±15 m above the chamber. CentauroIV has larger spread, thus the production height is higher than Centauro I.

Those Centauro-type events were found as a group of many parallel showers in the lower chamber, clustering in a small region of a few cm in size. Measuring the arriving directions, we are able to find their continuation in the upper chamber. Relative position of the event in upper and lower chamber is determined accurately with reference to the position of several showers of Pb-jets upper, which are recorded in both chambers.

Up to now we found five events of Centauro-type and four are completely analysed. Table I gives a summary of those Centauro-type events.

Table I Summary of Centauro Events

Event number		I	II	III	IV
Chamber number		15	17	17	17
Number of showers in upper chamber	unidentified(atmospheric γ-rays, electrons or Pb-jets upper)	1*	5	26	61
	identified Pb-jets upper	6	9	16	15
lower chamber	continuation of Pb-jets upper	7	6	5	3
	C-jets	29	9	8	13
	Pb-jets lower	7	8	8	7
Observed energy sum in TeV	upper chamber	28.1	57.6	150.1	195.5
	lower chamber	202.5	145.8	119.8	90.1
	Total	230.6	203.4	269.9	285.6

* It has double core structure, showing not Pb-jet upper but pair of γ-rays.

A) BUNDLE OF PARTICLES ARRIVING AT THE TOP OF THE CHAMBER

The distribution of depths of interaction points in the chamber is constructed for the local interactions, C-jets and Pb-jets. The observed distribution is consistent with an exponential function with the nuclear collision mean free path, value of which is commonly observed for cosmic-ray hadrons. Therefore, knowing the thickness of the chamber and using the number of observed Pb-jets and C-jets, it is possible to estimate the number of arriving hadrons at the top of the chamber for each Centauro event.

As showers in the upper chamber are produced by atmospheric electrons and gamma-rays, and by Pb-jets upper, we are able to estimate expected number of Pb-jets upper, knowing number of the arriving hadrons. Subtracting the number from the total observed number of showers in the upper chamber, we obtain the number of atmospheric electrons and gamma-rays arriving at the top of the chamber. The results of such estimation are presented in Table II for hadrons and Table III for gamma-rays and electrons.

Table II Multiplicity of Hadrons

Event number		I	II	III	IV
Height of main interaction (meters)		50	80	230	500
Observed multiplicity	Pb-jets	20	23	29	25
	C-jets	29	9	8	13
	Total	49	32	37	38
Calculated multiplicity	arriving at the chamber	71	66	63	58
	interacting in the air	3	5	13	32
	Total	74	71	76	90

Table III Multiplicity of γ-rays and Electrons

Event number	I	II	III	IV
Arriving at the chamber	1*	none	17	51
Produced in A-jets	4	13	30	47
Difference (γ-rays in the Centauro interaction)	none	none	none	4

* See note in Table I

B) MULTIPLICITY OF HADRONS AND γ-RAYS AT THE PARENT INTERACTION

In order to estimate the number of hadrons and γ-rays produced in the parent atmospheric interaction, we must know its height in the atmosphere. The height measurement through the geometry is possible only for the event Centauro I, where we are able to measure the divergence of direction of its showers*. The production heights for the other three events are estimated comparing their lateral spread with that of Centauro I. The distance R_N of a hadron from the energy weighted center of each event is connected to its P_T

$$P_T = E_N R_N / H \qquad (1)$$

H being the height of production. For application of the formula, the hadron energy E_N and the hadron P_T can be substituted by the visible interaction energy ΣE_γ and $k_\gamma P_T$.

Footnote * The geometry measurement was made in two ways, one is the comparison of distances between shower spots in X-ray films of the upper and lower chamber for those Pb-jets upper which have continuation in the lower chamber. The other is the comparison of distances between shower cores in nuclear emulsion plates at different depths in the lower chamber. The two give a consistent result of 50 ± 15 m.

Once we have an estimation of interaction height H and of the number of hadrons arriving at the top of the chamber, we can estimate the multiplicity of hadrons produced in the parent Centauro interaction and the number of their secondary atmospheric interactions (A-jets). The results are presented in the Table II, too.

From the expected number of secondary interactions in the air (A-jets) and average γ-ray multiplicity above the detection threshold per interaction, we obtain the expected total number of γ-rays which are produced in those A-jets and arrive at the chamber accompanying the hadron bundle. Subtracting the number from the observed number of atmospheric electrons and γ-rays at the chamber we have the number of γ-rays and electrons produced in the parent Centauro interaction. As presented in Table III, the number of γ-rays or electrons produced in the Centauro interaction is negligibly small, that is, the number of neutral π-mesons must be zero or negligible among the produced hadrons in the interaction*.

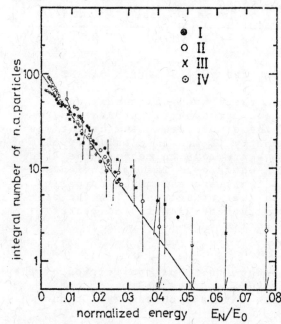

Fig. 2 Integral distribution of fractional energy of hadrons in four Centauro events, I, II, III and IV. The number of hadrons is after the correction of those penetrating through and those interacting in air.

Footnote*   See D) in next sub-section for another way of height estimation. Suppose the height estimation here be not right by some reason. If it would be significant over-estimation, the geometry method could have detected the case beyond the noise. More probable case would be under-estimation of the heights. Then, we should have more secondary A-jets, which would result even less production of γ-rays at the parent Centauro interaction.

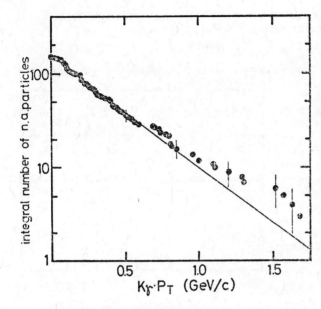

Fig. 3 P_T distribution of hadrons from Centauro interaction. Interaction height of Centauro II, III and IV are determined to give the same $<P_T>$ as Centauro I, and then the data of all four are put together.

C) CHARACTERISTICS OF CENTAURO INTERACTION

The distribution of fractional energy of hadrons, E_N/E_o, in the interaction can be constructed replacing the hadron energy E_N by the visible interaction energy ΣE_γ and the total energy E_o by sum of visible energies with correction factor for penetrating through. Fig. 2 shows the integral distribution where the fraction of particles non-interacted and interacted in the air has been corrected for. One sees that the spectrum is well represented by a single exponential function with the hadron multiplicity $N_o = 100 \pm 20$. It is expressed as,

$$N(\geq E_N/E_o) = N_o \exp(-N_o E_N/E_o) \qquad (2)$$

This exponential distribution with multiplicity of about one hundred gives good agreement to all of the four Centauro-type events. It makes us conclude that the four events belong to one and the same interaction type, i.e. Centauro interaction.

The transverse momenta of hadrons are measured with respect to the direction of motion of the energy-weighted center of the event. Fig. 3 shows the integral distribution of P_T of hadrons for the four Centauro-type events. Since we determined the height of production H to make the average $<P_T>$ equal among the four, the discussion should be only on the spectrum shape. There is no significant differences in the shape of P_T distribution among the four events, and here again we find an exponential function gives a good fit. If we assume that the γ-ray inelasticity k_γ is 0.20 ± 0.05, the average transverse momentum of hadrons is, from Centauro I, as

$$<P_T> = 1.7 \pm 0.7 \text{ GeV/c} \quad (3)$$

where a large error comes from uncertainity in the height measurement.

D) CENTAURO FIRE-BALL

The fire-ball hypothesis for the Centauro-type interaction is supported by the experimental fact that both distribution of fractional energy and of transverse momentum are well represented by an exponential function, which reminds us of the thermal emission. Under the fire-ball hypothesis, the average $<P_T>$ gives a measure of particle motion in the fire-ball system. If we assume, as an example, that the hadrons produced in the Centauro interaction are nucleons and anti-nucleons, their average energy in the fire ball system is found as $<E_N>^* = 2.3 \pm 0.8$ GeV.

Then, multiplying by the hadron multiplicity, $N_o \simeq 100$, we obtain for the rest energy of Centauro fire-ball as,

$$M_{cen} \simeq 230 \text{ GeV} \quad (4)$$

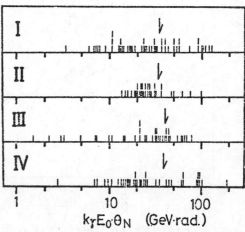

Fig 4A Angular distribution of hadrons in Centauro events in energy normalized scale $k_\gamma E_o \theta_N$.

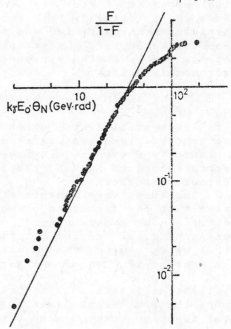

Fig. 4B F-plot of angular distribution under the normalization by hadron multiplicity expected from observed number of C- and Pb-jets.

Fig. 4 presents the angular distribution of hadrons in the Centauro interaction in the form of logtanθ plots and of F-plots. The emission angles θ_N are normalized multiplying them with $k_\gamma E_o$. Isotropy is manifested in the F-plot. Now, the total multiplicity of hadrons from the Centauro fire-ball is estimated as about one hundred from the extrapolated number of hadrons in their fractional energy distribution. This allows one to estimate the position of half angle of the decaying Centauro fire-ball, by counting particles from the forward direction until half of the multiplicity correcting for the detection efficiency, and the arrow in Fig. 4A indicates the half-angle. This gives us estimation of motion of the fire-ball and its rest energy. The result is $M_{cen} \simeq 200$ GeV which is consistent with the value given above in (4).

MINI-CENTAURO-TYPE EVENTS

During the systematic survey for the Centauro type interactions in Chacaltaya emulsion chambers, we found a number of examples of another new type of events. They show rich abundance of hadrons in arriving bundle of cosmic-ray particles just as Centauro type, but they are with much smaller multiplicity. We name them as "Mini-Centauro" events and we have thirteen events of such type up to the present. The characteristics of the Mini-Centauro type interaction is summarized as an atmospheric nuclear interaction which produced about fifteen hadrons in an average without any sign of accompanying neutral π-meson emission. Beforegoing into the analysis of observed thirteen events of Mini-Centauro type, we will present some of typical Mini-Centauro events as an example.

A) EVENT MINI-CENTAURO I

This event consists of 27 showers in the upper chamber and 10 in the lower chamber, all detected by naked eye scanning over the X-ray films. Three showers of the upper chamber continue into the lower chamber, thus the relative geometrical position in the both chambers is exactly determined. All the shower spots in X-ray films were traced into the nuclear emulsion plates in both upper and lower chambers, and the microscopic observation was made for all showers to study the shower core structure in detail.

Among 10 showers detected in the lower chamber, three are continuation of Pb-jet upper, two are Pb-jet lower and five are C-jets from the target layer. Out of 27 showers in the upper chamber, two are far from the center and considered not to belong to the concerned main interaction. All the others, 25 showers are located within a radius 5 cm from the center. Among them, 5 showers are identified as Pb-jet upper because of the criterion on the shower transition curve. Three big showers are found near the center, all of which have multi-core structure with a spread of about 1 mm and contain a Pb-jet upper within the spread. Thus it is considered that they are A-jets near the chamber but not atmospheric cascades. The same argument can be applied to two clusters of showers with lateral spread of 0.5~1.0 cm, because there is a Pb-jet associated to each cluster. Thus, eleven showers are left to be possible atmospheric gamma-rays. All of them are with low energy, their total energy being 31.8 TeV. Summary of such identification is given in Table IV, from which one sees large unbalance between hadron component and gamma-ray component in the event. Atmospheric gamma-rays are occupying such small fraction of the event that all of them can be assigned to secondary A-jets. The parent interaction is very likely producing only hadrons without emission of gamma-rays just as in the case of Centauro events.

Table IV. Hadrons in Mini-Centauro I

Lower chamber			Upper chamber		
shower No.	type	ΣE_γ in TeV	shower No.	type	ΣE_γ in TeV
101	C	126.3	15	A	23.4
102	C	85.4	16	A	40.0
103	C	23.7	17	A	80.4
104	C	14.1	6	Pb	2.0
105	C	4.1	7	Pb	3.2
107	Pb	3.2	11	Pb	8.2
108	Pb	2.1	13	Pb	2.5
			14	Pb	6.7
			9-9'-10	A	12.9
			a1-2-3	A	7.2

Associated atmospheric γ-rays are 11, with total energy sum 31.8 TeV.

B) EVENT MINI-CENTAURO XIII

This event shows more clearly the unbalance of charge states in its original interaction. Ten showers are observed spreading about ~3 cm in the both upper and lower part of chamber nº 17 with the total effective thickness of 1.7 nuclear collision mean free path. Five showers are identified to be Pb-jet upper, two C-jets and one Pb-jet lower. Two showers are left unidentified.

Thus, total visible energy of hadrons is 53 TeV, with 6 TeV for the two unidentified showers.

C) HADRON BUNDLE IN MINI-CENTAURO EVENTS

First we will present summary of observed thirteen events of Mini-Centauro in Table V. Arriving hadrons are detected through their secondary interactions as Pb-jet upper and lower, C-jet and A-jet.

Table V Summary of Mini-Centauro Events

Event n⁰	Hadrons visible energy sum(TeV)	number	Electrons,γ's energy sum (TeV)	number	Total obser. energy (TeV)	Estimated interaction height (m)
I	446	17 (8)	32	11	478	600
II	267	13 (4)	168	38	435	1600
III	270	14 (6)	167	27	437	900
IV	130	21 (6)	52	27	182	1000
V	70	10 (9)	17	9	87	800
VI	108	9 (5)	130	35	238	1500
VII	63	9 (4)	125	41	188	600
VIII	78	10 (6)	84	23	162	650
IX	61	5 (4)	76	18	137	1900
X	48	11 (7)	30	9	78	400
XI	77	5 (4)	133	19	210	800
XII	482	7 (4)	25	8	507	650
XIII	53	8 (7)	6	2	59	300

( ) gives number of hadrons with $f \geq 0.03$

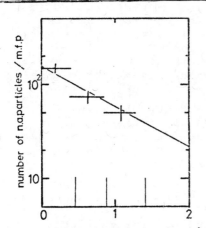

Fig.5 Distribution of points of nuclear interaction in the chamber from C- and Pb-jets in 13 Mini-Centauro events, expressing the depth in nuclear collision mean free path.

Fig. 5 gives the distribution of interaction points in the chamber for Pb-jets and C-jets of all the Mini-Centauro events. The distribution agrees well with the exponential distribution with the usual nuclear collision mean free path, as in the case of Centauro type. A-jets are identified when the shower cluster in the upper chamber contains at least one Pb-jet or C-jet within itself. Thus we are picking up only a part of such atmospheric secondary interactions as A-jet that occur very near the chamber, say << 100 m, in requiring them to have well recognized cluster structure.

Fig. 6 shows the distribution of fractional energy of

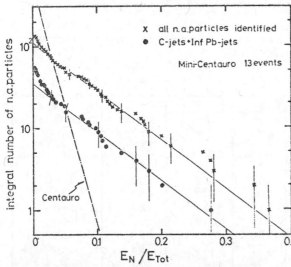

Fig.6 Integral distribution of fractional energy of hadrons E_N/E_{Tot}, for 13 Mini-Centauro events.

hadrons, E_N/E_{Tot}, for which is substituted by the ratio of visible interaction energy ΣE_γ and the total sum of visible energies of the event. There is no appreciable difference among the thirteen Mini-Centauro events, so that the results are presented summing up all the thirteen. The figure presents the two distributions separately, one for C-jets and Pb-jets lower where the detection method is straightforward, and the other for all, including Pb-jets upper and A-jets, too. Absence of appreciable difference between the two indicates absence of effects due to the difference in detection method. Now one sees that the fractional hadron energy spectrum follows an exponential form as in the Centauro case, while the slope is less steep. It shows that the Mini-Centauro events have much smaller multiplicity. In the region of small fractional energy, $E_N/E_{Tot} < 0.03$, the exponential distribution shows a rise over the exponential function, which is considered to be due to contribution of tertiary hadrons from the secondary atmospheric interactions. The extracted value to zero energy by the straight line shows that the average number of secondary hadron interactions detected as C-jets, Pb-jets and A-jets is 8.3 ± 0.8 for one Mini-Centauro event.

D) PARENT MINI-CENTAURO INTERACTION

The height of parent Mini-Centauro interaction in the atmosphere can be estimated through the diagram on the average of energy-multiplied lateral spread of secondary hadrons $<R_N \Sigma E_\gamma>$, and the fraction of energy retained by hadrons $\Sigma E_N/E_o$. The former quantity is the product of $k_\gamma <P_{TN}>$ and the interaction height H. Under the assumption of absence of γ-ray emission in the parent Mini-Centauro interaction, the latter quantity will follow the form $\exp(-H/\lambda)$ with λ the collision mean free path.

For application of this argument to the height estimation the hadron must be restricted only to the secondary ones directly produced from the parent interaction, which can be picked up by the criterion $E_N/E_{Tot} > 0.03$ as discussed above. Fig. 7 presents the diagram with experimental points. Here, black dots express the cases including the identified A-jets, near the chamber, and crosses those excluding them. Since the identified A-jets are all very close to the chamber, the difference between the two shows effect of fluctuation which is large for low multiplicity Mini-Centauro events. Centauro type events are also shown by open circles for comparison. The upper scale of the diagram for the production height H is obtained with use of average P_T value obtained from Centauro I ($<k_\gamma P_T>=0.35$ GeV/c^2), and the straight line express an exponential decrease with the nuclear collision mean free path. One finds that the experimental points are distributed along the straight line. This distribution of experimental points is indicating absence of γ-rays among the secondary products of Centauro and Mini-Centauro interaction. Seeing the consistency, the height estimation for each event can be carried out now by the upper scale in the figure.

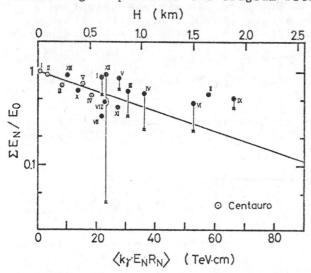

Fig.7 Diagram of energy fraction retained by secondary hadron component E_N/E_0 and average lateral spread of secondary hadrons $k_\gamma <E_N R_N>$ for Mini-Centauro events.

The energy attenuation of hadronic component gives another way of estimating the height of production for Centauro events because of its small fluctuation due to the larger multiplicity. The agreement with the previous estimation gave encouragement for applying the diagram to the height estimation of Mini-Centauro events.

Knowing the production height for each event, the hadron multiplicity at the parent interaction is estimated for ten Mini-Centauro events excluding the three wich travel more than one nuclear collision mean free path,

1.2 km at Chacaltaya. Number of all secondary hadron interactions in the chamber has already been obtained from the fractional energy distribution as 6.2 ± 0.6 per event. It gives average number of secondary hadrons arriving at the chamber as 8.7 ± 0.9. Since the average height of production for those ten events is 650 m above the chamber, it gives the hadron multiplicity at the parent interaction as,

$$N_o = 15 \pm 1.5 \tag{5}$$

with estimated number of atmospheric secondary interactions as 6.2 per one event.

For the transverse momentum distribution, we have already seen that the average value for secondary hadrons is the same as the Centauro case, i.e. $k_\gamma <P_T> = 350$ MeV/c. Fig. 8 gives the distribution of $k_\gamma P_T$ which is again of the exponential type.

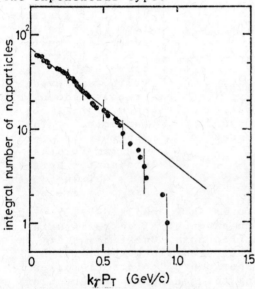

Fig. 8 P_T distribution of hadrons in Mini-Centauro events in integral form.

E) MINI-CENTAURO FIRE-BALL

We have seen that the distribution of fractional energy and transverse momentum follow exponential type in Mini-Centauro interaction too, as in the case of Centauro. It suggests the existence of a fire-ball of a new type, called Mini-Centauro fire-ball, as an intermediate product decaying into about fifteen hadrons, in an average, without emission of neutral pi-meson.

Rest energy of Mini-Centauro fire-ball can be estimated for the example in which all the produced hadrons are nucleons and anti-nucleons. The average energy of hadrons in the fire-ball rest system is obtained from $<P_T>$ as $<E_N^*> = 2.3$ GeV like the case of Centauro fire-ball. Multi plying the average multiplicity N = 15, we have

$$M_{Mini-Cen} = 35 \text{ GeV}/c^2 \quad (6)$$

If the γ-ray inelasticity k_γ is assumed as 0.3, it comes down to 26.5 GeV/c^2.

Angular distribution for secondary hadrons are given in Fig. 9 for ten events with production height lower than one nuclear mean free path. The logtanθ plots are composed jointly with use of the energy normalized angular scale $\theta_N E_{Tot}$. The estimated half angle of the Mini-Centauro fire-ball is indicated by the arrow, which was determined by counting secondary hadrons from the smallest angle up to the point of a half of estimated total multiplicity corrected for the detection efficiency. It gives an estimation of Mini-Centauro fire-ball as $M_{Mini-Cen}$ =31 GeV/c^2 with use of k_γ=0.2 which is consistent with the mass estimation given above.

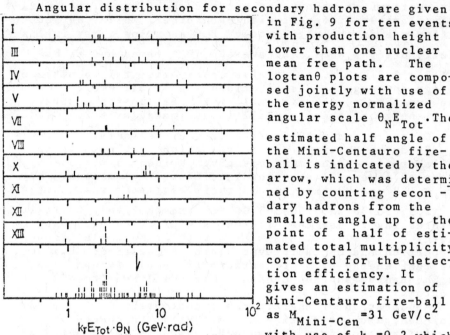

Fig.9 Angular distribution of hadrons in Mini-Centauro events in energy normalized scale $k_\gamma E_o \theta_N$

SUMMARY AND DISCUSSION

Experimental evidences are presented for existence of two new types of nuclear interactions, Centauro and Mini-Centauro. A decisive character of these new interactions is that neutral pi-mesons, i.e., γ-rays, can not be found among the produced particles. In the Multiple Production of Hadrons of these new types, emission of pi-mesons must be absent or only negligible fraction.

A bundle of hadrons from Centauro and Mini-Centauro interactions is detected as a group of parallel secondary hadron interactions, i.e., C-jets, Pb-jets and A-jets. Observation of those secondary interactions finds no difference between them and ordinary cosmic-ray nuclear interactions of comparable energy, in all available aspects including the collision mean free path, the γ-ray multi -

plicity, the transverse momentum and others. Thus we may conclude that secondaries of Centauro and Mini-Centauro interactions can not be much different from ordinary cosmic-ray hadrons, i.e., nucleons.

Hadron multiplicity in Centauro and Mini-Centauro interactions are about one hundred and fifteen, respectively, largely different with each other. While, we see no difference in the magnitude of P_T. Assuming γ-ray inelasticity being $k_\gamma \simeq 0.2$, the average $<P_T>$ turns out to be $\simeq 1.7 \pm 0.7$ GeV/c. All available experimental results on energy and angular distributions are in agreement with the hypothesis of a fire-ball with the isotropic decay. Rest energy of the fire-ball is estimated as $\simeq 200$ GeV/c^2 for Centauro and $\simeq 30$ GeV/c^2 for Mini-Centauro.

Study of Multiple Meson Production by Brasil-Japan collaboration experiment has given evidences for three types of fire-ball decaying into cluster of mesons. They are H, SH and UH-quantum and their estimated rest energy is 2.5, 20~30 and 200~300 GeV/c^2, respectively[4]. It is interesting to see the discreteness of fire-ball rest energy both in Multiple Meson Production and Centauro-type Hadron Production. Besides, the rest energy of Mini-Centauro fire-ball is near that of SH-quantum and the Centauro fire-ball near to UH-quantum.

Application of the fire-ball concept to Centauro and Mini-Centauro interaction is further strengthened by new information from Centauro V, which is now under analysis. The event contains a huge Pb-jet lower of $\Sigma E_\gamma \simeq$ 90 TeV, which carries about one third of the total observed energy, the rest being distributed among 65 showers. In Mini-Centauro XI, a large A-jet of $\Sigma E_\gamma \simeq$ 340 TeV is observed which shares about 3/5 of the total observed energy. Existence of such distinctly high energy jet must be due to the leading hadron which is likely to be a surviving particle after the interaction moving isolated from the rest of particles, the fire-ball. This indicates that the concept of partial elasticity still applies to interactions of such violent type.

From this information we believe that the Centauro or Mini-Centauro type interactions are not the break-up of an incoming cosmic-ray hadron into its constituents.

Acknowledgement

The collaboration experiment is financially supported in part by Conselho Nacional para o Desenvolvimento Científico e Tecnológico and Fundação de Amparo à Pesquisa do Estado de São Paulo in Brasil and Institute for Cosmic-Ray Research, University of Tokyo in Japan.

References

1. Brasil-Japan Collaboration, Conf. Paper, XIII Cosmic-Ray Conf. Denver, $\underline{3}$, 2227 and $\underline{4}$, 2671 (1973).
2. Brasil-Japan Collaboration, Conf. Paper, XIV Cosmic-Ray Conf. Munich, $\underline{7}$, 2393 (1975).
   M. Tamada, Nuov. Cim. $\underline{41B}$, 245 (1977).
3. Brasil-Japan Collaboration, Conf. Paper, XV Cosmic-Ray Conf. Plovdiv, $\underline{7}$, 208 (1977).
4. Brasil-Japan Collaboration, "Multiple Meson Production in the $\Sigma E_\gamma > 2 \times 10^{13}$ eV Region" presented at this conference.

Large-scale Emulsion Chamber Experiment at Mt. Fuji

M.Akashi, E.Konishi, H.Nanjo and Z.Watanabe (Hirosaki
Univ., Aomori), I.Ohta (Utsunomiya Univ., Tochigi),
A.Misaki and K.Mizutani (Saitama Univ., Saitama),
K.Kasahara, S.Torii and T.Yuda (ICR, Univ. of Tokyo,
Tokyo), I.Mito (Shibaura Inst. of Technology, Tokyo),
T.Shirai, N.Tateyama and T.Taira (Kanagawa Univ.,
Yokohama), M.Shibata (Yokohama National Univ.,
Yokohama), H.Sugimoto and K.Taira (Sagami Inst. of
Technology, Fujisawa)

Characteristics of high energy family events, which were observed with large-scale emulsion chambers exposed at Mt. Fuji (3776 m above sea level), are discussed in connection with multiplicity and transverse momentum. In particular we lay great emphasize on a detailed description of special events, i.e. big families with energies over 1000 TeV (F4-589 and FC-31) and families with peculiar structures (TITAN and FA-11) , in order to clarify what happen in very high energy region over 10^{14} eV.

1. — Introduction.

A bundle of high energy γ-rays that come from the same direction in the atmosphere is generally called as a γ-ray family, which contains photons, electrons and positrons being electromagnetic descendants of several —— genetically connected —— nuclear interactions occurring in the atmosphere. For the present, observation of high energy family events with a large-scale emulsion chamber at mountain altitude is considered to be only means for getting direct information on particle production in ultra-high energy region over 10^{14} eV as other detectors are more or less inferior to emulsion chamber in space resolution and energy determination especially of high energy showers.
Generally speaking, essential features of γ-ray families, i.e. energy spectrum and lateral spread of constituent γ-rays etc, are mainly governed by the energy spectrum and transverse momentum of energetic hadrons emerging in very forward or limiting fragmentation region despite electromagnetic cascade process succesively occurring in the atmosphere. In studies of those phenomena, however, we must always take into account the energy and the nature of primary initiating γ-ray family and the height of interaction, both of which are inaccessible to direct measurement. In particular, large fluctuations of the latter cause considerable deformation on most observables so that relevant features of the first or most energetic interaction might be strongly masked. But most of these difficulties may be overcome by performing the extensive Monte Carlo calculation based on various interaction models and primaries to compare with experimental results.
Furthermore, by emulsion chamber experiments we can often observe

a " pure " family event that come from a single interaction almost
without cascading in consequence of large fluctuations of the
interaction height mentioned above ([1]).   Such families are clearly
produced at the altitude very low in the atmosphere, say far lower
than 1 km above the chamber.   These events bring us various
interesting information that can never be   obtained only from the
statistical approach.
    Thus one can derive a lot of information about very high energy
interactions through genetic analysis of family events in the energy
intervals as wide as possible.   For example scaling breakdown,
large P_t production, new phenomena and so on will be crucially tested
by the emulsion chamber experiments.
    The energy range of γ-ray families so far observed at Mt. Fuji
(3776 m) is from ΣE_γ = 30 TeV to about 6000 TeV.   This may
approximately correspond to the interaction of primary particles
ranging from 10^{14} to 10^{16} eV, though it depends upon the model and the
nature of primaries.
    In this paper, most of the content will be assigned to describing
details of several pure family events so far observed in order to
emphasize the indications of high multiplicity and/or large P_t
production in our concerned energy region.

2. — Experimental procedure.

A basic structure of the
emulsion chamber used in Mt.
Fuji experiments is a sandwich
of lead layers (thickness of
each layer is 1 cm) and
several interposed layers of
photographic sensitive material.
Photographic layer generally
contains two types of X-ray
film with different sensitivites
i.e. high sensitive type-N
X-ray film and low sensitive
(fine grain) type-RR X-ray
film.   Particulaly the type-
RR film is powerful to determine
the shower energy or to locate
its position in very crowded region.

Table 1.

Exposure List of Chambers

Chamber	Exposure	Area	Thickness
F-4	'71.8 - '72.7 (317 Days)	137 m²	6 c.u.
F-5	'72.8 - '73.7 (324)	134	6
F-6	'73.8 - '74.7 (356)	120	8 (10)
F-A	'75.8 - '76.7 (360)	16	70
F-B	'76.8 - '77.7 (365)	190	7 (6)
F-C	'77.8 - '78.7 (365)	50	28
F-D	'78.8 -	140	10

Total Exposure	Thin type F-4,5,6,B	545 m².year
	F-D	140 m²
	Thick type F-A,C	65 m².year

    Energy of each shower is determined by measuring the optical
density of shower spot registered in the X-ray films.   The
correlation between the optical density and corresponding shower
energy is carefully checked in every exposures by use of showers
detected in the chamber which include the nuclear emulsion plates
together with X-ray films , where the shower energy is determined by
electron counting method using microscope.   Detailed procedures
will be found in the paper of ref. ([2]).
    In our experiments, showers with energy greater than 1.5 TeV
are detectable without scanning bias of the X-ray films.   Also,

shower energies can be determined up to 100 TeV or more with feasible accuracy.

Up to date, two types of the chamber have been exposed on Mt. Fuji, as listed in Table 1. One is of thin type with thickness 6 ~10 c.u. and the other of thick type with thickness 28 and 70 c.u.. The latter is of great advantage to investigate the behaviour of hadronic components belonging to same family members.

3. — Results and Discussions.

From five series of exposures (F4-FB) which amounts to 545 m^2·year are detected 80 γ-ray families fulfiling the condition of $\Sigma E_\gamma \geq$ 100 TeV and $N_\gamma \geq$ 10, where ΣE_γ means the sum of all shower energies exceeding E_m in the family, E_m being 2 TeV. Also 86 γ-ray families in the energy range from 30 TeV to 100 TeV come from the chamber FB. In addition to this, by the thick chamber FA we observed several family events accompanied by energetic hadrons such as TAITAN and FA-11 which are of peculiar interest in connection with large P_t, as described later on. The chamber FC is now under analysis so that only preliminaly results of the event FC-31 being the biggest family event so far observed will be presented.

In this section, first we briefly summarize the average behaviour of γ-ray families and continuously discuss details of the f'-spectrum and some of pure family events in connection with high multiplicity and/or large P_t.

3·1 Average features of γ-ray families.

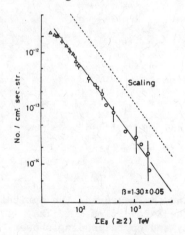
Fig.1. Integral ΣE_γ-spectrum at Mt. Fuji.

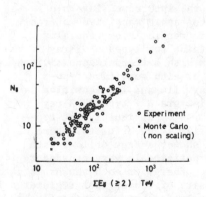
Fig.2. $N_\gamma - \Sigma E_\gamma$ correlation. Minimum energy is fixed to 2 TeV.

Fig. 1 shows the integral ΣE_γ -spectrum obtained at Mt. Fuji. The spectrum is well expressed as $(\Sigma E_\gamma)^{-\beta}$, where β = 1.30 $\pm$ 0.05.

This spectrum is closely connected with the primary spectrum as well as the nuclear interaction over 10^{14} eV.

In Fig.2 is presented the correlation between N_γ and ΣE_γ, which indicates approximately linear dependence. This gives us measures of the size of γ-ray families in our concerned energy region.

According to the Monte Carlo calculation ([1]), γ-ray families reach, on average, some equilibrium state through successive cascade processes and nuclear interactions before they arrive at an observation level deep in the atmosphere. Therefore, the power of ΣE_γ -spectrum and the $N_\gamma - \Sigma E_\gamma$ correlation is expected to become almost independent of the interaction model and the nature of primaries at mountain altitude. This may be confirmed by comparing our data with Chacaltaya one ([3]), which shows good agreement with each other. However the absolute intensity of the ΣE_γ -spectrum is fairly influenced by the model ([1]). As far as the Ryan's primary spectrum holds up to the energy region around 10^{16}eV, the simple scaling model gives too high intensity compared with experiment as demonstrated in Fig. 1. Scaling allow us a plausible explanation only if the primaries contain a significant fraction of heavy primaries.

Fig. 3 show the normalized lateral and ER distribution for all constituent γ-rays of energy in excess of 2 TeV in the families mixed together in the energy intervals between 100 and 300 TeV.

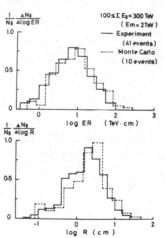

The lateral distribution basically depends on the transverse momentum of the secondary neutral pions as well as multiplicity in energetic p-air and π-air interactions, while the ER distribution strongly depends on the transverse momentum ([1]). As seen from this figure , experimental data seem to be consistent with the Monte Carlo calculation ([1]) based on the high multiplicity (two fire-ball) model with the normal transverse momentum, i.e. $<P_{t\pi}> \simeq 400$ Mev/c, though the statistics is yet insufficient to get definite conclusion. Here we would like to comment that scaling gives narrower lateral spread compared with non-scaling as energetic γ-rays are effectively produced. The difference is about two times([1]).

Fig.3. Lateral and ER-distribution of γ-rays of energy in excess of 2 TeV. Monte Carlo calculation is shown by dotted line (see text).

Thus, from the analysis of average features we may be able to draw conclusion that high multiplicity model with normal P_t is compatible with experiment, or in other words that the Feynman scaling in the fragmentation region breaks in our concerned energy region if the primaries over 10^{14} eV are almost same with that generally accepted in low energy region.

3·2 f'-spectrum.

As our concerned γ-rays are all high energy over 1 TeV, almost all members of γ-ray family would be originated from energetic neutral pions produced in the fragmentation region. Therefore, in general the production spectrum would be reflected on the energy spectrum of constituent γ-rays in the family in despite of cascading in the atmosphere. To examine this effect, Pamir group ([4]) introduced the variable defined as

$$f' = E_\gamma / \Sigma' E_\gamma \quad , \quad E_\gamma \text{ stands for energy of a given}$$

family member and summing in the denominator runs over all γ-rays fulfiling the condition of $f' \geq f'_{th}$ (fixed). As easily proved,

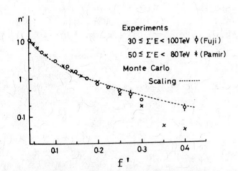

Fig.4. f'-spectrum in the energy intervals of $\Sigma' E_\gamma$ between 30 and 100 TeV. $f'_{th} = 0.04$.

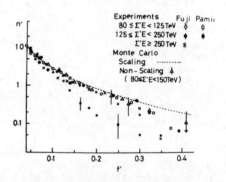

Fig.5. f'-spectrum in the energy region over $\Sigma' E_\gamma = 80$ TeV. $f'_{th} = 0.04$.

$\Sigma' E$ is proportional to the primary energy E_o under the assumption of scaling. In this case f' will correspond to the Feynman scaling variable defined as $x = E / E_o$. Therefore, this analysis allows us to compare conveniently the data differing in $\Sigma E'$ in order to examine whether the scaling breaks or not in our energy region over 10^{14} eV.

In Fig. 4 and 5 are shown our experimental data together with Pamir group's, devided into four different energy region of $\Sigma' E$. Pamir data show good agreement with ours in the energy region lower than 125 TeV, while theirs start to shrink drastically over 125 TeV in contrast with ours showing still scaling behaviour.

To clarify whether their result is caused by some physical processes or not, Monte Carlo calculations based on different models and primaries have been carried out by members of Pamir group ([5]). In spite of their efforts, however, calculations proved only that

f'-spectrum is insensible to both the model and the nature of primaries, indicating the asymptotic scaling behaviour even if the Feynman scaling completely breaks. This may be clear from the fact that drastic changes at the production are strongly masked by successive cascade processes with scaling behaviour in the thick atmosphere. In the figures experimental data are compared with the Monte Carlo calculations in both cases of scaling ([5]) and non-scaling ([1]). Calculations are consistent with our data as was expected.

3·3 High multiplicity ? Big family events over 1000 TeV.

We have already observed 6 big family events with total energy sum greater than 1000 TeV as listed in Table 2. These

Table 2. Big events

Event	n_γ	ΣE_γ (TeV)	$<E_\gamma>$ (TeV)	$<ER>$ (TeV·cm)	$<R>$ (mm)
F4-589	247	1880	7.6	4.4	5.8
F4-89	183	1115	6.1	5.4	8.9
F6-118	435	2135	4.9	11.4	23.2
F6-345	366	1503	4.1	6.3	15.4
FC-104	~400	~1800			
FC-31	~800	~5500			

$E_\gamma \geq$ 1.5 TeV.

families may be, on an average, originated from the interaction of primaries around 10^{16} eV. Such big events are generally composed of a large number of high energy γ-rays over 100 so that one can analyze on each event to get more direct information about particle production in ultra-high energy region.

In the formation of γ-ray families, there is one interaction responsible for the largest energy contribution to family members. We call it the main interaction. Monte Carlo calculation ([1]) teaches us that energies liberated from the main interaction, in most cases, account for over 50 % of the total energy sum. Also the production heights of main interaction distribute widely around 2~4 c.u. above the chamber. Furthermore simulations allow us to observe pure family events with a fairly large fraction of 10~20 % in the energy region over ΣE_γ=100 TeV. By examing the form of energy spectrum and ER distribution of family members, it may be possible to judge whether the family concerned is pure or not. The event F4-589 is considered to be pure or genuine. Following is the details of this event.

F4-589 (Pure family event).

First we present the ER distribution of this event together with the event F6-118. In this figure the dotted line is the average ER distribution of γ-rays per event for all families in the energy intervals of ΣE_γ from 500 to 1000 TeV, or in other words it means the equilibrium ER distribution of family with average energy $<\Sigma E_\gamma> = 640$ TeV.

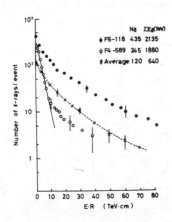

Fig.6. Integral ER-distribution of constituent γ-rays of energy in excess of 1.5 TeV. Dotted line is an eye guide.

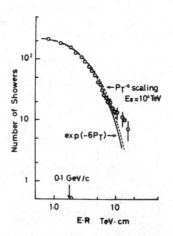

Fig.7. Comparison of data with the P_t-distribution. see text.

The form of F6-118 is close to the equilibrium one, while F4-589 shows distinct feature, i.e. is of the beutiful exponential form up to about 10 TeV·cm. This may imply that the event F4-589 was produced at the altitude very low in the atmosphere (see discussion about the energy spectrum). The slowly decreasing tail composed of about 10 % of total γ-rays over 10 TeV·cm may be a small contamination from the first interaction in higher altitude.

The product ER is nothing but the product of the transverse momentum P_t and the height H if the family come directly from a single interaction (main interaction only) without cascading. In low P_t region, the P_t-distribution of γ-rays is well expressed as the form of $\exp(-6P_t)$. Thus if the event F4-589 is pure and the P_t distribution still holds up to very high energy region, then the steep part of the ER distribution should fit to $\exp(-6P_t)$ as shown in Fig. 7. From this the production height is roughly estimated as H = 180 m. In this figure our data is also compared with the P_t distribution that include the large P_t contribution of the P_t^{-8} - type scaling ([6]). In this case comparison gives almost same production height. However the distribution including the P_t^{-4} - type contribution by no means fits to data as this slope is too gentle in the energy region around 10^{16} eV.

Next we discuss the energy spectrum of constituent γ-rays. In Fig. 8 is shown the energy spectrum of γ-rays of F4-589 and F6-118 in the fractional form, respectively. F4-589 also shows the exponential form in contrast to F6-118. The exponential form cannot be expected without assuming that the production height of γ-ray family is very near to the chamber. That is to say, in proportion to suffering cascade, the energy spectrum would come near to the power form as F6-118. The situation is well

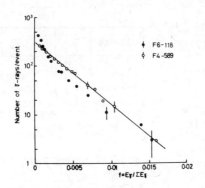

Fig.8. Energy spectrum of γ-rays expressed in the fractional form. Minimum energy is fixed to 1.5 TeV.

understood in the following discussions.
First we assume that the family concerned is produced by a single interaction (main interaction only) and next that the neutral pions are generated with the energy spectrum of

$N_{\pi^0} \exp(-X) dX$ and

$X = N_{\pi^0} E / k_\gamma E_0$

where E_0 is the primary energy and k_γ the inelasticity to neutral pions and N_{π^0} ($= 1/2\, N_\gamma$) the multiplicity of neutral pions. On this assumption we can easily calculate the energy spectrum of constituent γ-rays at any depth t , using the cascade theory under the approximation A. Calculated results, which are shown in Fig. 9, may be displayed by the diagram of n_γ / N_γ and $N_\gamma E_\gamma / 2k_\gamma E_0$

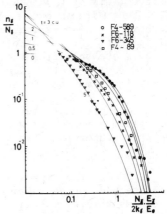

Fig.9. Comparison of the energy spectrum of γ-rays in four big families with calculation.

with parameter t where n_γ and E_γ is the observed multiplicity and energy of γ-rays, respectively. From this figure, we can easily understand that the family coming from very high altitude has an energy spectrum close to power type. By fitting the data to this diagram as shown in Fig. 9, the production height of F4-589 is estimated as $t \lesssim 0.5$ c.u. (about 250 m above the chamber), which is consistent with the one derived from the ER distribution. $k_\gamma E_0$ and N_γ at the production is also estimated from this figure. In Table 3, the quantities thus obtained are summarized together with other three big families, derived under the same assumptions.

Table 3.

Event	N_γ	$k_\gamma E_0$ (TeV)	H (c.u.)	F
F4-589	350	2200	0.5	0.85
F4-89	300	1800	1.5	0.64
F6-118	500	4000	2.5	0.53
F6-345	470	5600	≳ 5	0.27

$F = \Sigma E_\gamma / k_\gamma E_0$

H: Height of main interaction

This table may demonstrate that big four families have very similar structures one another. Estimated multiplicity of neutral pions for F4-589 is 100 - 200, and others take almost same value.

Our big family is compared with other cosmic ray ([7]) and accelerator ([8]) data in Fig. 10. Cosmic ray data should be compared with the accelerator one taken from proton-light nucleus (C,N or O) to match the target effects one another. Also estimated multiplicity for Centauro([9]) and ours are only from the forward region so that real multiplicity may be higher than this. The figure seems to prove that the multiplicity increases with energy stronger than log s or $\log^2 s$.

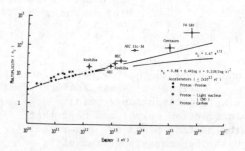

Fig.10. Energy dependence of the multiplicity. Multiplicity of F4-589 is estimated using the relation of $n_c = 2n_{\pi^0}$.

FC-31 (Biggest family event).

FC-31, the biggest family event so far observed at Mt. Fuji, was detected in the chamber FC developed in this August. The thickness 28 c.u. of this chamber is sufficient to get almost all information on the lateral and longitudinal behaviours in the chamber. Followings are only preliminary because the detailed analysis is now in progress.

Fig. 11 is the photograph of this event registered in the type-N X-ray film at the depth of 10 c.u.. As is clearly seen from this photograph, the event is clearly composed of two parts.

That is to say, (1) Big halo with deep-black nucleus and (2) High energy showers accompaning the halo.

(1) Big halo with deep-black nucleus.

Almost all energies of this event is concentrated in this part. This part has clearly two structures on the type-N film. One is the deep-black nucleus where X-ray film of this part became completely black and the other the halo surrounding this nucleus.   At the maximum of 14 c.u., the radius of nucleus on the type-N film reaches to about 1 cm and that of halo extends up to about 2 cm or more.

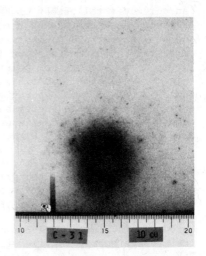

Fig.11. Photograph of FC-31 on the type-N X-ray film at the depth of 10 c.u..

By microscopic scanning of the type-RR films at 4 and 6 c.u., we can find a large number of high energy showers even in the black nucleus. Also about 60 high energy showers ($E_\gamma \gtrsim 2$ TeV) are locally observed in the halo between the radii from 1 to 2 cm.   Thus large amounts of electrons associated these high energy showers would make this halo and nucleus parts.

The optical density of the uniformly blackened part on the X-ray film is approximately proportinal to the number of electrons passing through this area.   The relation is calibrated by use of the uniform electron beam extracted from the electron synchrotron of INS ([2]). Therefore we can construct the lateral electron density distribution making the halo and nucleus part at each depth, measuring the optical density by the microphotometer.   The result is shown in Fig. 12.   It may be noted that the lateral distribution can be roughly expressed by the exponential form as $\exp(-r/r_0)$ except the core region, where r_0 takes the value of 0.45 - 0.55 cm around the maximum of 14 c.u..   The value r_0 is smaller than the one expected from the minimum attenuation length of photon in the case of pure cascade shower , i.e.~1.2 cm([10]).

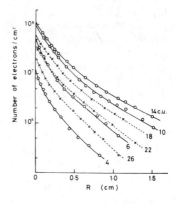

Fig.12. Lateral distribution of electron density at several depth.

Integrating this lateral distribution multiplied by $2\pi r$, we

can get the total number of
electrons passing through the halo
and nucleus part.   The transition
curve thus obtained is shown in Fig.
13.   From the maximum point of
this curve, i.e. $N_{max} \sim 5.5 \times 10^7$
and $t_{max} \sim 14$ c.u., we can roughly
estimate the average energy and
number of γ-rays incident upon the
chamber as follows,

$<E_\gamma> \sim 10$ TeV and $<N_\gamma> \sim 460$.

In Fig. 13 is shown the comparison
between the data and the theoretical
curve of 10 TeV.   Multi-γ-rays
case gives almost same transition
curve with a single γ-ray case if
its average energy is fixed to 10
TeV.

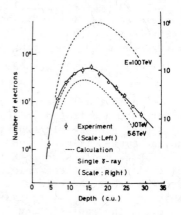

Fig.13. Transition curve of FC-31.

Total area under the transition
curve would correspond to the total
track length Z_0 in cascade theory,
which gives the measure of the total
energies released into the halo and nucleus part, i.e.

$$E_c \cong \varepsilon_0 \cdot Z_0 \sim 4800 \text{ TeV}.$$

Transition curve seems to attenuate exponentially in the deep
depth far beyond the shower maximum as $\exp(-t/\lambda_0)$ with $\lambda_0 \cong 5.0$
c.u..   This λ_0 is of course larger than the one expected from the
pure cascade shower, in which it is approximately equal to the
minimum attenuation length of photon,
$\lambda_{min} = 3.7$ c.u. for lead[10].   This
slow decrease is clearly reflecting
the effect of successively
interacting hadrons contained in this
part.

Also, the value λ_0 might
correspond to the attenuation length
measured in air shower experiment.
$\lambda_0 = 5$ c.u. in lead is equivalent to
~ 190 g/cm² in air.

In Fig. 14 the lateral
distributin at the shower maximum
is compared with the theoretical
curves of age parameter s=1, 1.4 and
2.0 under the Approximation B,
calculated by Nishimura and Kamata
([11]).   At larger radii over 1
Moliere unit (about 1.6 cm in lead)
experimental data seems to fit the
curve $s \sim 1.2$, where s = 1.2 is

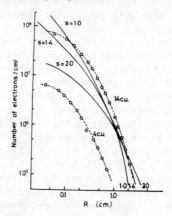

Fig.14. Comparison of the lateral distribution at 14 c.u. with the calculation under the App. B.

estimated from the tail of transition curve by equating $\lambda_0 = \lambda_1(s)$, $\lambda_1(s)$ being the well known cascade function. Deviation in smaller radii is due to the lateral spread of incident γ-rays (average spread at 4 c.u. is about 0.7 cm), which may reflect the transverse momentum.

(2) High energy showers accompanying the halo and nucleus.

Many high energy showers are locally found in the inner and outer region of halo and nucleus.
(a) Outer region $R \geq 2$ cm. 100 showers are detected with energy greater than 1.5 TeV and its total energy sum is 370 TeV. The energy spectrum is of the power form as $E^{-\beta}$, where $\beta = 1.6 \pm 0.1$.
(b) Middle region between R=1 cm and 2 cm. Only high energy showers over 3 TeV are detectable with no bias because of the high background electron density covering this region. The slope of energy spectrum is slightly gentle compared with the one in (a). Total energy sum is about 430 TeV (≥ 3 TeV). It may be remarkable that several high energy hadrons are found in the neighborhood of the nucleus (maximum shower energy is about 65 TeV).
(c) Inner region $R \leq 1$ cm. In this region there are so many showers that it is almost impossible to determine the energy of each shower separately. About 400 showers (≥ 2 TeV) are roughly perceptible in this region.

Adding the energies of all high energy showers detected to that of the halo and nucleus, we can get the total energy of this event, which will exceed 5500 TeV.

Unfortunately we have no direct information about the production height of this big family event. However, if we pay attention to the fact that the average energy of γ-rays making the halo and nucleus is rather high, i.e about 10 TeV, and also that energetic hadronic showers are found mainly near the margin of the nucleus, this event seems to be generated by the main interaction not so high in the atmosphere. This would reduce the possibility that it may be from the interaction of a heavy primary, and consequently make likely very high multiplicity. Here it should be emphasized that the whole behaviour of our event in the chamber bears a striking resemblance to that of " Andromeda " (^{12}) except the total energies. [Andromeda was $\sim 1.5 \times 10^4$ TeV – ed.]

Also, as pointed out in this section, similarities of the behaviour between our event and air shower phenomena would be very suggestive to guess the interaction mechanism dominantly occurring in air shower energy regions.

3˙4 Large P_t ? TITAN and other similar events.

Family events accompanied by hadronic showers as their family members are of peculiar interest as it bring us more direct information about particle production without passing through cascade process. In this section we will describe details of such interesting events in connection with large P_t production.

TITAN, which was detected in the thick chamber FA of 70 c.u., is a family with total energy sum ΣE = 630 TeV ($\geq$ 1.5 TeV), composed of 36 high energy showers. The target diagram is presented in Fig. 15.

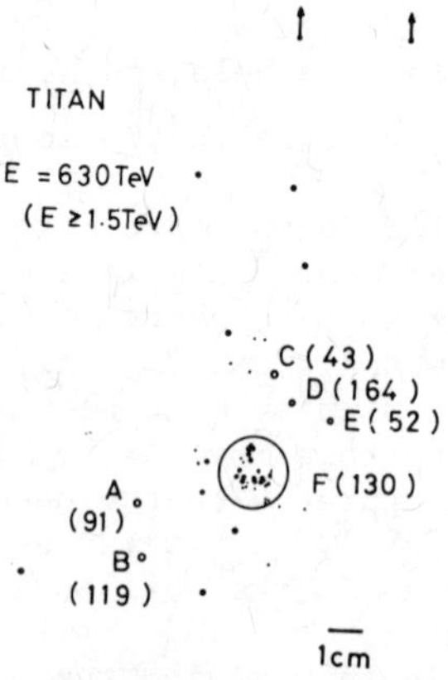

Table 4.

Shower	E (TeV)	Δt (c.u.)	$k_\gamma P_t$ (GeV/c)	
A	91	33	1.1	S
B	119	2	1.8	S
C	43	44	0.5	S
D	164	4	2.1	
E	52	2	0.8	
F	130		0.5	
Total	599		6.8	

S : Successive int.
H = 1 km (maximum estimate)
Δt; starting depth of shower

Fig.15. Target diagram of TITAN. Numeral in the bracket is energy in unit of TeV.

The structure shows conspicuous differences in comparison with γ-ray families so far discussed. That is to say : 5 energetic showers (A - E, ΣE = 470 TeV) lie in the outskirt of a normal γ-ray family (F) with large lateral spread from the center. As discussed in ref. ([13]) and learned from Table 4, three showers (A, B and C) must be hadronic because of their successively interacting development or the deep starting point in the chamber. However, showers D and E are delicate and may be produced by a γ-ray or an electron. If that is the case, their production height may be limited within 1 or 2 c.u. at most, because they come into the chamber without any associated visible showers. Also, provided that both are decay product of neutral meson such as π^o or η^o, then their production height is estimated as 4.4 km for π^o or 1 km for η^o, respectively. But height 4.4 km is too high to investigate decascading of both D and E in the atmosphere, while 1 km may be accessible.

Next we discuss the family F. This is composed of 18 γ-rays with energy greater than 2 TeV. In Table 5 various mean quantities of this family are compared with the one averaged for all γ-ray families in the energy intervals between 100 and 200 TeV ($<\Sigma E_\gamma>$ = 133 TeV). Distinct differences are seen in $<R>$ and $<ER>$.

Table 5.

	ΣE_γ (TeV)	n_γ	$<E_\gamma>$ (TeV)	$<ER>$ (TeV·cm)	$<R>$ (cm)
F	128	18	7.1	0.9	0.3
"Average"	133	23	6.0	13	2.2

$E_\gamma \geq$ 2 TeV.

In Fig. 16 is shown the energy spectrum and the ER distribution of constituent γ-rays in F. The ER distribution is of the exponential form and also the slope of energy spectrum is fairly gentle compared with the equilibrium one (in this case the spectrum is roughly expressed as $E^{-\beta}$, where β = 1.5-1.6).

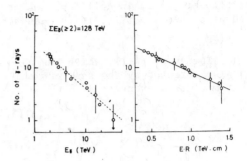

Fig.16. Energy spectrum and ER distribution of γ-rays in the family F. R is the distance from the energy weighted center of F.

All these properties can not be explained by the family F being the pure cascade. Following three cases are probable for the origin of family F. (a) secondary interaction produced by one of several energetic hadrons coming from the same interaction, (b) fairly pure γ-ray family coming from the same interaction point with other energetic hadrons, and (c) normal γ-ray family coming from the first interaction in higher altitude. The case (c) may be excluded from the above discussins. If the family F is pure, then the height of about 40 m would be roughly obtained as the minimum estimate using $<P_{t\gamma}>$ = 160 MeV/c. Therefore, if (b) is the case, the production height of TITAN would be very low as a logical result. Also if 5 energetic showers are all hadronic and (a) is the case, most probable production height of this event would be less than 500 m. Anyway, all discussions done above may support that the production height of TITAN. is to be lower than 1 km.

Fig. 17 shows the ER vector balance, showing clear jet structure. Estimated minimum $k_\gamma P_t$ values are summarized in Table 4. In general the inelasticity k_γ takes a value between 1/3 - 1/6 so that real P_t values will be several times as large as that in this table.

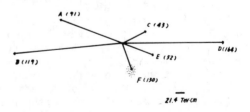

Fig.17. ER vector balance of TITAN. Numeral in the bracket is energy in unit of TeV.

As discussed above, TITAN seems to manifest the production of very high energy hadrons with large P_t in forward region. Its jet structure and also approximate energy balance between (A + B) and (C + D + E) may suggest that this interaction is ascribed to the hard scattering of constituent parton in colliding nucleons ([14]) or the hadronic decay product of heavy particle with mass greater than 10 GeV.

FA-11

The event FA-11, which has almost same structure with TITAN, was also detected by the same chamber FA. Fig. 18 shows the target diagram of this event.

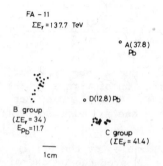

Fig.18. Target diagram of FA-11. Numeral in the bracket is energy in unit of TeV.

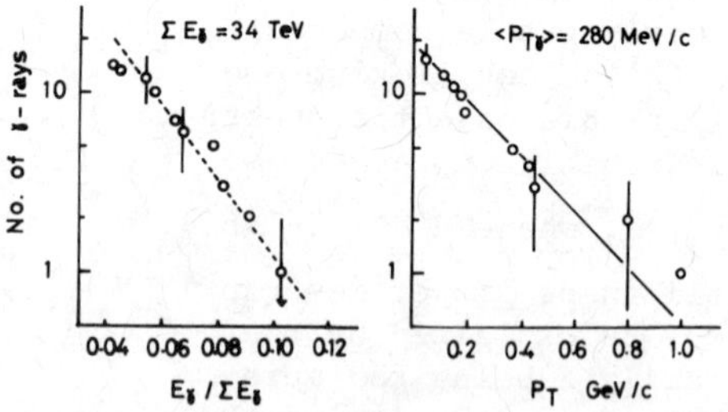

Fig.19. Energy spectrum of γ-rays in a fractional form and P_t distribution of γ-rays in cluster B.

Conspicuous structure is being composed of two hadronic showers (A and B) and two well separated γ-ray families (B and C). The cluster B includes a hadronic shower in their family members.
Fig. 19 is the energy spectrum of γ-rays in cluster B, being of the exponential form. As the multiplicity is not so high, we have tried the $\pi^0 \to 2\gamma$ coupling method to estimate the production height of the cluster B. 5 pairs of γ-rays give quite consistent height with an average value of 30 m. The $P_{t\gamma}$ distribution of γ-rays is presented also in Fig. 19, which are obtained by use of H = 30 m. Average value of $P_{t\gamma}$ is 280 MeV/c being fairly large owing to the detection bias of γ-rays.

Cluster C is not so pure as cluster B because of few cascade processes in air. However the energy spectrum and also ER distribution may be compatible with cluster C coming from the height lower than 1 c.u..

Thus it would be very natural to assume that clusters B and C are γ-ray families made by secondary interaction of energetic hadrons produced at the same altitude with hadrons A and D. In this case, the height of main interaction would be lower than 1 km if we take into account this probability.

Estimated $k_\gamma P_t$ values are summarized as follows,

A : 1.9 , B : 1.9 , C : 1.5 , D : 0.3 GeV/c.

Thus we can find that the event FA-11 resembles to TITAN.

Double or multi-cores

Another peculiar family events which may be closely connected with TITAN and also FA-11 are with double or multi-cores structure. That is to say : γ-ray families composed of two or more clusters mutually separated with a great distance larger than 5 cm, each of which has total energy sum greater than 30 TeV. The criterion defined here is rather artificial, but would be reasonable for clustering if one took into account that the mean lateral spread of γ-ray families is of the order of 2 cm. Such families are observed with a fraction of several % in our concerned energy region.

If several energetic hadrons are produced with large P_t at the altitude high in the atmosphere, then some of those hadrons will develop into γ-ray families with detectable energy until they reach the emulsion chamber. Thus one can have a chance to observe such family events under appropriate conditions. Observation of TITAN and FA-11 may support this picture. Of course the possibilities that each cluster in a family may come from the different interaction point of primary particle, or that large fluctuations may make such structures in a family would be never disregarded. Extensive Monte Carlo calculation will give more reliable answer to us.

Double or multi-cores in air shower experiments would be closely connected with our phenomena. In connection with this, it may be very suggestive that the average P_t and observation rate of multi-cores in air shower are more or less consistent with our emulsion chamber experiment.

4. — Summary.

In spite of the various limitations discussed above, large-scale emulsion chamber experiment can crucially test the gross features of particle production at very high energy region over 10^{14} eV. Following is the summary of our results and discussions.

(1)   The average behaviour of γ-ray families in the energy region over 10^{14} eV may be incompatible with the simple scaling model as far as the primaries are mainly composed of protons up to very high energies. There are several evidences that the interaction mechanism changes with energy, in favour of a more rapid increase of multiplicity at very high energy. See (3).
The average transverse momentum in the fragmentation region would increase very slowly with energy, if any.

(2) f'-spectrum of γ-ray families clearly shows the scaling behaviour in our concerned energy region. However, this does not always mean that the Feynman scaling in the fragmentation region is still maintained up to very high energy.

(3)   The structure of the big family F4-589 shows a fair indication of very high multiplicity. FC-31 and other big families seem to be generated by the same interaction mechanism with F4-589. Especially striking similarities between FC-31 and air shower phenomena are very suggestive to know what happens in ultra-high energy region over

10^{16} eV.

(4) TITAN seems to manifest the production of very high energy hadrons with large P_t in forward region. Particularly the jet structure of this event shows marked distinction compared with other similar events. FA-11 and other family events with double or multi-cores structure would be closely connected with TITAN.
Thus the observation of such events would give us a clue to clarify what mechanism is superior to the large P_t production at very high energy region.

The authors would like to express our thanks to Sengen Shrine, Kawaguchiko Office of The Ministry of Enviroment Maintenance and Gotenba Meteorological Observation Office for extending every facility necessary for carying out the experiment at Mt.Fuji to us.
We are also indebted to the technical staff of ICR, Univerity of Tokyo, Mms E Mikumo, K Sato, M Tsujikawa, S Toyoda and Mr T Kobayashi, and to Miss Y Sato of Hirosaki Univeristy for the scanning of the events or their co-operative efforts.
This work was carried out in part under the support of the Scientific Research Fund by the Ministry of Education.

References.
1. Kasahara K, Torii S and Yuda T, 1977 Proc. 15th Int. Conf. on Cosmic Rays, Plovdiv 7 236.
   Torii S, J. Phys. Soc. Japan, 44 , 1053 (1978).
   Kasahara, see Fujimoto Y et al, Prog. Theort. Phys. Suppl.32 246 (1971)
2. Ohta I et al, 1975 Proc. 14th Int. Conf. on Cosmic Rays,Munich 9, 3145.
3. Konishi E, Shibata T and Tateyama N, Prog. Theort. Phys. 56 (1976).
4. Pamir Collaboration, 1975 Proc.14th Int. Conf. on Cosmic Rays, Munich 7, 2370.
5. Dunaevsky A M et al, 1977 Proc.15th Int. Conf. on Cosmic Rays, Plovdiv 7, 354.
6. Halzen F, Nuc.Phys., B92, 404(1975).
7. Koshiba M, 1963 Proc.8th Int. Conf. on Cosmic Rays, Jaipur 5,293
   Sato Y, Sugimoto H and Saito T, J.Phys.Soc. Japan, 41, 1821(1976).
   Hoshino K et al, Proc. Int. Conf. on High Energy Phenomena, 168 (CRL, Univ. of Tokyo, 1974).
8. Thomé W et al, Nuc. Phys., B129, 365 (1977).
   Hebert J et al, Phys. Rev., 15, 1867 (1977).
   Vishwanath K A et al, Phys. Lett. B53, 479 (1975).
9. Brazil-Japan Collaboration, 1973 Proc.13th Int.Conf. on Cosmic Rays, Denver 3, 2227. Also, these Proceedings.
10. Yuda T et al, Nouvo Cimento, 65A, 205 (1970).
11. Kamata K and Nishimura J, Prog. Theort. Phys. Suppl.No.6, 93(1958)
12. Brazil-Japan Collaboration, 1971 Proc.12th Int. Conf. on Cosmic Rays, Hobart 7, 2775.

13. Akashi M et al, 1977 Proc. 15th Int. Conf. on Cosmic Rays, Plovdiv $\underline{7}$, 184.
14. Bjorken J D, Phys. Rev. D8, 4078 (1973).
    Berman S M, Bjorken J D and Kogut J B, ibid $\underline{4}$, 3388 (1971).
    Field R D and Feynman R P, ibid D15, 2590 (1977)

High Energy Interactions at Cosmic Ray Energies

S.Miyake
Institute for Cosmic Ray Research,Univ.of Tokyo

ABSTRACT

As a summary of super high energy events and exotic phenomena observed in cosmic ray experiments, some trends in very high energy physics are discussed in this paper. The results show that (1) violation of scaling law in interaction character at above 100 TeV, (2) increase of P_t at above 100 TeV, and (3) existence of exisotic phenomena observed deep underground.

INTRODUCTION

Although the cosmic ray experiments cannot provide the kind of detail that is routine with the accelerator experiments, one can glean from cosmic ray experiments some trends in high energy physics using the superiority over the machine in terms of energy. The variety of cosmic ray observations are ; (1) jet showers are observed by emulsion chambers (ECC) at balloon,aircraft and mountain altitudes, (2) extensive air showers (EAS) are observed at mountain altitudes and at sea level, and (3) muons and neutrinos are observed at deep underground. By the use of all those data, one can study hadron-hadron, lepton-hadron and lepton-lepton collisions. Thus,cosmic ray experiments extend to super high energies far above the existing accelerators. The subject to be studied will be to show how super high energy events differ from a simple extrapolation of the interaction characteristics in accelerator energy region.

I. Interaction Characteristics at Cosmic Ray Energies

(1) A Mild Violation of Scaling in 10 - 100 TeV Energy Range

The data of direct observation of nuclear interactions are available in this energy region by ECC. ECC is a device that can detect high energy gamma rays and also hadron jet showers generated in ECC itself provided the ECC is thick enough. Therefore, one uses normally a variable X' ($E_\gamma / \Sigma E_\gamma$) similar to the Feynman variable, frequently used in the accelerator experiments. In order to transform X' to X (Feynman variable),however, one has to know the relation $E_\gamma \rightarrow E_{\pi^0}$ and $\Sigma E_\gamma \rightarrow E_0$. I have estimated these transformations as multiplication factors of about 1.8 for the former and about 0.45 for the latter. In estimating the factor for the lattercase, we have taken into account the selection biases for the events and also the steep nature of the energy spectrum of their parents. As a result, one gets X''= 1.2 X.
For the sake of further discussion,we will take the following empirical relation for the X distribution ;

$$dn = A\, E_0^{2\alpha} \exp(-B E_0^{\alpha} X)\, dX \qquad (1)$$

which implies $\langle n_s \rangle \propto E_0^{\alpha}$, $\langle X \rangle \propto E_0^{-\alpha}$, and that scaling is valid if $\alpha = 0$.[1,2]

Using the above exponential function for X' distribution in the energy region of a few tens TeV, we get $\sim \exp(-7X')$ which is the same as $\sim \exp(-8.4X)$ in terms of X. The coefficient of the exponent is about 6.5 in the accelerator experiments and there is not much difference between these two X distributions. Nevertheless, direct comparison of both will not be correct because the one for cosmic ray experiments is biassed in favour of large inelasticity events. The coefficient of the exponent for the cosmic ray data increases about 10 % with each decade of increase in the primary energy[3]. This, then, suggests that there is a mild violation of scaling characterised by $\alpha = 0.1$ in Eq. (1).

(2) Diffusion of Cosmic Rays in the Terrestrial Atmosphere

In order to study the mild violation of scaling in greater detail, we will use the cosmic ray data on the diffusion of high energy particles in the atmosphere. There are many observations on high energy hadrons and gamma rays using ECC at various altitudes from the top of the atmosphere down to the mountain altitude. Fig.1(a) and (b) compare what is expected on the basis of the scaling model and an energy independent inelastic cross section with the observed results on gamma rays[4] ; (a) for 1 TeV at various depths in the atmosphere, and, (b) the energy spectrum at mountain altitude (650 g/cm²). Deep in the atmosphere, the differences are large not only in the absolute value of the intensities but also in the exponent of energy spectrum. In order to get a reasonable fit between the expected and the observed results, I like to intro-

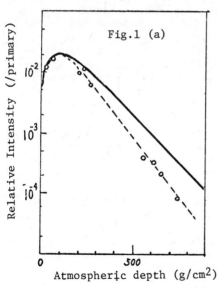
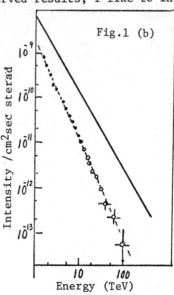

duce following three factors ;
(i) The exponent in the energy spectrum of primary cosmic rays increases smoothly with energy.   As an example, I am using here the form ;
$$I(>E) = 1.0 (E + 2)^{-1.55 - 0.02 \log E} /cm^2 sec\ sr \quad (2)$$
where E is in unit of GeV, for total energy spectrum.   Assuming "normal" chemical composition, the energy spectrum of nucleons will be the form ;
$$I(>E) = 0.6 (E + 1)^{-1.55 - 0.02 \log E} /cm^2 sec\ sr \quad (3)$$
In this case, about 80 % is a proton component and rest of them are nucleons bounded in the nuclei of alpha particle or heavier nuclei of the primary cosmic rays.   The spectrum implies a steepening of the exponent by 0.04 for each decade of energy.   The exponent of the relevant primary energy region becomes steeper, the deeper one goes in the atmosphere as given by ;
$$\Delta \gamma = 0.04 \log [1/(1-k)]^z \quad (4)$$
where k is inelasticity and z is the depth in the atmosphere in units of nuclear mean free path.   This is because the primary energy responsible for the events is higher at observations deep in the atmosphere.
(ii) Inelastic cross section increases slowly with energy as ;
$$\sigma = \sigma_0 (1 + \delta \ln E) \propto \sigma_0 E^\delta \quad (5)$$
which gives a change of the exponent $\Delta \gamma = \delta z$, for instance $\Delta \gamma = 0.08$, at mountain altitudes (650 g/cm^2) if we use $\delta = 0.01$ including the effect of energy dependence on k (inelastisity).
(iii) Scaling is violated mildly with increasing energy as shown in the previous section.   This results in a steeper energy spectrum for secondary particles than that of primaries ; $\Delta \gamma = (\gamma - \alpha)/(1-\alpha) - \gamma$

Taking the above three factors into account, one can get nice fits to the experimental observations in the atmosphere ; thus the problems shown in Fig. 1 are solved.   Nevertheless, it must be pointed out that such a trend especially the factor (iii) in the above, must be started at around a few tens GeV to get a nice fit in the absolute intensity of observations.   The most plausible values for the parameters from the above exercise are ; $\alpha = 0.1$ or slightly higher value but less than 0.2, and, $\delta = 0.01$ or slightly higher value but not more than 0.02.   The former agrees well with the conclusion of the previous section, and the latter also seems to be reasonable in the energy region above 1 TeV.   Although the experimental results at low energies around 1 TeV suggest, rather rapid increase of cross section[5], the increase obtained here for high energy region must be a very slow one -- a conclusion forced upon us by detailed studies of diffusion of cosmic rays in the atmosphere.

In the analysis of this section, the limiting fragmentation hypothesis had been employed, which means that there is no distinction between the collisions proton-air nucleous and proton-proton as far as concerned the character of secondaries in the fragmentation region which is major component observed in cosmic ray experiments because of very steep energy spectrum.

(3) Strong Violation of Scaling Law at more than 10^{14} eV

In the energy region above 10^{14} eV, it is difficult to get direct information similar to the lower energy region because of the low rate of events. Therefore, people use gamma ray families in ECC exposed mountain altitudes which are the results of high energy hadron-air collisions in the atmosphere (called as A-Jet), and also the data of extensive air shower (EAS) to estimate the nature of super high energy interactions. The energy spectrum of gamma ray families observed at mountain altitude (total energy of the group of Gamma rays produced A-jet) is a form of $E^{-1.35}$ in integral[6]. The difference of this energy spectrum from ordinary energy spectrum of the exponent about 1.7 in the upper atmosphere, will be explained by the difference in average production height which is energy dependent as can be seen in Fig.2. The mean production height, $E^{0.35} \propto \exp \Delta h/\lambda$ h, is higher by about 80 g/cm^2 (attenuation length in this energy region) for every decade of higher energies. For example, the mean production height for 100 TeV is about 1.5 km above the observation level, and it will be about 3 km and 4.5 km for 1000TeV and 10000 TeV respectively. Because of this different mean production height, one can not compare with each other straight; however, there are wide distribution in production height for each energy range, so that, one can get some idea to find out some trends of energy dependency of interaction characteristics with energy.

Looking at some typical examples obtained by ECC as shown in Fig.3 (a) to (d), with total energies of gamma rays of about 100 TeV, 500 TeV, 1500 TeV and 5000 TeV respectively, one will find that each of these figures has particular features ; (a) the event consists of several secondary particles of energy of the same order, as expected from scaling law. (b) The energy in this events is much higher than in the case (a) ; besides the secondaries similar to those in (a) there is a group of very low energy (pionization like) gamma rays near the center of the event. (c) The group of low energy gamma rays near the center grows up in size as if robbing away the energy from the scaling type high energy particles. Finally, (d) at the highest energy region, one can not see any superior energetic particles at all in the event but finds only a homogeneous distribution of innumerable particles of small energy (lateral

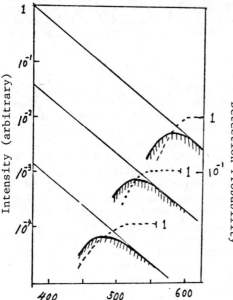

Fig.2 Distribution of production height for 3 energies.

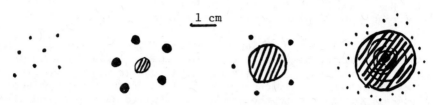

~100 TeV   ~500 TeV   ~1500 TeV   ~5000 TeV

Fig.3 Schematical diagram of ECC A-jets (at 10 c.u.) distribution has a peak at the center of the group).

From these examples, one sees, therefore, a clear trend of change in the hadron interaction characteristics with energy ; the scaling law may be valid in the low energy region up to around 10^{14} eV, and it is strongly violated in the very high energy region above 10^{15} eV.

There may be another type of explanation for above events saying that since the production height is higher for larger energy events, successive collisions of secondary particles with air nuclei may make such features explained above.   However, it is difficult to explain the lateral distribution of energy flow of the gamma ray group.

In the energy region higher than 10^{15} eV, the only way to study such high energy phenomena is by the observation of extensive air showers (EAS).   The EAS phenomenon is complicated and one observes only the tail which is quite far from the first interaction near the top of the atmosphere.   Nevertheless, from a comparison of observations with various expectations, the generally accepted view is that the interaction at very high energies is very hard at the earlier stages of the development of EAS and the total energy of primary particle is split into innumerable secondaries of rather low energy.   Thus, the maximum in the development of EAS is high in the atmosphere and the height does not come down so fast towards the ground with increasing EAS energy.   To understand these features of EAS, one adapts one of the three following views : (i) high multiplicity model $n_s \propto E^{1/2}$ at energies higher than 10^{14} eV, (ii) primaries are changing their chemical composition from proton to heavier nuclei like iron, then, the fast dissipation of energy may be understandable even in the frame work of scaling law, and, (iii) not only the violation of scaling but the nature of secondaries produced also changes from mesons to baryons at very high energy region.

The maximum of the transition curves of EAS in the atmosphere is at around 300 - 400 g/cm^2 and moves down slowly with increasing primary energy.   The shape of the curve is similar to that of single electro-magnetic cascades of 10^{12} - 10^{13} eV even for energies of EAS of 10^{16} - 10^{17} eV.   The arguments (i) and (iii) can explain the characteristics but (ii) will be difficult although it may explain only for a narrow band of energy.   Without going into a detailed discussion, I believe that there is a general agreement among EAS researchers high multiplicity model close to $E^{1/2}$ is one of the best solution to understand EAS phenomena[8].   The fast degradation of energy into large number of particles is thus supported by many groups studying A-jet by ECC and EAS in the energy range from 10^{14} -

eV to 10^{18} eV.

(4) Average Multiplicity Vs Energy

All the discussions in the previous sections on the average multiplicity vs energy are summarised in Fig.4. In the lower energy region in the figure, $< 10^{14}$ eV, the interactions are more or less similar to scaling type though the onset of a mild violation of scaling can be seen in the energy region close to 10^{14} eV.

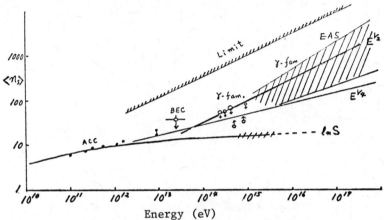

Fig.4 Average multiplicity vs energy

The continuation of scaling type interaction into higher energy region will have to be terminated at around 10^{15} eV. In the higher energy region, $> 10^{14}$ eV, fluctuations in n_s progressively increase with energy ; also the average number increases as $E^{1/2}$ starting from 10^{14} eV. In the super high energy region, 10^{15} eV - 10^{18} eV, it is difficult to show exactly the experimental points on this diagram ; so, probable range are shown by the shaded area along the line of $E^{1/2}$ with some width. A limitation line in the figure is just showing $E^{1/2}$ line start at 1 GeV which means that total available energy has been spent to produce large number of low energy pions.

(5) Average Transverse Momentum, P_t, Vs Energy

Fig.5 shows the relation of average transverse momenta vs primary cosmic ray energy, together with some accelerator data. In the figure, cosmic ray data tend to show smaller P_t values compared with accelerator data ; for, P_t in cosmic rays is estimated with respect to the energy center of the secondary particles which leads to a systematically low P_t. In general, P_t rises very slowly (logarithmically) with energy from 0.3 GeV/c at 10^9 eV to 0.4 GeV/c at around 10^{12} eV. Nevertheless, it increases with energy rather fast at energies beyond 10^{14} eV. The points plotted in this higher energy region are all from cosmic ray observation ; four points are taken from the observation of parallel muons deep underground[9]. The average muon energies are estimated from range-energy relation. The average distance from the central axis of the distribution is estimated from the distribution of decoherence curve which have a form of function of $\exp(-r/r_0)$. And, the production height is assumed to be 20 km above the ground level. The average energy of primaries is taken as $50 \times E^{0.2}$ (TeV) times of the average muon energy arriving at the depth. The energy estimated may be different by what model has been used to estimate primary energy of these parallel muon events.

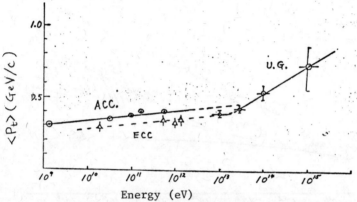

Fig.5 Average transverse momentum vs energy

(6) Production Mechanism of High Energy Muons

Energy spectrum of cosmic ray muons can be measured directly by using magnetic spectrometers up to about 10 TeV. Somewhat indirect method on super high energy muons up to 100 TeV, is based on underground observation of muon ranges ; such as India-Japan collaboration in Kolar Gold Field.[10] The energy spectrum of muons can be deduced from Depth-Intensity relation using an energy loss equation and a correction for range fluctuations by bremsstrahlung. The energy spectrum of muons thus obtained for vertical direction is expressed as $5.0 \times 10^{-8} E^{-2.7}$ cm^{-2}sec^{-1}sr^{-1} (E in TeV).

Upto energies of nearly 1000 TeV, one can calculate primary energy spectrum starting from the muon energy spectrum under the assumption of limiting fragmentation and Feynmen scaling. One finds that the primary spectrum so derived almost coincides with direct observation of primaries[11] in the absolute intensity as well as the slope. (mild violation of scaling is not important at low energies where direct observation exists).

It is well known that there is a direct production of muons in hadron interactions with a ratio of about 10^{-4} with respect to pion production in the accelerator energy region[12]. If such process dominates muon production in cosmic rays, there must be a change of exponent in muon spectrum from 2.7 to 1.7 at energies where such process dominates, because muons produced directly will have the same exponent of primaries, 1.7, whereas muons resulting from the decay of pions or kaons will have a steeper exponent by - 1 which is

brought out by the competition between the decay and interaction of
the parent mesons.   However, in the observed muon energy spectrum
in the vertical direction up to 100 TeV, there is no appreciable change in slope.   From this fact, one can set an upper limit for the
direct production of muons as 2×10^{-3} of pions at primary energy of
several times of 10^{14} eV.   It means that the ratio of direct production mechanism does not increase with energy faster than $E^{1/2}$.

(7) The Other Possible Change of Interaction Character at 10^{17} eV

Fig.6 shows the average age parameters of EAS determined
from lateral structure function, vs size of EAS observed at mountain
altitude (2770 m.).   The value of s is expected to decrease (EAS
are younger) with increasing energy, analogous to a single electromagnetic cascades.   One finds from the experiments, however, two
breaks in the EAS size dependence of s, one at the shower size 10^5
and the other at close to 10^8 (corresponding to about 2×10^{14} eV
and 6×10^{16} eV respectively[13]).   The trend of the change is that
the age become older all of a sudden and then starts becoming young
again with increasing size.   The first change at the smaller size
supports the change of interaction at the energy resulting in high
multiplicity as discussed in the previous section.   It is not
clear why the second change at about 10^{17} eV is occuring ; but,since
the trend in the change is similar to the first one, a large number
of heavier particles may be produced at this energy region.

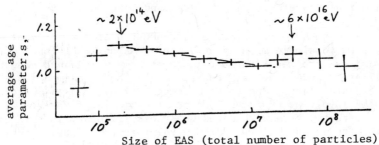

Fig.6 Average age parameter vs size of EAS

II. Exotic Phenomena Observed Underground

(1) Possible Example of Heavy Lepton Events[14]

In the deep underground observation for neutrino detection
in K.G.F.(India), six events in which two or three penetrating particles have a vertex in the air have been reported.   Fig.7 shows
one of these and the most plausible interpretation will be as follows ; (a) cosmic ray neutrino collides with iron nucleous in the magnetic iron core producing an unstable charged particle among others
(b) the unstable charged heavy particle decays in flight into three
charged particles, and (c) one decay product, a charged particle,
decays again into three charged particles in the air.   It may be

written as ;
$$M_1^\pm \to M_2^\pm + \mu^+ + \mu^- \quad \text{and} \quad M_2^\pm \to \mu^\pm + \mu^+ + \mu^-$$
Mass and lifetime of these particles are estimated as ; M_1 is ~ 8 MeV and $\sim 10^{-10}$ sec, and M_2 is ~ 3 GeV and $\sim 10^{-9}$ sec, respectively. If these are heavy leptons, the original neutrino is also probably a heavy leptonic neutrino.

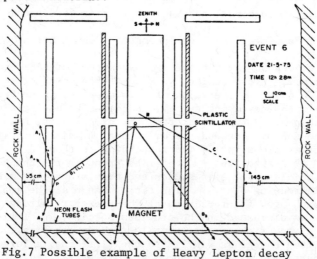

Fig.7 Possible example of Heavy Lepton decay

(2) Anomalous Showers in Deep Underground Observations[15]

In the K.G.F. experiments conducted at a variety of depths, in addition to single muon events, electromagnetic cascades of different sizes were recorded in telescopes comprising lead absorbers, scintillators and neon flash tubes. Most of these cascades are small and are clearly due to electromagnetic as well as photonuclear processes involving muons, but there are some large showers with energy content $\geq$ several hundred GeV whose frequency is much higher than the predicted values. General features of the events are ; (i) high energy of the cascades: the order of TeV, (ii) broad angular distribution (large zenith angle), and , (iii) frequency of shower generation is independent of depth. These features are difficult to understand in terms of conventional knowledge on muons which are vertically collimated deep underground.

The most plausible interpretation of the cascades may be $\tilde{\nu}_e + e^- \to ?^- \to e^- + \tilde{\nu}_e$, because of the high threshold energy of the events. If this is the case, the production cross-section of the events is estimated as the order of 10^{-32} cm^2/Nucl.

(3) Bundle of High Energy Muons Observed Underground

In K.G.F., the observation of muons by means of muon detector of 12 m^2 in area which comprised of proportional counters, neon flash tubes and iron absorbers is now operating at a depth of 3375 hg/cm^2, corresponds to about 2 TeV. There are many events recorded as multi-parallel-muons which are the tail of high energy interactions near the top of atmosphere. The number spectrum of such parallel muons observed has a steep distribution and most of these events are less than 6 muons. However, there are 4 events which contain 10 - 12 muons for each, and the rate of these events is quite high compared with the extrapolation from lower muon numbers in the spectrum.

The speciality of those 4 events, although the statistics is still not enough, are the followings ; (a) zenith angle is small $< 20°$, (b) lateral spread is small and essential part of the event is within a circle of 2 m diameter, and (c) parallelity of muon tracks is good and it can not be considered as local products in rock. The rate of such muon bundles corresponds to the rate of primary cosmic rays of the order of 10^{16} eV, therefore, there may be a special production mechanism of such muons above the energy.

The underground experiments will be continued still at various depths in K.G.F. as the collaboration project of India and Japan. And, the exotic phenomena reported in this paper will be studied in detail near future.

REFERENCES

1) J.Benecke et.al. Phys. Rev. Vol. 188 No.5 2159 (1969)
2) R.P.Feynman Phys. Rev. Lett. 23 1415 (1969)
3) see ECC papers in this symposium
4) K.Kasahara and Y.Takahashi Prog. Theor. Phys. 55 1896 (1976)
5) Y.Takahashi et. al. in this Symposium
6) Mt. Fuji group in this Symposium
7) T.Yuda Private communication
8) T.K.Gaisser Rapporteur paper 15th ICRC Plovdiv (1977)
9) M.R.Krishnaswamy et. al. 15th ICRC Vol.6 161 (1977)
10) M.R.Krishnaswamy et.al. 15th ICRC Vol.6 95 (1977)
11) M.J.Ryan et.al. 12th ICRC 1 173 (1971)
    N.L.Grigorov et.al. 12th ICRC 5 1746 (1971)
12) L.M.Lederman Phys. Reports 25c 151 (1976)
13) S.Miyake et.al. 15th ICRC Plovdiv (1977)
14) M.R.Krishnaswamy et al. Pramana Vol 5 No.2 59 (1975)
15) M.R.Krishnaswamy et.al. 15th ICRC Vol.6 137 (1977)

SPECULATIONS CONCERNING POSSIBLE NEW PHYSICS FROM EXPERIMENTS ABOVE 10TeV[*,†]

G.L. Kane
Randall Laboratory of Physics
University of Michigan
Ann Arbor, MI 48109

We can classify the kinds of new physics which we can hope to learn from experiments at higher energies -- from cosmic rays and from colliding beam accelerators -- into two categories.

(A) <u>Standard experiments.</u> These include a number of important areas which have been well studied and reviewed already, such as (a) measurement of σ_{TOT}, scaling and scaling violations, and related "$\ell n s$" physics; (b) production of gauge bosons $W^{\pm}, Z^{\circ}$; and (c) large p_T data resulting from basic constituent interactions. Since these have been well covered I will on the whole not discuss them here.

(B) <u>Other</u>. There are a number of possible new kinds of effects. Of course, one could speculate without constraint, but today one has available viable gauge theories of strong, weak, and electromagnetic interactions and it seems preferable to speculate within the context of these new effects. It may be that none of the following speculations will be realistic and we are only free to discuss them at the moment because our understanding of the theory is minimal. Even so, some good, either theoretical or experimental, may come from such considerations.

The topics I will discuss are
(1) Possible large "weak interactions" at very high energies, with an associated cross section that is a significant part of σ_{TOT}; what characteristics would enable us to detect such new effects?
(2) Can we learn some things about QCD and gluons from small p_T, large cross section experiments?

†Research supported in part by the U.S.D.O.E.

(3) At large p_T but very high energy, cross sections in QCD get very large.

(4) Some new heavy meson excitations are suggested by increased understanding of the gauge theories and could appear experimentally.

LARGE "WEAK INTERACTIONS" AT VERY HIGH ENERGIES

What happens at very high energies in the unified gauge theories of weak and electromagnetic interactions? First, consider the Weinberg-Salam theory,[1] with a single Higgs boson doublet to provide the masses of the gauge bosons ($W^{\pm}$, Z^0) and the fermions, via spontaneous symmetry breaking. Suppose the Higgs boson masses are not large, i.e. $m_H \lesssim m_W$. Then it can be shown that no weak interaction effects get large. At very large p_T, for example, the weak and electromagnetic cross sections will be comparable in size. The weak interaction will be relatively isotropic, so it is a tiny part of the cross section at small p_T, and thus a tiny part altogether.

However, if $m_H \gg m_W$, some amplitudes in the boson sector become larger[2,3], because the coupling strengths are proportional to m_H. Then one can imagine obtaining large cross sections. It is interesting to consider ways to observe possible "weak" effects at very high energies. Before we do that, briefly consider how such effects could be imagined to arise.

Consider Higgs boson scattering via an s-channel Higgs,

giving an amplitude

$$M = \frac{9}{4} g^2 \frac{m_H^2}{m_W^2} \frac{m_H^2}{s-m_H^2} = 18 \, (G_F \, m_H^2) \, m_H^2/(s-m_H^2)$$

and a cross section

$$\sigma = \frac{81}{4\pi s} \, (G_F \, m_H^2)^2 \, \frac{(m^2 H)^2}{(s-m_H^2)^2} \, .$$

If we assume $m_H = 1$ TeV $= \sqrt{s}$, an energy which can be achieved by the colliding doubler at FNAL even for constituent subenergies, and put $s-m_H^2 \simeq m_H \Gamma_H$, then $\sigma \simeq 260$ mb$/\Gamma_H^2$ with Γ_H in GeV. Such a cross section greatly exceeds the unitarity limit, since it arises from a single partial wave. Similar results arise for other diagrams. Consequently they will be suppressed. Nevertheless, it is conceivable that the dynamics gives rise to amplitudes from the Higgs sector which are large enough to give observable effects, directly or through interferences with strong interactions.

Is this relevant to experiments? Here a subtle complication arises -- in the simplest Weinberg-Salam theory, with one Higgs doublet, even if large effects did arise the answer is no.[2,4] To do experiments one must begin with fermions; e.g. $q\bar{q}$ or e^+e^- collide (the $q\bar{q}$ in hadrons). Then all of the large amplitudes get multiplied by the tiny factor $(m_f/m_W)^2$, essentially because that factor appears in the Higgs-fermion coupling. One can make a strong statement: in the standard theory at high energies there will be no weak interaction effects not directly due to production of W's and Z's; observation of such effects would contradict the standard theory.

Is this general? No. One can construct theories[5,6,7,8] which differ only in the Higgs sector, in which the Higgs-fermion coupling is no longer suppressed by the factor m_f/m_W. In essence one needs two Higgs doublets, one having vacuum expectation value v and giving mass to fermions, with $v \sim m_f$, and the other having vacuum expectation value V and giving mass to the gauge bosons, with $V \sim m_W$. No contradiction with experiment arises[8] for Higgs-fermion couplings up to about two orders of magnitude times that of the standard theory. While an extension of the standard theory along these lines is not required by any data, it is of interest to have the boson and fermion mass scales set separately in the Higgs sector. If nature were like this extension it would be relatively easy to detect Higgs bosons and their effects, and -- most relevant here -- it is not prohibited from giving large cross sections associated with the Higgs sector at high energies.

So far I have argued that even within the context of the conventional gauge theories it could happen that large cross sections arise from effects associated with the Higgs sector. The arguments are meant to stimulate our thinking rather than to lead to definite conclusions; they certainly do not imply that such effects will be present. More important is the question: how could we detect large Higgs-related phenomena if they were present?

To give some answers to this, consider the characteristics of normal strong interactions at present experimental energies:
   (a)   Hadrons have a certain size, and differential and total cross sections are qualitatively fixed by geometrical arguments, $\sigma_{TOT} \simeq \pi R^2$, $d\sigma/dt \simeq \sigma_{TOT}^2 \exp(R^2 t)$ $d\sigma/dt$ also has a diffraction minimum.
   (b)   Isospin is a good symmetry
   (c)   Parity is a good symmetry
   (d)   Large p_T events are very improbable
   (e)   Increased energy input goes mainly into relative motion, not into particle masses and increased multiplicity. Consequently, multiplicity grows as $\ell n s$.

If a new interaction (e.g. related to the Higgs sector as above) becomes important at very high energies there is no guarantee that it will respect any of these results. Consider them in turn.

   (a)   If a new interaction is present it could have a distance scale which is not connected to normal hadronic sizes. Then σ_{TOT} could change[9] fairly abruptly to a new value; presumably only an increase will occur as a different distribution in partial waves would be involved and cancellations are unlikely. Similarly, $d\sigma/dt$ could broaden significantly if a part of the cross section involved scattering on a shorter distance scale. (In this context it is amusing that recent data[10] shows a relatively isotropic $d\sigma/dt$ for pp elastic scattering at 400 GeV/c, with $d\sigma/dt \sim \exp(0.7t)$ for $-t \simeq 10$ GeV^2, while for small t, $d\sigma/dt \sim \exp(12t)$.)

   (b)   Since Higgs-quark couplings are proportional to the quark mass, and[11] $m_u \neq m_d$, the Higgs-quark interaction does not conserve isospin. If it led to large high energy interactions, one might find large isospin violations at such energies. For example, one could find

$$\sigma_{TOT}(pp) \neq \sigma_{TOT}(np) \neq \sigma_{TOT}(pd)/2$$

or

$$n(\pi^\circ) \neq (n(\pi^+) + n(\pi^-))/2$$

with similar results for other resonances. At present there is no way to estimate what effects to expect (if any); the key point is that such effects could occur and we should be alert to find them.

(c) In a similar way it could happen that large parity-violation effects occur at high energies. For example, to give an illustrative mechanism, in the model of Ref. 8 the charged Higgs can be quite massive and it couples only to left-handed fermions. There are several ways[12] to detect such effects.

-- If a Λ is detected, it has a decay angular distribution

$$W_\Lambda(\theta,\phi) = N \left\{ 1 + \alpha \sin\theta \sin\phi \, P_y + \alpha \sin\theta \cos\phi P_x + \alpha \cos\theta P_z \right\},$$

where θ, ϕ are the decay angles in the Λ rest frame, $P_{x,y,z}$ the components of the Λ polarization, and α the measured asymmetry, $\alpha \simeq 2/3$. If parity is conserved in the production of the Λ, then $P_x = P_z = 0$. Thus the presence of the $\sin\theta \cos\phi$ or $\cos\theta$ terms would require parity violation in the production. A non-zero P_z could occur from direct "weak" production of the Λ, while a non-zero P_x would have to arise from interference of a parity-violating and a parity-conserving amplitude of non-zero radative phase. Similar results[12] hold for production of ϕ, K^*, ρ, Δ, ... In general it is possible to learn from the strong decay angular distribution of a resonance if its production was parity violating.

-- One can form parity-violating observables such as

$$\langle \vec{p}_{\pi^+} \times \vec{p}_{\pi^-} \cdot \vec{p}_{beam} \rangle .$$

and look for non-zero values.

-- Consider a large p_T quark jet, which fragments into two (π_a, π_b) or more pions. Then the angular correlations of $\vec{p}_a$, $\vec{p}_b$ may allow the determination of longitudinal and transverse polarizations of q. (13)

(d) A new mechanism associated with Higgs or "weak" interactions could presumably lead to different behavior at large p_T. This could get confused with the changes in large p_T behavior expected from QCD when hard scattering of constituents takes over (see below), and it will be important not only to look for changes at large p_T but to check whether the usual strong interaction symmetry laws apply, as expected in QCD.

(e) Finally, if a new mechanism enters there is no reason for the energy dependence of the multiplicity to stay the same. There could also be different multiplicity growth in different regions. Of course, it could happen that accidently two different mechanisms gave the same growth, or, multiplicity changes could be an easily detected clue to new phenomena.

How likely is any of the above scenarios? Although they are plausible, presumably they are not very likely. But the important thing is that the absence of any new effects at very high energies would also tell us a great deal about how the theory develops. Since good theories exist now, it is of great value to confirm their predictions, and to help guide their application in areas where calculations are difficult.

Low p_T QCD

Next turn to possible ways of learning about gluons and QCD from small p_T information. This is not as good as looking directly at perturbative QCD arguments and experiments directly related to qq or qg collisions. However, there may be useful physics to be learned, and a good deal of the total cross section is relevant

to this region so it merits serious study.(14,15).
I'll mention three possibilities.

(a) It may be that single gluons give rise to the central region clusters.(14,15) An important test of this is their flavor neutrality if they originate solely from gluons. Theoretically, it will eventually be possible to calculate multiplicities, mass distributions, rapidity gap distributions in terms of QCD quantities. These will have some s-dependence -- at higher energy, are there more clusters with energy-independent characteristics, or do the characteristics change too?

(b) An important possibility may be the opportunity to directly measure the gluon distribution function in the most important region of x, around $x \simeq 1/3$. To see this, consider particle ratios in central region production. Since hadrons from gluons will be flavor neutral, ratios such as π^+/π^-, K^+/K^-, $\bar{p}/p$, $\bar{\Lambda}/\Lambda$ must go to one if the hadrons come from gluons.

Normally it is very hard to measure the gluon distribution function, i.e. the probability of finding gluons with a fraction x of the hadrons momentum $G(x)$, because weak and electromagnetic currents do not interact with gluons. Consider the curves in Figure 2,

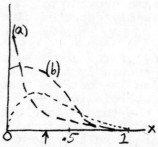

where the dashed lines show two possible gluon functions (a), and (b), and the dotted line the valence quark distribution. Look at $x \simeq 0.3$ (the arrow). Then for distribution (a) the quarks dominate over the gluons, and the particle ratios will not be unity, nor will they get closer to unity as s increases and the gluons have more opportunity to dominate the central region. But for distribution (b), the gluons dominate over the quarks, and as s increases the particle ratios will get closer to unity at this x value. At each x, we can determine whether $G(x) \gg q(x)$, $G(x) \simeq q(x)$, or $G(x) \ll q(x)$.

Since $q(x)$ is measured in lepton reactions, and the area under $G(x)$ is known from the momentum sum rule, this effectively allows a useful determination of $G(x)$. Data at current energies suggests that such a procedure may work at higher energies.

(c) An important prediction of QCD is that gluon-gluon interactions will be strong, and that probably resonant states -- called glue-balls -- will be formed. Although this subject is not well understood[16] most work indicates that σ_{gg} will be large, with several low-spin, low-mass resonances. Perhaps it will look as in Figure 3, where the dashed line indicates the unitarity limit.

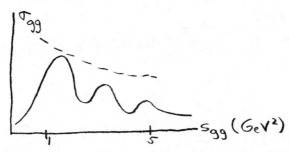

Solutions to QCD will have definite predictions for σ_{gg} as well as the gluon distribution functions. One possible indirect way to probe these is via the behavior of the very high energy total cross section. If we assume part of σ_{TOT} is due to gluon-gluon interactions, we can write that part as

$$\Delta\sigma_{TOT}^G = \int dx_1\, dx_2\, G(x_1)\, G(x_2)\, \sigma_{gg}(x_1 x_2 s)$$

Where $G(x_1)$ gives the probability of finding one gluon with momentum fraction x_1 in one hadron, and similarly for $G(x_2)$. One can argue[15] that the observed rise in σ_{TOT} is due to such a mechanism. For example, at low energies and typical x values, σ_{gg} is small. At higher energies σ_{gg} is in the region of its unitarity limit and $\Delta\sigma_{TOT}^G$ is large. If $G(x) \sim 1/x$ for small x, as is expected, then $\Delta\sigma_{TOT}^G$ grows as ℓns ($\ell n^2 s$) if σ_{gg} falls (becomes constant) at large s. The absolute size calculated from the above formula is consistent with what is needed to explain ISR data, assuming a σ_{gg} with a few low energy resonances. Thus eventually, as we understood QCD better, it may be possible to predict the size and s-dependence of σ_{TOT} at very high energies in terms of low energy gluon properties.

LARGE p_T QCD

Finally I will give a few remarks from a conventional large p_T QCD viewpoint. These are based on calculations from R. Field[17]. The main point is that at very high energies QCD predicts that the large p_T cross section will be rather large. FOr example, the inclusive π° cross section has the property

$$E\frac{d\sigma}{d^3p} (\pi^\circ, p_T = 30 \text{ GeV/c}, 90^\circ, \sqrt{s} = 500 \text{ GeV})$$

$$\simeq E\frac{d\sigma}{d^3p} (\pi^\circ, p_T = 6, 90^\circ, \sqrt{s} = 19.4).$$

Similarly, the cross section for pp → a jet with $p_T > 2$ GeV has risen from of order 1% of σ_{TOT} at (say) $\sqrt{s} = 15$ GeV, to as much as 40% of σ_{TOT} for $\sqrt{s} = 800$ GeV.

These cross sections are large because the cross section grows and approaches a p_T^{-4} fall off as the energy increase. This is due in part to the asymptotically free nature of the theory. It is also kinematic in part, with interactions occuring at parton x values more toward the maxima of the parton distributons as more energy is available and the kinematic restrictions matter less.

It will be of great importance to check these rather dramatic predictions.

HEAVY MESONS

Just as glueballs arise from the interactions of the quantum in QCD, analagous excitations will be found in Weinberg-Salam theory. Recently, Nambu[18] has shown that classical solutions representing monopole-antimonopole configurations may be stable or nearly so and may give particle or resonance solutions. Einhorn and Savit have shown[19] that stable configurations of Z°-flux may exist as particle states (Z°-balls). These could decay into several $Z^°$'s plus whatever states $Z^°$ is strongly coupled to, such as fermion-antifermion pairs.

So far, the arguments used by the above authors have been made for classical solutions, which correspond to highly excited states. However dimensional arguments alone indicate that the masses would be large, of order 1 TeV. Whether production cross sections will be large enough to lead to observation of such states is not known, but is not excluded. Both theoretically and experimentally, the existence of such states in the spectrum of gauge theories must be taken seriously.

CONCLUSIONS

Even within the framework of the conventional gauge theories it could happen that rather surprising behavior is in store at very high energies, including significant changes in σ_{TOT} and $d\sigma/dt$, and large parity or isospin violations. On the other hand, perhaps no new kinds of effects occur. The latter alternative should not be considered boring; it has implications for the development of the gauge theories that are just as interesting and important as if some dramatic new behavior is found.

ACKNOWLEDGMENTS

I would like to thank the organizers, especially T. Gaisser, for their hospitality and for the opportunity to speak on this subject, and M. Einhorn for helpful conversations.

REFERENCES

1. S. Weinberg, Phys. Rev. Lett. $\underline{19}$, (1967) 1264; A. Salam, in *Elementary Particle Physics*, ed. by N. Svartholm. (Almquist and Wiksells, Stockholm 1968).
2. M. Veltman, Acta Physica Polonica $\underline{B8}$, (1977) 475.
3. B.W. Lee, C. Quigg, and H.B. Thacker, Phys. Rev. $\underline{D16}$, (1977) 1519.
4. H. Haber and G.L. Kane, Nuc. Phys. B, in press; UM HE 78-19.
5. F. Wilczek, Phys. Rev. Lett. $\underline{39}$, (1977) 1304.
6. J.E. Kim and G. Segre, University of Penn. Preprint UPR-0094T.
7. P. Ramond and G.G. Ross, Cal Tech preprint LACT-68-674.
8. H.E. Haber, G.L. Kane, and T. Sterling, preprint UM HE 78-45.
9. Through dispersion relations there will be constraints on how rapidly and how much σ_{TOT} can change. These have been considered by Pomeranchuk. See, for example Nuc. Sci. Abs. 30490; "Bounds on the Range

of Increase of Weak Interaction Cross Sections", Pomeranchuk, I. Ya., Yadem. Fiz:, $\underline{11}$, 852 (1970).
10. S. Conetti et al., Phys. Rev. Lett. $\underline{41}$, 924 (1978).
11. S. Weinberg, "The Problem of Mass", A Festschrift in honor of I.I. Rabi, (N.Y. Acad of Sc., 1977).
12. H.E. Haber and G.L. Kane, Nuc. Phys. B, inpress; UM HE 78-11.
13. G.L. Kane, J. Pumplin, and W. Repko, UM HE 78-29, submitted to Phys. Rev. Lett.
14. L. Van Hove and S. Pokorski, Nuc. Phys. $\underline{B86}$, (1975) 243; L. Van Hove, Acta Phys. Pol. $\underline{B7}$, $(\underline{1976})$ 339.
15. G.L. Kane and Y.P. Yao, Nuc. Phys. $\underline{B137}$, (1978) 313
16. For a recent summary of work, see G.L. Kane, Proc. of the XIII Annual Rencontre de Moriond, March 1978.
17. R. Field, Private communication, and invited plenary session talk at the XIX International Conference on High Energy Physics, Tokyo, 1978, CALT-68-683.
18. Y. Nambu, Nucl. Phys. $\underline{B130}$, 505 (1977).
19. M.B. Einhorn and R. Savit, Phys. Letters $\underline{77B}$, 295 (1978).

HIGH ENERGY COSMIC RAYS, PARTICLE PHYSICS AND ASTROPHYSICS

A. M. Hillas
Physics Department, University of Leeds, Leeds LS2 9JT, England

ABSTRACT

At energies of 10^{15} to 10^{18} eV particles degrade their energy more rapidly in traversing the atmosphere than do low-energy particles — i.e. there is a departure from simple scaling — but if the rapid rise in hadronic cross-sections continues so far one need assume no other change in the average characteristics of interactions seen at accelerators to explain a wide range of characteristics of air showers, though discordant sets of data still permit different interpretations. If one such discord, between shower size spectra obtained in the (old) Chacaltaya experiment and experiments at lower altitudes, were to be resolved in favour of the latter, one would obtain a cosmic-ray energy spectrum fitting very well to the low-energy data, and the primary composition might also fit smoothly. A small shift in the energy scale might have significant consequences for the 10^{20} eV cut-off. The "knee" in the spectrum near 3×10^{15} eV remains, and appears not to be an effect of magnetic trapping, but more probably a source effect. The muon flux fits this composite spectrum, but there are problems with gamma-rays. It is remarked that emulsion chamber experiments are probably not usually selecting nucleon interactions.

INTRODUCTION

At energies above 100 TeV we are not able to make very direct observations from which we can unambiguously separate the information on the energy spectrum of cosmic rays, their composition, and the detailed nature of particle interactions. Nevertheless, there are several recent trends in this area which are of interest to those who use cosmic rays as an accelerator beam, to astrophysicists, and to other particle physicists.

The energy domain of extensive air showers will be considered first, where the wide range of available energy ($\sim 10^{15}$ to $>10^{19}$ eV) would be expected to allow even crude measurements to distinguish between very different laws of partition of energy amongst secondary particles. Many authors (e.g. [1], [2], [3]) have deduced that above 10^{14} eV interactions depart radically from Feynman scaling, and typically call for a variation of the multiplicity of secondary particles rising as $E^{\frac{1}{2}}$ (and typically assuming no leading pions). However, recent observations at Tien Shan [4,5], and recent calculations [6] considerably weaken the case for any serious departure from scaling, provided that hadronic interaction cross sections rise considerably at high energy.

After considering air showers, one turns to the 10-1000 TeV region of the cosmic-ray spectrum, which is intermediate between the domains well studied with balloons or through air showers: how well can one fit together the work in these two areas?

EXTENSIVE AIR SHOWERS

The atmosphere may be regarded as a large calorimeter, in which most of a primary particle's energy is deposited in the form of ionization (including atomic excitation in this term), apart from a small fraction taken away by neutrinos and by particles stopping in the earth. The incident particle makes interactions with nuclei in the atmosphere: mesons are generated; these make further collisions with air nuclei; and at each collision a fraction of the energy is carried off by π^os, which decay to photons and initiate electron-photon cascades. Most of the energy is dissipated in electron-photon cascades. Schematically, the number N of fast charged particles varies with depth t in the atmosphere as shown in Fig. 1. As the primary energy increases the depth of maximum, t_{max}, increases, and N_{max} increases nearly proportionately to E.

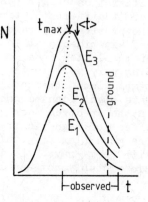

Fig. 1.

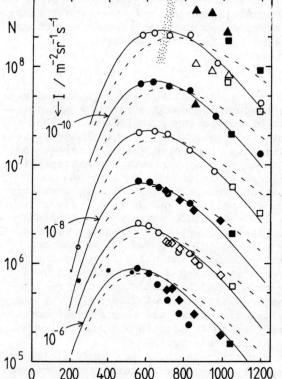

The area under the curve would give the primary energy E — N being deduced in practice from the rate of ionization per $g.cm^{-2}$ — after making an allowance for the energy taken by neutrinos, etc..

There has been much discussion of the way in which t_{max} increases with E: t_{max} is smaller than was expected by simple scaling of particle interaction products from lower energies, indicating a more rapid degradation of energy than expected; but before discussing its interpre-

.. Fig. 2.

tation, some problems with the data will be pointed out.

Fig. 2 shows the variation of N with atmospheric depth obtained by the method of "constant intensity cuts" in the size spectrum, where showers occurring at the same rate at different depths (vertical or slanting) are identified with the same energy. (In the presence of fluctuations in shower development, this procedure more nearly reproduces the root-mean-square shower size $N_{rms}(t)$ for fixed E [7], but the difference from $N(t)$ is thought generally to be only a few percent.) Various experiments at sea level, and that at Tien Shan (covering the range 700-1000 g cm^{-2}) agree well amongst themselves, but there are systematic differences between these and the results reported by two separate experiments at Chacaltaya (530-1100 g cm^{-2}). The recent Chacaltaya experiment [8] gives N about a factor 3 larger than most other experiments in the region of overlap, a discrepancy arising from differing estimations of the fraction of particles at small distances from the shower axis. However, even the older Chacaltaya experiment [9] gives N systematically higher than the Tien Shan and sea level results by a factor 1.5, a fact which has not been emphasized before. It is not clear at present who is right, but it appears as though a relative shift of this magnitude is necessary to combine the data, and for the purpose of presenting the data, the scale adopted by the majority is used in Fig. 2: the old Chacaltaya points are scaled down by a factor 1.5 in N. (This is far out of line with the new experiment, though the shape of the $N(t)$ curves, which is of particular interest here, is much the same.) The Durham group[10] have reported preliminary estimates of the depth t_{max} from Cerenkov radiation: these (indicated by stippled band) are in accord with the other observations of $N(t)$.

In Fig. 2, the squares represent data taken at sea level (1030 g cm^{-2}, from Moscow, Yakutsk, MIT and Kiel), triangles are for Volcano Ranch data, diamonds Tien Shan, large circles Chacaltaya (N reduced as stated above), and small circles are aircraft data from Antonov et al.[11,12]. Each set of points refers to showers arriving at the same integral flux (rate I of showers of size $\geqslant N$) filled and open symbols alternating for successive factors of 10 change in I. The lines are obtained from model calculations, to be introduced later.

Various characteristics of showers observed near sea level, such as the shape of the lateral distribution of particles, and the proportion of muons, are to a large extent governed by the shower "age" at the observation level, and hence on t_{max}, so the position of the peak in $N(t)$ is the most important single characteristic of the shower. At the risk of over-simplification, it is instructive to examine the main factors determining the depth of shower penetration.

MEAN DEPTH OF ENERGY DEPOSITION

The depth at which N reaches a maximum comes only indirectly out of calculations. More amenable to discussion is the mean depth $\langle t \rangle$ of energy deposition,

$$\langle t \rangle = \int t\, N(t)\, dt \Big/ \int N(t)\, dt ,$$

(energy deposition rate being $\propto$ "N"), and it is not very different from t_{max}: typically $t_{max} = (0.87-0.93)\langle t \rangle$.

Consider a cascade model simplified to the extent of ignoring kaon and antinucleon production, so all the secondary particles are pions or surviving nucleons; and examine the fate of one parcel of the incoming energy to the point where it is deposited as ionization:

Fig. 3: [diagram showing t_N, t_π, t_{EM} segments with nucleon, pion, $\pi^\circ \to \gamma$ legend]

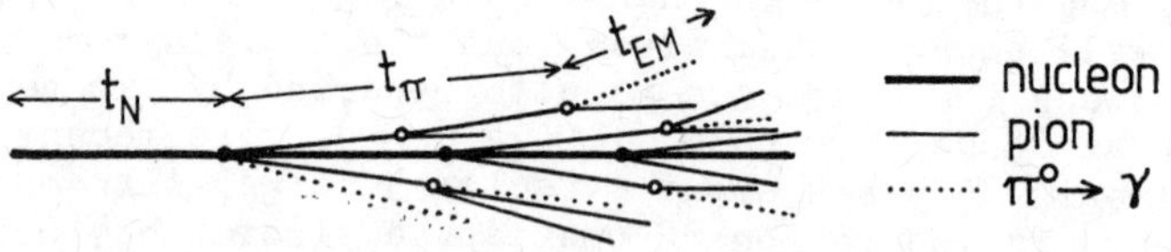

This parcel travels a distance t, carried first by a nucleon, then by a pion, say, and finally in the electromagnetic cascade, so that

$$t = t_N + t_\pi + t_{EM} ,$$

adding the distances travelled in each form. Averaging over all the parcels of energy in the primary particle, and over many primaries, one has for the mean depth of energy deposition,

$$\langle t \rangle = \langle t_N \rangle + \langle t_\pi \rangle + \langle t_{EM} \rangle .$$

In crude terms, all of the energy travels with the nucleon to its first collision, then a fraction $(1-K)$, where K is the inelasticity, travels along a further path with the nucleon, etc., so that if the mean free path of the nucleon is λ_N,

$$\langle t_N \rangle = \lambda_N + (1-K)\lambda_N + (1-K)^2 \lambda_N + \ldots$$
$$= \lambda_N / K \approx 2.\lambda_N .$$

Then, 2/3 of the energy travels along the first flight path of the charged pions, a fraction K_2 remains in a second path .. etc..
$K_2 \approx 2/3$, but may be more at the highest energies, depending on the charge ratio of leading pions, and will be less at lower energies, because of $\pi-\mu$ decay, so

$$\langle t_\pi \rangle = (1.5-2).\lambda_\pi .$$

The mean depth of energy deposition in an electromagnetic cascade generated by the photons from a decaying π° of energy E_{π° is $L_{rad} \cdot \ln(2.3\, E_{\pi^\circ} / \epsilon_{crit})$, from which

$$\langle t_{EM} \rangle = L_{rad} \cdot \ln\left(\frac{2.3\, E_{\pi^\circ} \text{typ}}{\epsilon_{crit}} \right) ,$$

where L_{rad} and ϵ_{crit} are the radiation length and critical energy in air (about 35 g cm^{-2} and 77 MeV respectively), and $E_{\pi^o typ}$ is a typical energy of π^o generated in the cascade. This is not the energy of π^o produced directly by the primary particle, as only ~1/6 of the energy goes immediately into π^os: it is weighted over several early generations of π^os:

Approximately, $\ln(E_{\pi^o typ}) = \dfrac{\sum E_{\pi^o} \cdot \ln(E_{\pi^o})}{\sum E_{\pi^o}}$, summed

over all π^os in the cascade. For a scaling model, apart from complications due to $\pi-\mu$ decay (which breaks strict energy scaling features), $E_{\pi^o typ}$ increases in proportion to E_{prim}.

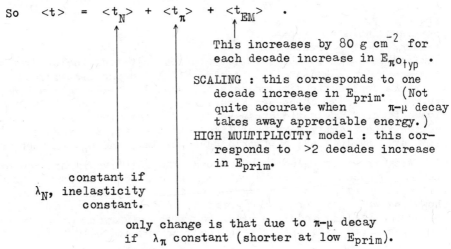

So $\langle t \rangle = \langle t_N \rangle + \langle t_\pi \rangle + \langle t_{EM} \rangle$.

- $\langle t_{EM} \rangle$: This increases by 80 g cm^{-2} for each decade increase in $E_{\pi^o typ}$.
  SCALING : this corresponds to one decade increase in E_{prim}. (Not quite accurate when $\pi-\mu$ decay takes away appreciable energy.)
  HIGH MULTIPLICITY model : this corresponds to >2 decades increase in E_{prim}.
- $\langle t_N \rangle$: constant if λ_N, inelasticity constant.
- $\langle t_\pi \rangle$: only change is that due to $\pi-\mu$ decay if λ_π constant (shorter at low E_{prim}).

But do λ_N, λ_π change?

A high multiplicity model, with rapid degradation of energy, produces a smaller depth $\langle t \rangle$ and a much smaller change with increasing energy, E_{prim}. The rather rapid increase in shower length with increasing energy expected from a scaling model can, however, be partly compensated by a decrease in the hadronic length $\langle t_H \rangle = \langle t_N \rangle + \langle t_\pi \rangle$ if λ_N and λ_π decrease with increasing energy.

The precise effect of decreasing λs cannot be obtained from this simple treatment, as t_N and t_π have to be averaged over a range of energies of hadrons in the cascade, but Ouldridge and the present author have made detailed calculations of shower development (and not ignoring kaons, etc.) assuming

$$\lambda = \lambda_o / \left\{ 1 + 0.039 \left(\ln(E/100\text{GeV}) \right)^{1.8} \right\},$$

this rather odd formula having been used by previous workers, to represent hadron attenuation in the atmosphere[13], and to calculate muon production (though this is actually insensitive to λ_N)[14]. Mean free paths of pions and kaons were assumed to have this same energy dependence. Fig. 4 shows this assumed λ_N as a function of energy, and the right-hand scale shows the equivalent inelastic proton-proton cross-section. For comparison, the p-p cross

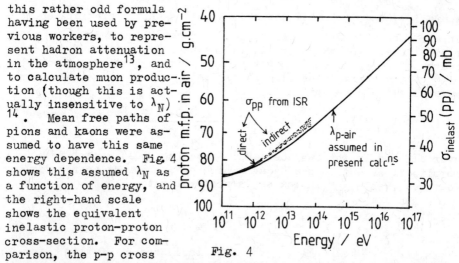

Fig. 4

sections measured at the CERN ISR are also plotted, including the high-energy values deduced by Amaldi et al.[15] using dispersion relations applied to measurements of forward scattering amplitudes. (Their total cross-sections have been multiplied by 0.825 to give estimated inelastic cross-sections.) The true rise in cross section is evidently very much like that assumed, at least to ~10^{14} eV. By 10^{17} eV, σ_{pp} would have tripled, and σ_{p-air} doubled.

From these detailed calculations, the relative importance of the hadron cascade and the electromagnetic cascade in contributing to the mean depth $<t>$ of energy deposition were found to be as shown below. Apart from the variation in the interaction cross section, a scaling model is used, (a) with λs varying as shown in Fig. 4, and (b) with the λs becoming constant above 10^{14} eV.

	(a)			(b)			
	σ rises indefinitely;			σ rise stops at 10^{14} eV;			
E per nucleon	$<t_H>$	$<t_{EM}>$	$<t>$	$<t_H>$	$<t_{EM}>$	$<t>$	
10^{15} eV	243	402	645	257	402	659	g cm^{-2}
10^{18} eV	225	588	813	314	587	901	g cm^{-2}

Qualitatively, this table confirms the effect which rising cross sections can have. Quantitatively, however, it underlines the danger of relying on simple arguments for anything beyond a qualitative understanding of what is happening in a cascade. The effect of π-μ decay in shortening the pion cascade is found to be large, especially at 10^{15} eV, so that in the energy range considered, $<t_H>$ remains almost constant (in case a), as this effect almost balances the changing λs, which are tending to shorten $<t_H>$ at higher energies. (Also the electromagnetic cascade does not lengthen as much per decade as the simplest prediction gives.)

The same calculations (scaling model, with energy-dependent mean free paths as indicated in Fig. 4 — see Appendix for more details) gave for the shower development $N(t)$ the curves drawn in Figure 2, assuming a conventional mixture of nuclear masses to be present (see Appendix). If the mean free paths were assumed instead to be constant, the dashed curves were obtained. The latter show much too small an attenuation rate at sea level (as has long been known), whilst the former are quite reasonable.

There still remain considerable problems in fitting all the data on shower size development, but some of these arise because of discrepancies between different experiments. A few years ago, the data on shower development curves near 10^{15} eV indicated much too great a fall-off from N_{max} to $N_{sea\ level}$ to be explained by a scaling model, with t_{max} very high in the atmosphere. The data were heavily weighted by the shape of the lowest Chacaltaya curve, which showed fast attenuation, and by aeroplane measurements of Antonov's group. The Chacaltaya data were inserted on Fig. 2 reduced by a factor 1.5, and they do not then suggest an anomalous steepness of the showers, but the question remains as to whether it is right to align the different sets of data in this way, and how to account for the discrepancy. Earlier attempts to reconcile scaling with the air shower data had ignored the lowest energies, and had called for an iron-dominated composition to ensure the smallest possible t_{max} (with a low energy-per-nucleon), but if the cross-section rises as assumed, the composition does not have to be unduly iron-rich. (The composition will be discussed again later.) Even so, although recent re-analyses[11] of Antonov's aircraft data, and preliminary new experiments have produced points which are not quite so high as before, the points at 200 g cm^{-2} on Fig. 2 are still well above the curves. It will be hard to produce such flat-topped shower curves unless there is a component present which makes some showers develop very much earlier than the normal ones.

Overlooking the possibility of such an additional anomalous component, for the present, the shower development curves $N(t)$ should not differ greatly from those drawn in Fig. 2 for $t < 500$ g cm^{-2} if they fit at $t > 500$ g cm^{-2}. Hence the curves may be used in determining the total energy deposition of showers occurring at each of the chosen rates, I, thus building up an energy spectrum over the range $3 \cdot 10^{15}$ eV to $3 \cdot 10^{17}$ eV, omitting the lowest curve which is much more uncertain. These energies, derived from Fig. 2, and corresponding to rates 10^{-7} to 10^{-11} m^{-2}s^{-1}sr^{-1} are used in Fig. 6, but before discussing the energy spectrum, and the nature of the primary particles, some further comments are called for on the relevance of other properties of air showers to the nature of high-energy interactions.

TRENDS IN PARTICLE INTERACTIONS AT E $>10^{15}$ eV ?

Summarizing the conclusions which have been drawn from the rate of shower development, energy appears to be degraded more rapidly than expected on the basis of simple scaling from the TeV region with no increase in cross-sections. One may attempt to account for this either by

(a) "high multiplicity models", invoked particularly to explain rapid development of showers in the upper atmosphere (though only the aircraft experiment now appears very unusual), and having normal cross-sections in order to yield sufficient fluctuation in the stage of shower development;

or (b) a large rise in the frequency of interactions (Fig. 4), but no essential change in the nature of interactions when they occur (i.e. scaling: the structure function $f(x)$ in the Feynman or Yen scaling variable x remaining independent of beam energy).

Since the rise in cross-section does now appear to occur, the second seems the more natural explanation, and in any case forms the simplest situation which should be examined first in order to find what type of deviation does occur. It has been proposed by Kasahara and Takahashi[16] that a small shrinkage in $f(x)$ (i.e. lower mean x) at high energy, combined with a somewhat slower rise in σ, would be more acceptable than (b) at energies around 100 TeV, to explain their observations on gamma-ray fluxes and hadrons: this deserves further examination at air shower energies.

Surprisingly, the choice between the radically different solutions (a) and (b) has not yet been convincingly decided in the air shower domain, as both have their advocates. I believe that some of the objections to scaling are not valid. For example Vernov et al.[3] have argued that the number of muons in showers is almost an order of magnitude greater than expected from scaling. Fig. 5 shows their measurements of the average number N_μ of muons above 10 GeV in showers containing N_e electrons — see the points in the diagram — and the line represents the prediction from their scaling model. (N_μ/N_e can be low in a scaling model for two reasons: t_{max} is large, and hence N_e can be high at sea level, as compared with a high-multiplicity model; and in order for pions to have the energy — tens of GeV — at which decay competes well with interaction, many successive steps of cascading are needed in a low-multiplicity model, and the energy losses to π^0s leaves little energy for μs.) In principle, the ratio is sensitive to the multiplicity in π-air collisions. However, it turns out that one has to be careful

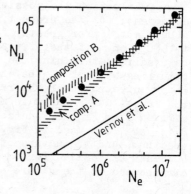

Fig. 5

about many details. Ouldridge and Hillas[6] found that many natural approximations in the paper referred to (and in some others) all affected the result in the same direction: none was vital but the cumulative effect on N_μ/N_e was up to an order of magnitude. (For example, Vernov et al. omitted kaon and antinucleon production, assumed proton primaries only, calculated for vertical showers rather than the observed range of slant depths, took a constant cross-section, and assumed that too many leading pions from pion collisions were neutral: all these approximations increased N_e or decreased N_μ.) Until one knows the true mass composition of the primary radiation, N_μ cannot be calculated exactly, and for the purpose of testing the plausibility of scaling, in Fig. 5, a conventional composition was assumed (see Appendix), together with a reasonable variant ("composition B") which was richer in iron near 10^{15} eV. Within these uncertainties, the results, shown by the shaded bands, match the data well.

It is of practical importance to be able to describe the lateral distribution of detector signal strength in 10^{17} eV showers at Haverah Park, and the same scaling model appears at present to give very reasonable results if the typical primary mass is ~12.

Another apparent problem for the scaling model — or any other model having a normal elasticity — is in the number of hadrons in air showers at mountain altitudes. Vernov et al. have quoted experimental results (most notably of the Tata group) having only a few percent of the predicted number of TeV hadrons; but there are experimental problems here : other experiments have given much higher hadron numbers, and the recent Tien Shan calorimeter data[5] agree well[6] with the predictions. There are evidently very great experimental problems here. Different workers appear to be in disagreement about identification of individual hadron cascades, measurement of hadron energies, and measurement of shower sizes.

In summary, the data on air showers, particularly above 10^{16} eV, do not appear to contradict a scaling hypothesis, provided that the inelastic hadron-air cross-sections have doubled by 10^{17} eV. This conclusion is not independent of assumptions about the primary composition, and would be hard to sustain with a nearly pure proton beam. (Fluctuations in shower development are observed, and are accounted for in scaling models largely by the spread in primary masses: in high multiplicity models they are accounted for by the fluctuations in starting-point of proton-induced showers.) Also, the present scatter in the data permits one, within reason, to ignore some of the data unfavourable to the hypothesis under examination. But it is necessary to make a hypothesis to start with, and then test the range of data which it fits. The alternative procedure of considering the parameters of an interaction model, the primary mass composition, and the energy spectrum all to be free parameters which are simultaneously selected to give the best fit to all the data is liable to give bad results if one tries to fit inaccurate data which has only a weak sensitivity to the parameters. Until the composition and the data discrepancies have been clarified, the strongest claim that can be made is that a wide range of data is consistent with scaling (modified by having rising cross-sections).

ENERGY SPECTRUM

Between $\sim 3 \times 10^{15}$ eV and $\sim 3 \times 10^{17}$ eV the shower development curves $N(t)$ in Fig. 2 are sufficiently well-defined that a large proportion of the energy deposition is observed, and the determination of the energy of the particles (corresponding to the selected arrival rates) is essentially calorimetric, rather than model dependent. If the cascade model were to be revised, the unobserved part of the curve in the upper atmosphere would not be expected to differ very much (leaving aside the aircraft data for the moment), and the energy lost in neutrinos would not differ significantly. The spectrum can be continued down to 10^{14} eV using the scaling model already mentioned to convert the actual shower size at Chacaltaya to primary energy. (At low energies this conversion depends on the effective mass of the primaries: a value of 12 is assumed.) The spectrum thus derived is plotted in

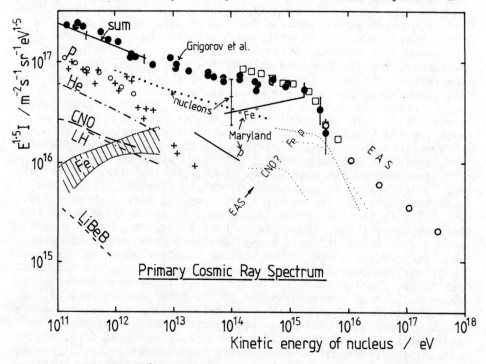

Fig. 6

Fig. 6, in the form $E^{1.5} \cdot I$, where I is the flux of particles of energy $\geq E$. Multiplication by $E^{1.5}$ disguises the steeply falling spectrum, but avoids the necessity for a highly compressed ordinate covering very many decades. The air shower results are shown as open circles (calorimetric points) and squares (derived from N at Chacaltaya). It must be repeated that the published sizes N from Chacaltaya have been reduced by a factor 1.5, to match data obtained in experiments at Tien Shan and sea level. If the truth lies between these two, the energies will have to be increased. The energy scale refers to the total energy of a particle, not energy-per-nucleon.

Sea-level experiments agree that there is a sharp bend in the shower size spectrum (See Fig. 7), as seen at Chacaltaya. The bend is seen also at Tien Shan (3340m altitude). (The oblique lines in Fig. 7 show the shift which was applied to the Chacaltaya sizes to yield the reduced sizes represented by the points.) The bend does not appear, however, in the Norikura spectrum[17] at 2770m. If all showers of a given size at one altitude have equal sizes when they reach a lower altitude, the bend in the spectrum should appear at the same integral rate, I, at different altitudes. The bend does not occur at identical rate at different altitudes, though, which means that individual showers follow different development curves, suggesting that a mixture of very different primary masses is present.

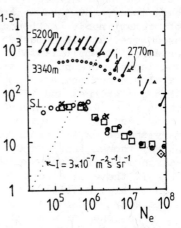

Fig. 7: shower size spectra.

Before discussing the composition, however, we should ask how well we can link this energy spectrum with that derived by other methods at lower energies.

COSMIC RAYS OF INTERMEDIATE ENERGY

The flux of particles up to ~3 TeV energy has been measured directly with balloon-borne apparatus, each nuclear type being separated. The fluxes of separate groups of nuclei[20] are shown in Fig. 6: protons (o from Ryan et al. , + from the Proton satellites of Grigorov's group — though above 1 TeV the nuclear charge identification has been questioned in this experiment), helium nuclei, the Li-Be-B group, C-N-O group, the heavier nuclei apart from the iron group ("LH"), and the iron group. Summing the fluxes of all groups, one obtains the line marked "sum", which is in good agreement with the spectrum of all nuclei obtained directly by the calorimeter in the Proton satellites, shown by the filled circles. These latter results of Grigorov et al. seemed very low in comparison with air shower data when first produced in 1965: they now fit the air shower data ("reduced" as noted above) remarkably well!

What other checks do we have on the spectrum in the region intermediate between the balloon and air shower domains? What happens to the composition here?

Apart from Grigorov's experiments, we rely here on measurements of the "incoherent" fluxes of secondary particles, and, given a model of nuclear interactions, these tell us about the incoming flux of nucleons rather than about how the nucleons were bundled into nuclei. The muon spectrum at sea level or underground allows us to deduce the spectrum of their parent pions at around 100 g cm^{-2} (after the appropriate allowance for kaon parentage has been made), and then we have to use a model to derive the spectrum of primary nucleons (bound or unbound). With the scaling model used here, and in agreement with other workers, the nucleon flux obtained is approximately the dotted line in Fig. 6 marked "nucleons", if one uses the Amineva [18] muon spectrum. The point labelled "nucleons" at 10^{14} eV was communicated by Takahashi at this conference, and is based on observations of gamma-ray cascades in emulsion chambers flown in aircraft. Gaisser, Siohan and Yodh [19] deduce the same spectrum slope in this region from the Maryland observations of unaccompanied hadrons deep in the atmosphere, though this deduction is somewhat more esnsitive to the hadron mean free path and elasticity.

If there is a serious departure from scaling, in the direction of high multiplicities, a higher nucleon flux will be required to account for the muons.

Postpone for a moment any discussion of the particle physics in this region. What are the primary particles?

PRIMARY COMPOSITION AND THE KNEE IN THE SPECTRUM

Many conflicting conclusions have been drawn by different experimenters about the composition near 10^{15} eV (listed, but not fully discussed in[20]). Only those points related to the cosmic ray spectrum, or raised at this conference, will be discussed here.

(i) From air shower size spectra. If one uses the scaling model already mentioned (see Appendix), the air shower spectra seem best described if the spectra of protons, iron and medium nuclei are somewhat as shown by the lightly dotted lines (labelled EAS⟶) in Fig. 6, but this solution is not necessarily unique. The different nuclei cannot be separated in any detail, and it is assumed that apart from protons the spectrum of each nucleus bends at the same energy-per-nucleon. The CNO nuclei are positioned only very roughly. With a spectrum not vastly different from this, one finds that the spectrum of electron-shower-size, N_e near sea level is dominated by the proton contribution, but the spectrum of muon size, N_μ, is dominated by the heavy nuclei. Unfortunately, data on the latter are still rather sparse.

(ii) From the incoherent muon flux.(i.e. the curve marked "nucleons"). If each nucleus of mass A and energy E is split up into A nucleons of energy E/A,

an integral spectrum $kE^{-\gamma}$ of nuclei gives a spectrum $k'E^{-\gamma}$ of nucleons, where $k' = k / A^{\gamma-1}$. Thus heavy nuclei are very inefficient producers of nucleons. In Fig. 6, if the total flux of all nuclei is no higher than shown (by Grigorov's points), the flux of primary protons should be about 75% to 90% of the flux of all nucleons. So if this "nucleons" curve is right, the spectrum of primary protons running just below it would join satisfactorily to the directly measured flux at 1 Tev and the level postulated to fit air shower spectra at 1000 TeV.

(iii) The Maryland group have presented to this conference an interpretation of their observations on delayed hadrons in small showers. Representing the primary radiation as a mixture of protons and iron nuclei, they required spectra of these two components as shown by the lines marked "Maryland" in Fig. 6 (drawn only in the energy regions to which their experiment was sensitive). If this result is right, the flux of nucleons must be much lower than drawn in Fig. 6.*

(iv) Whatever the detailed composition, the "knee" seen in the energy spectrum does not now appear to be explained as a simple consequence of more rapid leakage from a magnetic trapping region, which would have resulted in different nuclei having their spectra steepened at the same magnetic rigidity, as originally suggested by Peters. For if all nuclei had the same rigidity spectrum, the proton "knee" would occur at an energy 8 times less than the oxygen knee and 26 times less than the iron knee, and taking relative proportions of the nuclei from the extrapolated "source composition" obtained from low-energy observations (supposing fragmentation en route from the source to be small at this energy), the resulting total spectrum of all particles would have an extended break region, illustrated in Fig. 8, and not the sharper break that is observed. Either the spread in atomic masses is less than normal, or the bend in the spectrum occurs at a much higher rigidity for protons. In either case the rigidity spectra of different nuclei are quite different. The relative abundances used in Fig. 8 are probably iron-poor: this strengthens the conclusion drawn.

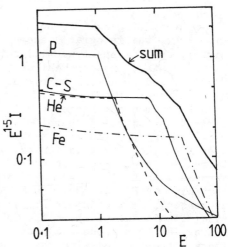

Fig. 8: superposition of similar rigidity spectra.

---

* In the original talk at the conference, I stated that there was no discrepancy, having drawn the "nucleons" line too low.
One may also note that, as Tonwar pointed out, the Tata[23] group concluded that the composition is "normal" at 10^{14} eV, from their observations on energetic muons in small showers.

GENERAL CONCLUSIONS

(i) For particle physics: I am impressed by the increasing success of scaling to account for properties of air showers, provided that hadron-air cross-sections continue to rise considerably (Fig. 4), though most air shower workers (excepting Gaisser, and Turver's group[21]) do not agree with this view. The same model and primary spectrum describes well the muon and hadron fluxes in the lower atmosphere.

But there are discrepancies. Four experiments give the numbers of TeV hadrons in air showers roughly as predicted, but three give much lower numbers, and two of the latter used cloud chambers (Miyake's group and the Tata group), in which the cascades could really be seen, so my choice of data could be wrong. There are large disagreements in shower sizes (or rates) found by different groups at mountain altitude. And the aircraft measurements of showers at 200 g cm^{-2} do not fit the predictions. (But if they are right, the integrated energy under the N(t) curves will be higher, and the spectrum will not agree with Grigorov's satellite measurements.) Measurement of the sizes of young showers is difficult, and more work here would be welcome.

There are problems with gamma-ray fluxes. The scaling model used here predicts fluxes of photons-plus-electrons which are too large by a factor ~3 near 10-30 TeV at mountain altitudes and higher. Paying particular attention to the gamma-ray fluxes (but also the other particle fluxes), Kasahara and Takahashi[16] have preferred a slower rise in σ and a slowly steepening distribution in Feynman x as E rises (say with the width of the distribution $xf(x)$ shrinking as $E^{0.15}$). This is a key area. I do not understand it at present, as the photon fluxes are still below prediction at aircraft altitudes, where they should tie in with sea-level muon fluxes, as both muons and photons arise largely from pion decay. These workers, analyzing the Japanese emulsion chamber experiments, also find little evidence for a large rise in cross-section, which should show up in a steepening of the hadron spectra with increasing atmospheric depth. The attenuation length for gamma-ray fluxes is expected to shorten— becoming less than 80 g cm^{-2} above 200 TeV— and although, according to private communication from Yodh, there is a preliminary indication of this in the Soviet emulsion chambers, it does not seem to be confirmed in the Japanese work. It is hoped that other groups will be able to contribute to this problematical area.

Finally, one should note that some high-multiplicity events are certainly observed in emulsion chambers: these would be surprising if the average multiplicity were described by scaling.

(ii) For users of the cosmic-ray accelerator beam: Near 1000 TeV the proportion of very heavy nuclei in the primary beam is probably considerably larger than at low energies, which may have some relevance for "air jet" experiments. But quite apart from the question of accompaniment by other nuclear fragments, the particle beam is complex at mountain altitudes. The collision models employed here show that most events releasing a given $\sum E\gamma$ are not due to nucleons: at 100 TeV, for example, the composition of hadrons might be 50% nucleons, 38% pions and 12% kaons, and although the treatment of "leading" kaons was inaccurate in this calculation, and the K/π ratio may hence be somewhat too high, $K\gamma$ is less for nucleons than for pions, so one would expect a minority of nucleons in any given sample. So the presence of a proportion of unusual events in any given sample of jets could possibly arise from an unusual initiating particle. Also, distributions in $(E\gamma/\sum E\gamma)$ or in $\log\tan\theta$ should not be compared simply with the corresponding data for proton collisions at accelerator energies.

(iii) The relevance of planned colliding-beam experiments at higher energies: It has been suggested that the advent of colliding-beam machines simulating 10^{14} eV collisions on static targets will leave us with no uncertainties in the hadron physics and thus permit such precise interpretation of (small) air shower experiments that the primary composition will be deducible quite unambiguously. Calculations of air shower processes show, however, that pion-nucleus collisions are the vital ones, particularly where muon production is concerned. Colliding beams will tell us nothing directly about these. We urgently need detailed information about particle spectra from pion-nucleus collisions.

(iv) Astrophysical implications: The form of the energy spectrum above 10^{15} eV is probably conveying information about the source region, rather than about propagation within the galaxy, as the "knee" in the spectrum does not seem to be a magnetic rigidity effect. It remains a possibility that the spectrum has a superimposed second component, richer in iron, in this energy region, rather like the "pulsar bump" of Strong et al.[22] (though less prominent and less iron-dominated than they originally proposed).

The (old) Chacaltaya experiment had served as a basic calibration of the energy scale for air showers, by calorimetry. If this experiment did indeed give N too large by a factor 1.5, the energies attributed to the largest showers would be reduced correspondingly. The promising scaling models of hadron collisions also seem likely to give a similar reduction in the energy attributed to large showers, as compared with older models. If this reduction were to be substantiated, the energy spectrum would not extend quite as far past the point at which losses due to reactions with the microwave photons commence; the maximum age since acceleration would not be so low and there need be little difficulty in accommodating particles from the Virgo supercluster, provided there was not extensive trapping.

However, we do not have a final energy calibration. The new Chacaltaya experiments would indicate a shift in the opposite direction. Cerenkov radiation from the sky provides the alternative calorimetric approach; and here the Yakutsk work had been taken to support the old scale. We shall soon have more data here[10].

APPENDIX: THE SCALING MODEL CALCULATIONS

Mean free paths for inelastic collisions of nucleons, pions and kaons in air were taken as 86, 114 and 128 g cm^{-2} below 100 GeV, and at higher energies (E GeV) varied as

$$\left[1 + 0.039 \left(\log_{10}(E/100)\right)^{1.8}\right]^{-1} .$$

The spectra of particles emerging from collisions were given in terms of $x = E_{sec}/E_{primary}$, $f(x)dx$ being the number in the range x to $x+dx$. For nucleon collisions (or antinucleons, annihilation being ignored), a leading nucleon emerged with the distribution given (as $x \cdot f(x)$) in Fig. 9a, pions emerged with

$$f(x) = A_\pi\, e^{-5x}(1-x)/x ,$$

kaons similarly (but with different factor A_k), and newly produced nucleons and antinucleons with

$$f(x) = A_n\, e^{-10x}/x .$$

A_n and A_k rose with increasing energy in such a way that the fraction of the incident energy taken by nucleon-antinucleon pairs and by kaons was

$$W_n = 0.033/(1 + 150/E), \quad \text{and} \quad W_k = 0.075/(1 + 40/E).$$

A_π was then adjusted so that pions took the energy not given to leading or produced nucleons, or kaons (typically leaving 45% for pions at very high energy, 44% for leading lucleons, 7½% for Ks). One third of these pions were neutral; and 4 types of kaon were produced in equal proportions.

Kaon collisions were treated in a similar way, though the leading particle was a kaon ($xf(x)$ as shown in Fig. 9a) of some kind: this was the only change (though this is not a very accurate description of kaon leading particles). Pion collisions created nucleon pairs and kaons exactly as above, but the pion spectra were as shown in Fig. 9b (at high energies; slightly enhanced when W_n or W_k fell). Variants 1 and 2 were taken to bracket the most probable spectra. (Best results were usually obtained closer to 2: π-p data from accelerators are closer to 1.)

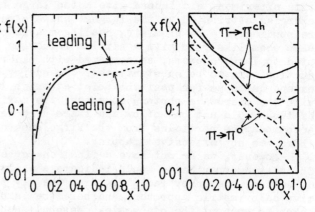

Fig. 9: (a) leading particles from N or K collisions; (b) π-nucleus $\longrightarrow \pi$.

A non-exponential atmosphere was used.

In order to check against errors, Ouldridge and Hillas made quite separate calculations. Ouldridge used a hybrid Monte-Carlo (for particles having energy $>0.001\ E_{prim}$) and layer-by-layer average treatment (for low-energy particles), in order to study fluctuation effects, particularly for the N_μ/N_e work, as well as average shower development. Hillas used numerical integration to solve one-dimensional diffusion equations to obtain the average shower development, and to examine the effects of composition mixtures in more detail.

In Fig. 5, "composition A" refers to a case in which iron hardly contributes at all to the "bump" in the energy spectrum between 10^{14} and 10^{16} eV: this is dominated by protons, with the medium elements becoming important beyond 10^{16} eV, and iron only near 10^{17} eV; whilst "composition B" supposed that iron dominated the "bump". In Fig. 2 a rigidity-dependent "knee" was used, as in the published paper[6], with a more conventional composition intermediate between the extremes mentioned above. The shaded band in Fig. 5 shows the range of uncertainty corresponding to varying the model for pion collisions between the extremes 1 and 2 in Fig. 9b, relating to the probability of having a very energetic pion emerge from a collision with a nucleus.

In reality, the various processes involving pions, kaons and nucleons are not represented by the very small group of functions $f(x)$ assumed here: in particular, the products of kaon interactions are very complex. Also, low-energy particles in the laboratory frame require more careful treatment. These improvements are now being studied.

A simple superposition model was used for interactions of nucleus projectiles. McIvor and Hillas are investigating the likely effect of greatly increased p-p cross-sections on nuclear fragmentation, as this may affect the amount of fluctuation in shower development.

REFERENCES

(The abbreviation PnthICCR means "Proceedings of the nth International Conference on Cosmic Rays".)

1. N. N. Kalmykov, G. B. Khristiansen and Yu. A. Fomin, P 12th ICCR, Hobart, $\underline{6}$, 2074 (1971),
   L. Popova and J. Wdowczyk, P 15th ICCR, Plovdiv, $\underline{8}$, 409 (1977)
   J. Olejniczak, J. Wdowczyk and A. W. Wolfendale, J. Phys. G, $\underline{3}$, 847 (1977).
2. P. K. F. Grieder, P 15th ICCR, Plovdiv, $\underline{8}$, 326 (1977).
3. S. N. Vernov et al., J. Phys. G, $\underline{3}$, 1601 (1977).
4. T. V. Danilova et al., P 15th ICCR, Plovdiv, $\underline{8}$, 129 (1977).
5. V. A. Romakhin, N. M. Nesterova and A. G. Dubovy, P 15th ICCR, Plovdiv, $\underline{8}$, 107 (1977).
6. M. Ouldridge and A. M. Hillas, J. Phys. G, $\underline{4}$, L35 (1978).
7. T. K. Gaisser and A. M. Hillas, P 15th ICCR, Plovdiv, $\underline{8}$, 353 (1977).
8. C. Aguirre et al., P 15th ICCR, Plovdiv, $\underline{8}$, 208 (1977).
9. M. LaPointe et al., Can. J. Phys., $\underline{46}$, S68 (1968).
10. R. J. Protheroe and K. E. Turver, P 15th ICCR, Plovdiv, $\underline{8}$, 275 (1977).
11. J. N. Stamenov and S. Z. Ushev, P 15th ICCR, Plovdiv, $\underline{8}$, 133 (1977).
12. R. A. Antonov et al., P 15th ICCR, Plovdiv, $\underline{8}$, 137 (1977).
13. R. A. Nam et al., P 14th ICCR, Munich, $\underline{7}$, 2258 (1975).
14. J. W. Elbert et al., Phys. Rev. $\underline{D12}$, 660, $\underline{D14}$, 314 (1975).
15. U. Amaldi et al., Phys. Lett. $\underline{66B}$, 390 (1977).
16. K. Kasahara, CRL-Report-62-78-6 (Cosmic Ray Laboratory, University of Tokyo), (1978),
    Y. Takahashi, J. Iwai and I. Ohta, ICR-Report-67-78-11 (Institute for Cosmic Ray Research, University of Tokyo), (1978).
17. S. Miyake et al., P 12th ICCR, Hobart, $\underline{7}$, 2748 (1971).
18. T. P. Amineva et al., P 13th ICCR, Denver, $\underline{3}$, 1788 (1973).
19. T. K. Gaisser, F. Siohan and G. B. Yodh, J. Phys. G, $\underline{3}$, L 241 (1977).
20. A. M. Hillas, Physics Reports, $\underline{20C}$, 59 (1975).
21. T. K. Gaisser, R. J. Protheroe and K. E. Turver, Rev. Mod. Phys., in press (1978)
22. A. W. Strong, J. Wdowczyk and A. W. Wolfendale, J. Phys. A, $\underline{7}$, 1489 (1974).
23. B. S. Acharya et al., P 15th ICCR, Plovdiv, $\underline{8}$, 36 (1977).

Chap. 7   New Cosmic Ray Experiments

WEAK-INTERACTION STUDIES WITH THE DUMAND DETECTOR

Arthur Roberts
Fermi National Accelerator Laboratory, Batavia, Illinois 60510[*]

ABSTRACT

The objectives of the DUMAND project (Deep Underwater Muon And Neutrino Detector) in high-energy physics, especially with regard to the weak interaction are reviewed and summarized in the light of recent work reported to the 1978 DUMAND Summer Workshop. Detailed Monte-Carlo studies are under way to determine the precision with which muon energy and nuclear cascade energy can be determined. These are relevant to the capability of the detector for measuring the y-distribution in neutrino-nucleon inelastic scattering in the energy range 2-50 TeV. Such measurements are expected to show the existence of the propagator factor provided there exists a charged boson in the mass range 30-100 GeV/c^2. Other aims of DUMAND are briefly described (excluding extraterrestrial neutrino detection, treated elsewhere).

BRIEF REVIEW OF DUMAND: THE 1978 SUMMER WORKSHOP

As of the fall of 1978, the DUMAND project has been in existence approximately five years; its start is usually taken as an ad hoc meeting at the 1973 International Cosmic Ray Conference in Denver. Since the first DUMAND Summer Workshop in the summer of 1975, events have moved at an accelerated pace, until now we can report on the seventh Summer Workshop, just concluded. It was a six-week meeting hosted by the Scripps Institution of Oceanography, La Jolla, and sponsored jointly by the DOE, ONR, NASA, and NSF. The proceedings will be published early in 1979. Plans for funding a feasibility study have already received tentative approval from one funding agency, and it is now our hope that such a feasibility study will be able to get under way in 1979 at the Scripps Institution of Oceanography in La Jolla, California.

The 1978 workshop comprised three two-week sessions, respectively on array studies, UHE physics and astrophysics with DUMAND, and ocean engineering problems. One of the aims of the first two weeks was to provide for the ocean engineering studies a model array to use for the purpose of the engineering studies; that array is shown in Fig. 2. Figure 1 shows an earlier, simpler,

[*]Operated by Universities Research Association Inc. under contract with the United States Department of Energy.

cubic-array concept now replaced by the hexagonal close-packed array of Fig. 2. The new array is wider and shallower than the cubic array, in order to improve the effective fiducial volume for the predominantly horizontal atmospheric neutrinos.

The first session was concerned, among other matters, with optical sensor design in which there is a choice between direct-view photomultipliers and wavelength-shifting sensor types in which the effective area for detection is much greater for the same photocathode area and allows smaller phototubes or larger effective cathode areas; however, this is done at the expense of considerable complication of design and construction. It also considered again the question of acoustic detection of nuclear cascades, about which the effective threshold has been in some doubt. A detailed examination of the problems of false triggering rates led to the conclusion that the effective threshold for acoustic detection would have to be set at a level for which the minimum energy for effective detection of a cascade at a distance of one km or more was 10^{16} eV. The rate of arrival of 10^{16} eV atmospheric neutrinos is very low and cannot even be estimated accurately, since it now appears that the principal component of such high-energy neutrinos is probably the prompt neutrinos, most likely originating in the very rapid decay of charmed particles. Thus the DUMAND array will be primarily optical.

A preliminary report on the 1978 Workshop has been prepared.[1]

PHYSICS OBJECTIVES

The major physics objectives of the 10^9-ton DUMAND array are presently thought of as follows:

1. A study of the neutrino spectrum $N(E)$ produced by cosmic-ray primaries in the atmosphere and determination of the variation with energy of the cross section $\sigma(E)$ with nucleons. Experimentally, what we can determine is $N\sigma$; that quantity must then be compared with calculated values of $N(E)$, which are derived from muon spectra and models of primary cosmic-ray interactions up to about 10^{15} eV, and from theoretical predictions of $\sigma(E)$ from IVB and quark models. The experimental observation provides an additional constraint.

2. A detailed study of the y-distribution (inelasticity) in the interaction of muon neutrinos and nucleons. We will show below why we think we can obtain a definite indication of whether or not the expected effect of the IVB propagator is present and how we can thus obtain a rough value for the IVB mass.

3. A search in the 10-100 TeV range for prompt neutrinos in the atmospheric spectrum. They can be recognized by their unique

(isotropic) angular distribution and by their composition which is expected to be half electron neutrinos.

4. A study of multiple-muon production in neutrino interactions in the TeV energy range.

5. A study of multiple-muon distribution in the cosmic-ray spectrum of muons, which is relevant to the problem of the composition of cosmic-ray primaries in the energy range 10^{14} to 10^{17} eV.

The extremely interesting capabilities of the DUMAND array for neutrino astrophysics will not be discussed here. Let us now examine the methods to be used in attaining the objectives above.

A. Measurement of muon and cascade energies

A detailed Monte-Carlo study program is under way whose purpose is to determine the precision with which the energies of nuclear cascades and of muons can be determined in a DUMAND array and the effect on such precision of the disposition and spacing of sensors. This information, together with accurate data on the absorption and scattering of light in the ocean, is necessary to design an array to achieve a desired performance level. In the present case, we are aiming for a precision of at least 50% in muon energy determination (above 1-2 TeV). It appears that an array with that accuracy for muons should be at least as good for cascades, although that remains to be verified. The 50% requirement is an estimate to be verified by further study (see the next section) of the experiment to be undertaken, the y-distribution in inelastic neutrino-nucleon scattering. In addition, it is important to obtain good directional information, especially on muons; the better the angular precision the more sensitive the apparatus to point sources of astrophysical neutrinos.

Most of the effort to date has been concentrated on muons, for which no similar program is known to us; a program on the source intensity of light from hadronic cascades is being written.[2] The procedure has been as follows:

1. We have adopted the formulae of Adair and Kasha[3] who have parametrized the four different modes of energy loss of fast muons. These equations yield the probabilities of an energy loss E in a given distance by each of the possible modes.

2. The trajectory of the muon is divided into 100-m segments, rather arbitrarily; the segments must be long enough to permit a determination of the loss in each segment.

3. Using the equations for the probability of an energy loss of given type and magnitude, energy losses are chosen at random with proper weighting to produce on the average the correct mean losses, with the correct distributions around the mean. A large

number of muons of different energies are so treated, so that mean energy losses in 100-m segments and their variances can be found. The total energy loss for each 100-m segment is calculated, one segment at a time.

At this point we have sufficient data to plot energy losses in tracks of various lengths for various energies; we have reproduced the energy straggling of a muon beam in sea water. Before we proceed further we examine the results to see if we can derive a satisfactory algorithm for the measurement of muon energy from the data.

The first thing we notice is that the energy lost in a 100-m segment by a muon of given energy can vary by a factor of about 30, and that averaging the total energy loss over, say, a kilometer path, gives poor results with enormous fluctuations. This is because there are occasional large energy losses, usually due to single bremsstrahlung or nuclear encounters, that increase the energy loss well above the mean value.

This situation is analogous to that encountered in the identification of charged particles by the measurement of dE/dx in a thin absorber, in which similar large energy losses can occur because of occasional close collision, thus giving rise to the so-called "Landau tail" to the energy loss distribution. We adopt the same procedure used in that case[4]: to measure the energy loss dE/dx in several segments, and discard the segments in which the largest energy losses occur, using the others to determine the unperturbed (by large fluctuations) energy loss and thus the particle energy. Figures 3-7 show how this is done. They represent the result of Monte-Carlo runs on 1000 muons of the same energy (in this case 10 TeV, but the result is true at all other energies of concern to us, down to at least 2 TeV).

We denote by Y_n^m the energy loss averaged over n 100-m segments of a track of m segments; the n segments are those with the smallest energy loss in the track. Figure 3 gives a plot of Y_5^{10}, i.e., the mean energy loss in the five 100-m segments of a 1000-m track which show the lowest energy losses. Figure 4 shows Y_8^{10} and Fig. 5 shows Y_{10}^{10}, i.e., the average over the entire track. It should be noted first that Y_5^{10} and Y_8^{10} are approximately gaussian, that the mean of Y_8^{10} is considerably higher than that of Y_5^{10}, and that Y_{10}^{10} is far from gaussian, with a long tail extending up 1.0 TeV, a value that corresponds to stopping the muon. Figure 6 shows the way this effect varies with muon energy, and finally Fig. 7 shows the variance of the energy loss distribution as a function of energy for various selections.

In the latter case, we note that the fractional error is essentially independent of energy, which follows from the equations used for energy loss; the form of the equations is unchanged until the muon energy drops below a few TeV. We note also that we cannot approach the desired 50% energy determination until the track is at least 400 m, preferably 600 m long. This is an important point; it affects the fiducial volume available for neutrino interactions, since neutrinos must interact at a point which allows a sufficient length of muon track to be visible. (This is the reason that the "standard" array adopted later in the session is a hexagonal array of larger diameter and lower altitude than the hitherto conventional cubic array.)

The next stages of the Monte-Carlo calculations are as follows:

1. Modify the program previously described, that generates muon tracks with randomly selected energy losses, to calculate as well the illumination produced at all points in the vicinity of the track. This has been done, using the angular distribution of the Čerenkov radiation calculated by Belayev et al.[5] and shown in Fig. 8. The calculation includes the attenuation of the water as well. Thus we end up with a map whereby we may find the illumination at any point in the vicinity of the track.

2. The next stage will be to set up an array of optical sensors in a lattice and to find the illumination produced at each detector. From this we will try to reconstruct both the trajectory of the muon and the energy loss at all points along the track, and thus by degrees reach the point of measuring the energy. This part of the calculation has not yet been done.

B. Detection of W-boson via its propagator in inelastic neutrino-Nucleon Scattering

A major part of our effort in the analysis of potentially interesting experiments with the DUMAND array has gone into a study of the feasibility of an experiment to detect the presence of the W in the interaction of neutrinos with nucleons. The experiment takes the form of studying the y-distribution in inelastic muon-neutrino scattering from nucleons. The existence of a W-boson with a mass in the vicinity of 50 to 100 GeV/c^2 has significant effects on both the cross section and elasticity of the interaction in the neutrino energy range 2 to 100 TeV, through the effect of the propagator due to the boson mass.

The study has two important aspects. First is a study of the interaction itself, to determine with what precision the energies of the hadronic cascade and the muon need to be determined to get significant results, what the effects of both random and systematic errors in energy determination will be, the influence of the steep neutrino spectrum on the results, and the procedures to be used in analyzing the data. These can all be investigated by Monte-Carlo methods.

The second part is a determination of the accuracy with which both cascades and muon energies can in fact be determined in a DUMAND array; this was discussed in the previous section.

Both of these studies are now under way; we give here a brief recapitulation of progress to date on the first one, the study of the interaction.

1 Procedure

First, the equations of Gaisser and Halprin[6] for the inelastic cross section, which depend explicitly on both x and y, were numerically integrated over x at a series of values, thus leaving predictions of the y-dependence of the cross section (see Fig. 9). These are model-dependent to the extent that they depend (but not strongly) upon assumed quark-structure functions taken from inelastic electron and muon scattering.

The predicted y-distributions for a series of energy bins for the incident neutrino were then calculated and tabulated. Storing such tables in a computer memory then makes it possible to choose a neutrino energy according to an assumed spectrum shape, pick a random number that assigns a properly weighted y value to the neutrino interaction, and thus obtain an event with known neutrino energy and known energy of the product cascade and muon.

The next step is to decide on an error variance for the energy of each of the two disintegration products, the cascade and the muon, and then to choose a random number, which together with the assumed variance, determines the energy of each of these two components. The neutrino energy as measured is then the sum of the two components, and the y-value the fraction of the total energy taken by the cascade. We can now sort out the resulting energies and y-values and compare them directly with the correct values represented by the originally selected particles.

Another parameter that can be treated by such means is the admixture of events due to muon antineutrinos. The y distribution for antineutrino events is sharply different from that of neutrino events, having a $(1-y)^2$ distribution at low energies. However, both neutrino and antineutrino y-distributions are equally affected by the IVB propagator. Fortunately, the interaction cross section of the antineutrinos is much less than the neutrino cross section (1/3 at low energies) and the abundance, from both the +/- ratio of cosmic-ray kaons (the main source), and direct calculations based on pion and kaon yields in primary interactions (which are not known too well) the actual contamination is readily corrected for.

The error in this procedure is the uncertainty in the fraction of the interactions due to the neutrino component. Figure 10 shows this, the y distribution being approximated by the ratio of events with y > 0.5 to those with y < 0.5.

In the case of a steeply falling spectrum like that of the cosmic-ray neutrinos (which we have taken as an integral spectrum varying as $E^{-1.5}$ including the variation of cross section with energy, there is a strong tendency for experimental errors to bias the energy distribution. At any given energy an experimental error in the energy of given magnitude will as often be positive as negative. But the positive error deposits the event in a bin of higher energy which is less populated than the parent bin; while a negative error of equal magnitude deposits in a lower energy bin which has a greater population. The effect is thus to shift the mean energy upward, and to bias the higher energy bins by whatever y-distribution the incorrectly assigned particles may end up with. Figure 11 illustrates this.

This process is readily simulated on the computer. Given the measurement accuracy, we can find the distortions introduced in the distributions from contributions from all sources, and unfold the distributions to correct for the contamination.

The final error will then contain not only the statistical errors, but the possible residual systematic errors introduced in making the corrections. However, the slow change of the y-distribution with energy, and the insensitivity of the shape of the distribution to energy errors lead us to believe that, even though this program has not yet been carried out, it offers no difficulties in principle; in practice, we will have to determine whether the unavoidable residual uncertainties in the precision of measurement and in the antineutrino component are sufficient to jeopardize the measurements. Since the presence of the propagator exerts so marked an effect on the y-distribution, it is difficult to see how uncertainties in the measurement errors can mask it, but the final results of the calculation will make that clear.

C. Multiple muons from primary cosmic rays

Before closing we should like to mention briefly the considerable interest attached to the study of multiple muons originating in primary interactions in the atmosphere, which are relevant to the highly interesting question as to the composition of cosmic-ray primaries in the energy range 10^{15} eV and above. The important study of the Utah group on this question[7] has now been supplemented by a calculation by Elbert[8] of the rates to be expected in a 1 km^2 DUMAND array which has 20,000 times the area of the Utah underground array. It is shown in Table I. Examination of the table indicates that for this purpose the resolution of the DUMAND

Table I. Multiple muon event rates (J. Elbert, 1978)

N	Rate	Percentage Contributions By Primaries				
		H	He	N	Mg	Fe
1	12.0 s^{-1}	81	14	2.7	1.4	0.8
2	0.28 s^{-1}	48	25	12	9.0	7.0
3	0.03 s^{-1}	24	22	18	17	19
4	0.01 s^{-1}	16	18	19	20	28
5	0.004 s^{-1}	13	16	18	21	33
6	6.6 h^{-1}	12	14	17	21	36
7	3.6 h^{-1}	11	14	17	21	38
8-10	4.5 h^{-1}	11	13	16	20	40
>10	3.5 h^{-1}	9	11	16	20	45
>100	0.4 d^{-1}	9	11	15	20	45
>1000	0.6 yr^{-1}	9	11	15	20	45

array needs to be closely examined to see whether the number of muons in a multiple event can be determined well enough. The mean spacing of the muons at the array is estimated by Elbert as about 5 m. If the sensor spacing of the array, now thought of as in the 30-40 m range, is inadequate for this purpose, as seems quite possible, it may be worthwhile to consider a single layer of sensors, probably at the bottom, with sufficient resolution, but perhaps not covering the entire area.

D. Detection of high-mass objects decaying into two or more muons

The size of the DUMAND array is sufficient to detect events in which two widely separated muons reach the array and to determine their point of origin. A massive neutral boson like the Z^0 is a candidate for such parentage, also pairs of charged particles that can each decay into a muon. We estimate that the array can detect particles with masses up to about 1 TeV formed in primary interactions, if they decay in such a mode. The background for such events (two muons) is random coincidences; with a counting rate of 15 sec^{-1} for the whole array, and a resolving time of 30 nsec, there will be six per day. These will of course be randomly distributed in energy, position, and direction, and the decay muons from a real event should be readily distinguishable on these counts. Events involving more than two muons will have much lower backgrounds, as e.g., the decay of a massive Higgs boson into two Z^0's that yield four muons.

REFERENCES

1. A. Roberts, J. Learned, F. Reines, R. Oakes, D. Eichler, D. Schramm, and G. Wilkins, DUMAND 78/5, Scripps Institution, UCSD, La Jolla, 1978.
2. V. W. Jones, Proc. 1978 DUMAND Summer Workshop, to be published.
3. R. K. Adair and H. Kasha, in Muon Physics, edited by V. W. Hughes and C. S. Wu (Academic Press, New York, 1977), Vol. 1, p. 324.
4. G. Igo and R. M. Eisberg, Rev. Sci. Instrum. $\underline{25}$, 450 (1954).
5. A. A. Belayev, I. P. Ivanenko, and V. V. Makarov, Proc. 1976 DUMAND Summer Workshop, Fermilab, 1977, p. 563; also ibid., Proc. 1978 DUMAND Summer Workshop, to be published.
6. T. K. Gaisser and A. Halprin, XV International Conf. on Cosmic Rays, Plovdiv, Bulgaria, 1977, Vol. 6, p. 265.
7. G. H. Lowe, M. O. Larson, H. E. Bergeson, J. W. Cardon, J. W. Keuffel, and J. West, Phys. Rev. $\underline{D12}$, 651 (1975); J. W. Elbert, J. W. Keuffel, G. H. Lowe, J. L. Morrison, and G. W. Mason, ibid. $\underline{D12}$, 660 (1975).
8. J. W. Elbert, Proc. 1978 DUMAND Summer Workshop, to be published.

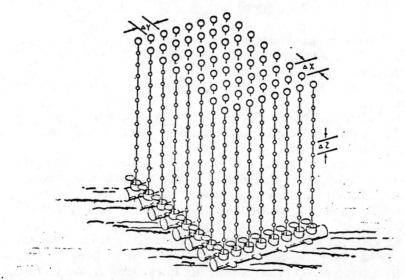

Fig. 1. Initial DUMAND concept: a simple cubical array of sensors.

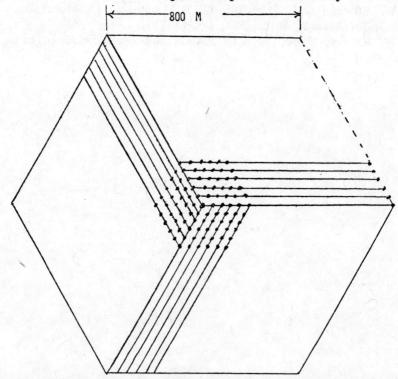

Fig. 2. 1978 Workshop concept: plan view of a hexagonal array with 62 **rows** containing 1262 strings. This is not a perspective view of a cube. Not all rows or string locations are shown.

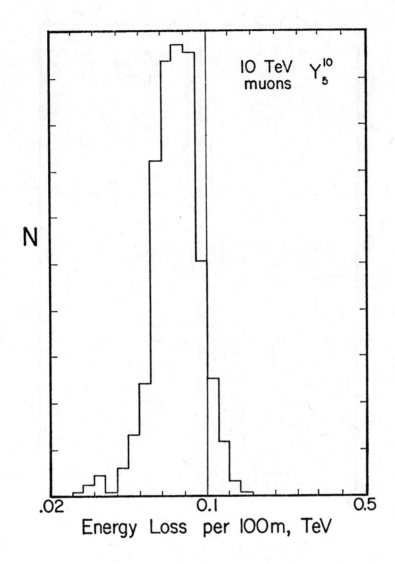

Fig. 3. Distribution of Y_5^{10}, the average energy loss in the five segments of a ten-segment muon track that show the lowest energy losses. Each segment is 100m long.

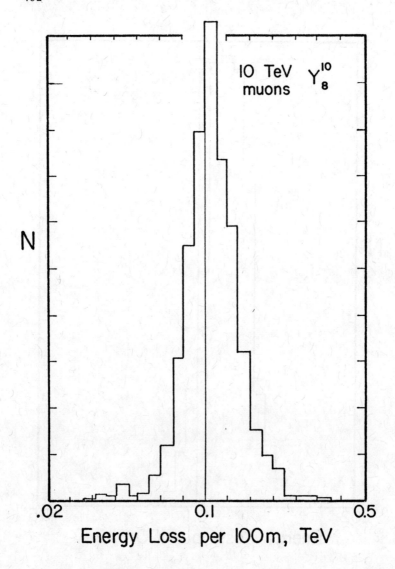

Fig. 4. The same as Fig. 3, except that we take the eight segments with the lowest energy losses; note the increase of the mean value.

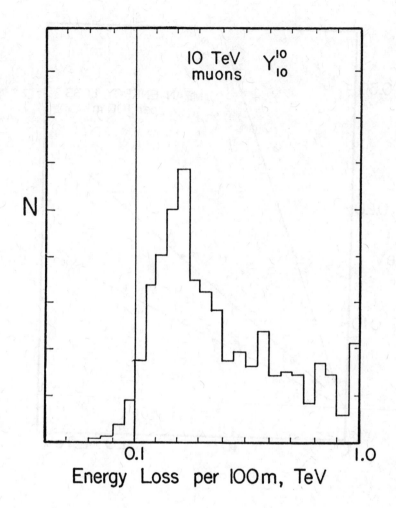

Fig. 5. Distribution of the average of all ten segments of the track; note now the long tail, going up to complete stopping of the particle, due to a few very large energy losses.

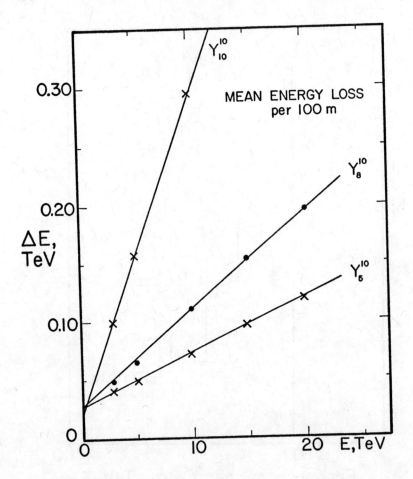

Fig. 6. The mean energy losses of the last three figures, as a function of particle energy.

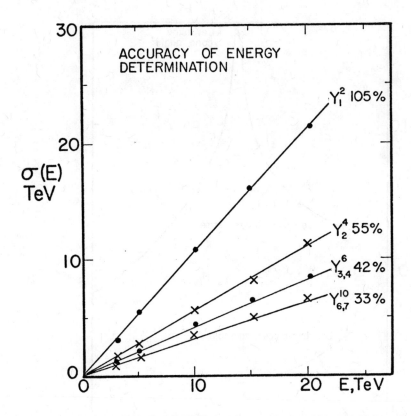

Fig. 7. The accuracy of the energy determination of the muon, as determined for several cases, with different numbers of segments. The value given is the variance of the energy loss distribution.

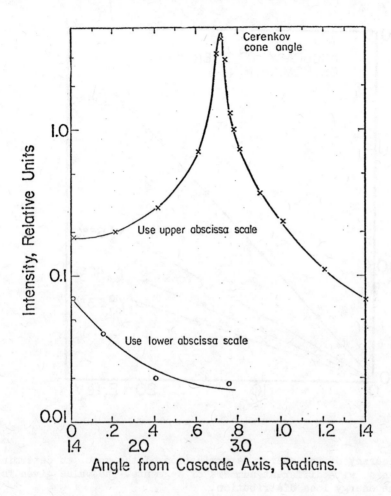

Fig. 8. Angular distribution of the Cerenkov light from a 10-TeV electromagnetic cascade at large distances, from Belayev, Ivanenko, and Makarov[5].

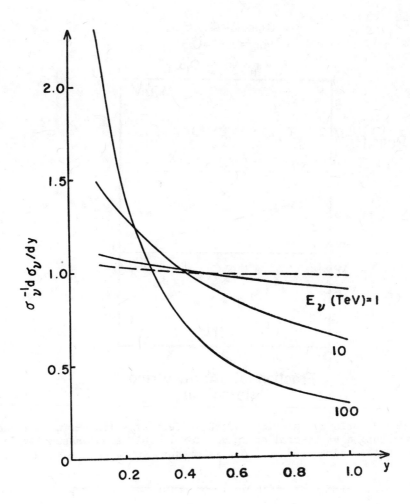

Fig. 9. The y-distribution for muon neutrinos, according to Gaisser and Halprin[6].

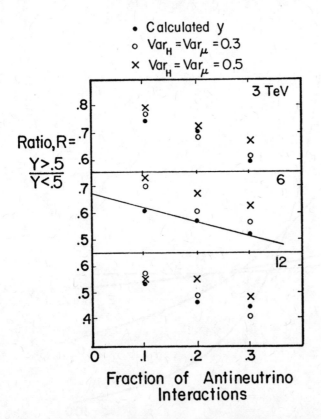

Fig. 10. Monte Carlo calculations of the effect of antineutrino component on the y-distribution at several energies, for two different assumed values of the variances of the hadron and muon energy measurements.

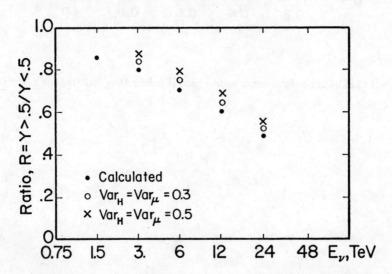

Fig. 11. Monte Carlo calculations of the y-distribution as a function of neutrino energy, for two assumed values of the energy variances.

THE HOMESTAKE NEUTRINO DETECTOR AS A COSMIC RAY
COMPOSITION ANALYSER*

M. Deakyne, W. Frati, K. Lande, C.K. Lee and R.I. Steinberg
Physics Department, University of Pennsylvania
Philadelphia, Pennsylvania 19104

ABSTRACT

The use of the Homestake Neutrino Detector in analysing the mass composition of the cosmic ray primaries by measuring the multiplicity of very high energy muons produced in air showers is described. It is likely to be of value for $E_o > 10^{15}$ eV.

I) INTRODUCTION

One of the major unresolved problems in the study of high energy cosmic rays is the composition of primaries for $E_o > 10^{15}$ eV. Most of the investigations in this regime have been carried out with extensive air shower arrays. Unfortunately, the EAS detectors are primarily sensitive to the energy of the cosmic ray primaries and not to their mass. One promising alternative approach to investigating the primary mass composition is to study the high energy meson multiplicity distribution produced in the first interaction length of the atmosphere. Several factors tend to give rise to a higher muon multiplicity from the interaction of heavy primaries such as Fe than from the interaction of protons. The larger cross section for heavy nucleus interactions results in these interactions occurring at higher altitudes thus providing a longer path for pion decay before absorption by nuclear interaction. It is also expected that a heavy primary behaves like A independent nucleons each with about 1/A of the energy of the primary.

Given the above scenario one experimental approach to mass analysing the high energy primaries would be to measure the multiplicity distribution of muons observed in a large area deep underground particle detector. The deep underground locale would impose a range and thus minimum energy requirement on the detected muons and the large area would insure that the muon shower is contained.

II) HOMESTAKE UNDERGROUND DETECTOR

We have recently put into operation a large deep-underground multi-purpose neutrino and muon detector[1]. This instrument, which is located at a depth of 4850 ft in the Homestake Gold Mine,

---

* This research is partially supported by the U.S. Department of Energy under contract no. EY-76-C-02-3071 and by the U.S. National Science Foundation under grant no. AST 78-08669.

Lead, South Dakota, is a multi-counter hodoscope in the form of a box 20 m long, 10 m wide and 7 m high, Fig. 1. The sides of the hodoscope consist of a set of water Cerenkov counters, while the top of the hodoscope consists of a series of liquid scintillation counters. In addition to detecting the interactions of low energy neutrinos the detector can also record muons that pass through the hodoscope. These muons arise either from the interaction of high energy neutrinos in the rock surrounding the hodoscope or from high energy cosmic ray production in the upper atmosphere.

For the purpose of this conference we wish to focus on the cosmic ray muons and to consider the possibility of utilizing the multiplicity distribution of these particles to investigate the mass composition of cosmic ray primaries with $E_o > 10^{15}$ eV.

Elbert[2], has recently described a model that predicts the multiplicity distribution to be expected from various mass primaries as a function of their energy. His model, which summarizes the results of detailed Monte Carlo calculations and lower energy observations[3], predicts that the multiplicity of muons above an energy E_μ due to a primary of energy E_o

$$N_\mu (> E_\mu) = \frac{A\, G(x)}{E_\mu \cos\theta}$$

where A is the mass of the primary, $x = \frac{E_\mu}{E_N}$ and $E_N = \frac{E_o}{A}$. The function $G(x) = ax^{-b} e^{-cx} K(x)$ where $a = 0.0145$, $b = 0.757$, $c = 5.7$ and $K(x) = 1.4 - 2x$ for $0.1 \leq x \leq 0.7$ and ≈ 1 for $x < 0.1$.

For our detector, which is at a depth of 4400 m.w.e., E_μ varies from 2.6 TeV at a zenith angle $\theta \approx 0°$ to 10 TeV at $\theta = 55°$. Using the above equations we have calculated the expected muon multiplicity vs. E_o for our depth assuming proton and Fe primaries, Figs. 2 and 3. It should be noted that for small zenith angles we investigate $E_o \gtrsim 10^{15}$ eV while for large zenith angles ($\theta \sim 55°$) we have an effective threshold of $E_o \gtrsim 10^{16}$ eV. Unfortunately, our apparatus does not provide a measure of E_o. In order to provide a basis of comparison with experimental data we have integrated the energy dependent multiplicity distributions over the cosmic ray primary spectrum. This provides an absolute muon multiplicity prediction, Figs. 4 and 5.

In order to provide a true multiplicity measure it is necessary that the detector be large compared to the typical separation between muons. Given that meson production occurs at a height of 15-20 km above sea level and that the typical $P_t \approx 0.4$ GeV/c while $<E\mu> \approx 4$ TeV for those muons reaching our detector, we find that muon separations will be 2-3 meters. Since this is small compared to the size of our detector (10 m x 20 m) we should be able to contain most of the muons in a shower that reach our depth.

III) ADDITION OF AN EXTENSIVE AIR SHOWER ARRAY

There is a risk in integrating over the cosmic ray energy spectrum in that it is quite possible that there is a different energy dependence for proton primaries than for heavy primaries. This may be especially important in the region around 10^{15} eV. One possibility for avoiding this problem and possibly shedding some light on the energy dependence of the primary spectrum would be to operate a surface extensive air shower array in coincidence with our underground detector. The surface array would provide a measure of the total energy of the shower associated with each muon observed by the underground detector.

An additional feature of this combined instrument is that it would precisely determine the direction of the incident primary particle. Since both the surface detector and the underground detector would locate the shower core to within a few meters and the lever arm between the two detectors is 1.5 - 2 km we would know the primary direction to $\pm$ 1-2 milliradians. Such a precise determination of primary direction would aid in the search for localized sources of cosmic radiation.

REFERENCES

(1) M. Deakyne, W. Frati, K. Lande, C.K. Lee, R. Steinberg and E. Fenyves, Conference Proceedings Neutrino - 78, p. 887, Purdue University 1978.
(2) J.W. Elbert, Proceedings of the 1978 DUMAND Summer Workshop. We are indebted to T.K. Gaisser for aiding in the application of this model to our detector.
(3) J.W. Elbert et al., Phys. Rev. D12, 660 (1975).

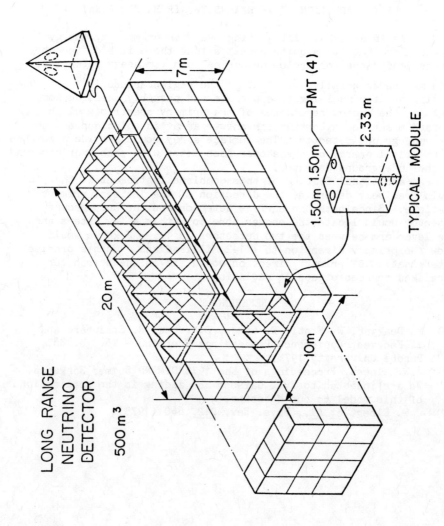

Fig. 1   Drawing of the Homestake Neutrino Detector showing the water Cerenkov counter modules along the sides and the liquid scintillator counters on the top.

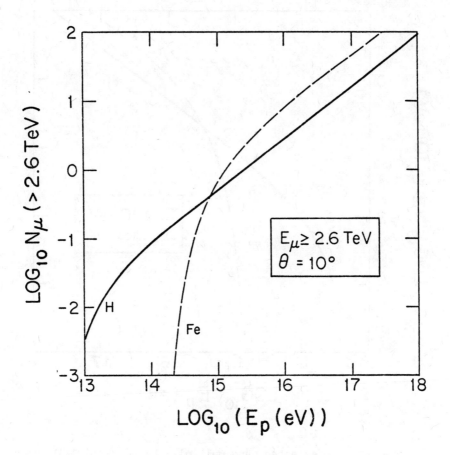

Fig. 2   The expected muon multiplicity (E_μ > 2.6 TeV) produced by H and Fe cosmic ray primaries.

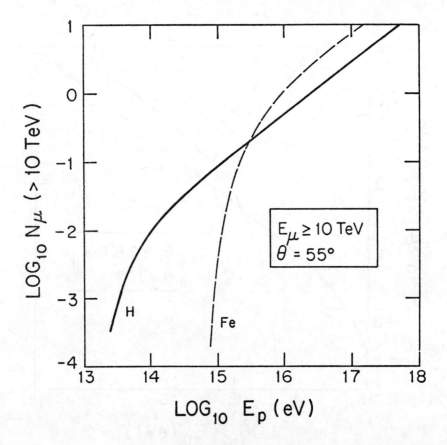

Fig. 3   The expected muon multiplicity ($E_\mu > 10$ TeV) produced by H and Fe cosmic ray primaries.

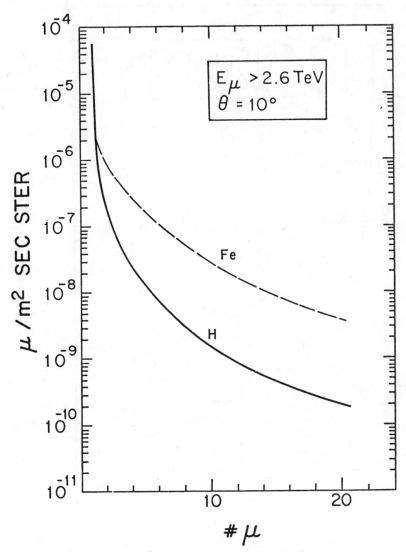

Fig. 4   The muon multiplicity distribution after integration over the cosmic ray primary spectrum for $E_\mu > 2.6$ TeV.

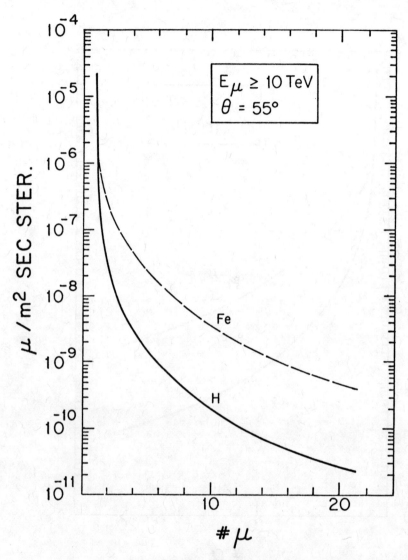

Fig. 5 The muon multiplicity distribution after integration over the cosmic ray primary spectrum for $E_\mu > 10$ TeV.

REPORT ON THE UTAH FLY'S EYE

G. L. Cassiday, H. E. Bergeson, E. C. Loh
J. W. Elbert and D. Steck
Department of Physics, University of Utah
Salt Lake City, UT 84112

ABSTRACT

Extensive air-shower trajectories and sizes (numbers of charged particles) have been measured using an optical detection system at Volcano Ranch Station near Albuquerque, New Mexico. Light produced by atmospheric scintillation and Cherenkov emission by shower particles was measured at distances of 0.7 to ~10 km. The shower sizes determined by the optical measurements are in satisfactory agreement (an average of 10% higher) with measurements by the ground-level scintillation-counter array at Volcano Ranch. Twenty-four optical units have now been constructed on top of Little Granite Mountain, Dugway, Utah, and are nearly operational. Important parameters of the optical system and electronic data acquisition system are presented along with data rate estimates based upon the calibration experiment performed at Volcano Ranch.

INTRODUCTION

Ultrahigh-energy ($>10^{16}$ eV) cosmic rays are potentially useful for studying high-energy nuclear interactions. Up to now, the combination of small incident flux and limited detector size has prevented the full exploitation of this potential. In 1965, Greisen[1] described the possibility of detection of remote air showers by optical methods, a scheme which would enable a detector to have an effective viewing volume of ~$10^{13} m^3$, but subsequent attempts[2-4] to use this method were unsuccessful. We report measurements of light production by distant air showers, using an optical system consisting of three "mirror units" described below.

Remote optical detection of air showers is possible because shower particles (primarily $e^{\pm}$) excite nitrogen molecules which subsequently emit light isotropically. Although the energy loss per meter in air increases with increased pressure, deexcitation by molec-

ISSN 0094-243X/79/49417-12 $1.50 Copyright 1979 American Institute of Physics

ular collisions rather than scintillation also increases with pressure. As a result, the light production per meter per particle is almost (within ~10%) independent of pressure.[5] Shower particles travel in a relatively small "packet" at practically the speed of light, thus forming well-defined trajectories. Consequently, shower distances and sizes can be found without undue reliance on calculations based on models.

CALIBRATION EXPERIMENT

a. Description

In this experiment an optical system consisting of three 1.5-m-diameter $f/1.0$ mirrors and 36 photomultiplier tubes (12 per mirror) was positioned 1.53 km from the center of the ground-level scintillation-counter array (hereafter denoted SA) at Volcano Ranch, New Mexico.[6] The optical system was pointed toward the SA so that its field of view passed just above the ground-level counters as shown in Fig. 1. Hence, most air showers falling within the SA passed through the aperture of the optical system. The optical system was operated in coincidence with the SA in order to determine (1) whether remote optical sensing of air showers is possible, and if so (2) whether the optical signals agree with expectations based on shower sizes and trajectories obtained with a conventional counter array.

The spherical mirrors were aluminized on the front surface. In each mirror's focal plane a cluster of twelve aluminized-plastic, hexagonal-faced funnels helped gather light onto the uv-sensitive, 90-mm diam cathodes of EMI 9861 B photomultiplier tubes. The maximum angle subtended by a funnel-phototube combination was 5.8^0. Before and after the experiment the cathode efficiencies (~23% at 3900 Å) and gains of the phototubes were measured in the laboratory. During the experiment the optical sensitivities were monitored for each phototube using a single, stable light pulser coupled to the mirror units by optical-fiber cables. The integrated pulse heights and signal-arrival times were stored, and later printed, for all showers which triggered the SA and also the optical system.

b. Measurements of Trajectories

A total of 44 showers was registered by both systems in twelve consecutive nights of operation (~100 h). Shower sizes and trajectories were calculated in the usual manner from data given by the SA.[6] The sizes ranged from 1.3×10^7 to 6.2×10^8 particles; the shower energies, from $\sim 5 \times 10^{16}$ to 2.5×10^{18} eV. The apparent brightness of each shower was calculated directly from the optical-pulse integrals. However, the intrinsic brightness could be obtained only for 15 showers with signals well above threshold and tracks well within the apertures of the top mirror unit and one of the bottom ones (see Fig. 1). The determination of a shower trajectory from the optical data involved a two-step fitting procedure. First, the location of the shower-detector plane in space was determined by fitting a straight line to the phototube data of Fig. 1. Next, the shower's impact parameter R_p and ground impact angle ψ (see Fig. 2) were determined from the arrival-time measurements plotted in Fig. 3. In order to determine both R_p and ψ from the optical data, it is necessary that there be significant curvature in the graph illustrated by Fig. 3. Since a given shower could pass through the aperture of no more than two of the three mirror units, the field of view was so small that little curvature was present; so fits were made possible by constraining the trajectories to pass through the point of impact of the shower core with the ground, as determined by the SA. Comparison of shower trajectories so obtained with those determined independently by the SA yielded average differences of magnitude 4.8^0 in the projected zenith angle α (see Fig. 1), and of magnitude 8.6^0 in the angle ψ. Both values are consistent with the estimated combined angular resolutions of the two systems.

c. Calibration Results

After reconstruction of the trajectories, shower sizes were evaluated from the optical data and compared to the SA results. Because the shower tracks were often more than one phototube-aperture wide, the signals from adjacent phototubes were combined. The ratios of "optical size" to SA size are given in Fig. 4(a), assuming that light production is constant, isotropic, 4.0 photons per meter per particle. Ratios above 1.0 imply that more light was received than was expected from atmospheric scintillation. The data of Fig. 4(a) determine a calibration curve for coverting light measurements to shower sizes. The systematic errors of SA size measurements are ±20% and instrumental errors in the optical measurements are estimated to be ±30%. Such errors are mainly due to uncertainties in the phototube gains, cathode efficiencies, and the mirror and funnel reflectivities.

A shower-development calculation estimated the effects of both Cherenkov light received directly and of light which was received after scattering out of the accumulated Cherenkov-light beam by atmospheric aerosols and air molecules (Rayleigh scattering). It also estimated the effects of shower-size changes in the distance interval between the optical emission point and the SA observation point. The calculation indicated that on the average ~38% of the light reaching the detector was lost because phototube signals generated by the lateral edges of the showers were below threshold. The optical sizes of Fig. 4(b), which include corrections for those effects, show very little angular dependence. The large peak at $\theta < 30^0$ in Fig. 4(a) was due to Cherenkov light scattered by aerosols as well as Cherenkov light received directly. The θ dependence of the direct Cherenkov light is caused primarily by the angular spread of the charged shower particles. In Fig. 4(b) the average ratio of the optical sizes to the SA sizes (1.10) changes by only ±0.07 as the atmospheric aerosol content is varied within the probable error estimates.[7] The rms deviation of the ratios about the average value is 0.39. The large corrections at small θ values are sensitive to shower-development fluctuations and errors in measurement of θ;

hence, the data fluctuate more at the smaller angles.

A sample of twenty showers which triggered the SA but did not trigger the optical system was also analyzed. In seven cases, the trajectory was outside the optical aperture. In the remaining thirteen, the expected optical signals, as calculated from the SA size of each event and the geometry, were below threshold in all channels.

We have demonstrated the ability of an optical system to detect, reconstruct, and measure sizes of remote air showers. We now discuss the progress of a 67-unit detection system currently under construction on top of Little Granite Mountain, Dugway, Utah.

DESCRIPTION OF THE FLY'S EYE

a. Introduction

Before discussing the Fly's Eye itself, we briefly mention the physics we hope to do with such a device--(1) measure the primary cosmic ray spectrum for energies $10^{16} \lesssim E \leq 10^{21}$ eV (2) measure the proton-air interaction length (3) separate protons and heavy nuclei in primary cosmic rays (4) directly measure the longitudinal development of EAS and thereby test various particle interaction models (5) search for primary cosmic ray anisotropies (6) look for high energy ($\approx 10^{20}$ eV) neutrino interactions by "upward-going" shower signatures and (7) search for anomalies. Most of these experiments are made possible only by the unique combination of resolution and data gathering power which the Fly's Eye exhibits.

The Fly's Eye consists of 67 1.5 meter-diameter spherical mirrors. Each mirror is viewed by a cluster of 12 (or 14) photomultiplier tubes and associated light collecting funnels. The optics is such that almost the entire sky (about six steradians) is covered by the detector. (Listed in Table I are the physical details of the detector). A shower induced by a high energy primary cosmic ray will be "seen" by a number of "eyes" in the Fly's Eye for a significant portion of its entire trajectory through the sky. This situation is schematically depicted in Fig. 5.

TABLE I

Specifications for Fly's Eye Optical System

Number of mirrors	67
Number of Photomultiplier Tubes (PMT)	873
PMT Type	75 mm EMI 9861 B
Total solid angle Ω	5.8 steradians
Diameter of mirror	1.59 m
Radius of curvature	3.05 m
Reflectivity of mirrors (in our angular acceptance)	85-92%
Intrinsic spot size for on axis parallel incident rays	0.02 m
Full cone angle/PMT	.089 radians
Solid angle/PMT	.007 steradians
PMT operating gain	$\sim 1 \times 10^4$
Average anode current (typical night sky)	10×10^{-6} amp
Amplifier gain following PMT	10^3
% area obscured by phototube light cone cluster	14%
Winston cone reflectivity	85%

As the shower progresses across the sky, it is "seen" successively by a sequence of photomultipliers which lie along a great circle on the Fly's Eye. The timing sequence plus the geometry of "struck" tubes permit the determination of the shower's trajectory. Once the trajectory is determined, the light yields from the struck photomultipliers can be converted into shower electron number. The measured light yields are obtained very simply by integrating the photomultiplier pulse if that pulse is above background noise. Thus, the Fly's Eye basically measures the size of the shower in electron number as a function of distance along its trajectory and it is this observable with which we do physics.

b. Physical Layout and Current Site Status

Each mirror and associated photomultiplier cluster is housed in a 2.13 m long x 2.44 m diameter corrugated cylindrical steel pipe. Each of these units is attached to two steel legs welded to steel plates which are rock bolted and cemented to Little Granite Mountain. Shown in Fig. 6 is a photograph of several of the units in place. Figure 7 is the field of view as seen from one of the units. (In general, visibility is quite good). Each corrugated housing has been correctly surveyed in an azimuth, but is free to rotate about a horizontal axis. Each unit faces the ground during the day and is manually swung into a locked position at night, thus pointing to an appropriate position in the sky. This technique protects the mirrors from the weather and sunlight even though the viewing end of each unit is always open. The phototube clusters are, of course, covered during the day to prevent light exposure. The physical disposition of these units is shown in Fig. 8. Currently, twenty-five detectors are completely installed on the hill. Their location is depicted by circles. The bases for the remaining 42 detectors are also installed. They are depicted by the small rectangular feet.

Also shown in Fig. 8 is the housing for the electronics and computer system. A fifty-foot trailer houses a tool storage area and workshop (used for wiring, etc.), a computer input/output workshop area and a room adjacent to a kitchen for eating and cooking while on site. In addition, four large Air Force packing crates were mounted on a foundation, insulated, made weatherproof, physically connected together to make one long trailer-like structure and then attached to the trailer itself via a passageway. This second structure was then wired for power to house the transducer electronics, data acquisition electronics and computer. The computer room, itself, has been air-conditioned.

As seen in Fig. 8, there is a service road which completely surrounds our central housing area. Furthermore, at the northwest corner of the hill is a communications structure used by Dugway. We were constrained by Dugway to (1) keep the service road free for at least a 20-foot width all along its length and (2) not mount any structures beyond their communications structure. This constraint forced us to deploy our mirrors in three main sectors: (1) north of the electronics housing on the other side of the access road; (2) around the housing and totally within the encircling access road; and, (3) on the southern side of the hill across the access road from the central housing. This mode of deployment resulted in rather long cable runs from the units to the electronics housing. In some cases, cables must extend about 300 feet. Furthermore, stringing the cable proved to be no small task because of the unforeseen complications of small animals such as kangaroo rats. These animals chew the cable. Consequently, we were forced to erect metal cable troughs, starting at the central electronics house and running in several spurs which service various segments of the mirror array. At each mirror location, a flexible metallic sheath starts at the horizontal axis of the mirror housing, runs down one of the housing support legs and then attaches to the adjacent cable service spur. In this way, cables are completely covered (except where they exit to the photomultipliers) by metal covering all the way into the electronics housing trailer. Moreover, a trench had to be dug across the service road and a small concrete causeway with removable iron plate tops had to be built to house two of the main cable trough spurs as they crossed the road. At this time, about 1/3 of the spur system has been installed and cabling for eight mirror units has been laid. This cabling has successfully resisted assault by both weather and rodents for several months.

The electronics housing for all 67 units is now complete. Power busses have been installed in the housing to handle the electronics for 32 units. Also, power for the computer, computer interface electronics, address generation electronics, programmable pulse generator, 20 MHZ timing system, and photomultiplier high voltage distribution system is complete. All internal patch panels for wiring in signals and high voltage for all 67 units is also complete.

c. Electronics Overview and Status

By far the most complicated problem about the Fly's Eye and the most difficult to successfully solve has been the electronics. The complexity and difficulty arise primarily from three sources: (1) the input signal from a single photomultiplier has a tremendous dynamic range...of the order of 10^5-10^6; (2) there are a tremendous number of photomultipliers...of the order of 1,000 (3) light sensors may "see" a signal well before any triggering decision has been made. Thus, analog storage of fast signals for times as long as 10 µsec is required. The first difficulty is essentially caused by the great variation in distance of optical sources from the detector (0.30-50km). This variation results in electronic pulse widths that vary between 50 nsec and about 15 µsec. (Some of this time variation is also caused by differences in angle between the optical line of sight and the shower axis). Moreover, the variation in size of visible showers at nearby distances is of the order of 100, i.e., pulse heights can vary between 50mv and 5v. Thus, pulse integrals can vary by 10^5-10^6. The second difficulty is obvious--a lot of complex electronics must be built, tested, and then installed. The third difficulty is most severe. As an EAS works its way across the field of view of the Fly's Eye, several light sensors will, in general, have received information before a "trigger" can be generated. This implies that signals must be "delayed" or stored in analog form for relatively long periods of time before signal processing can even commence, since an EAS might take as long as 100 µsec to develop. However, the nearby EAS develop in just a few µsec and individual pulse widths may be as short as 50 nsec, hence, the analog "resolution" must be one part in 10^3 of the storage "delay" time, and this places severe constraints on the parameters of any analog storage device. Such things as transient recorders or charge-coupled devices (CCD arrays), although nice in principle, prove far too costly for a light sensor array of this size and dynamic time range. One might also imagine the possibility of digitizing immediately every light signal above some threshold and then digitally processing the information. This, of course, would require parallel processing with a minimum of 1000 reasonably fast A/D converters! Again, three years ago both the CCD technique and the parallel processing A/D technique were outlandishly expensive. That situation is rapidly changing and could, today, represent a viable alternative to the data handling system we have, in fact, already implemented.

Shown in Fig. 9 is a block diagram of the electronics. The signal from each phototube is first amplified by a fast current preamplifier physically mounted on the base of each photomultiplier. This amplified signal is then sent into the central electronics housing area via an RG 174 coaxial cable. There it proceeds through a patch panel network to a centralized rack area containing the front-end signal processing electronics (which we call ommatidial boards in analogy with the operation of a Fly's Eye). One ommatidial board (schematically shown in Fig.10) is devoted to processing the signal from each photomultiplier. The signal is here split into four separate channels. Each of these channels is further split in two: one portion going through a filter so that only pulses of an appropriate rise time, height and width will trigger a discriminator. The other portion of the signal is delayed and then routed through a gated active integrator and peak finder. If the pulse triggers a discriminator in a given channel, the signal is integrated, its peak value is obtained and each of these values is held in analog form (in any or all of the four channels) for subsequent digitization. Furthermore, triggering a discriminator produces a logic pulse of an appropriate width which is sent to a "mirror" coincidence board; another logic pulse is sent to a latch board which records the time at which the input pulse crossed the threshold and finally a data present condition is asserted by setting a flip/flop. Thus, to recapitulate, each ommatidial board responds to pulses above a certain threshold (determined by noise levels), integrates and finds the peaks of the pulse, establishes a data present condition, generates timing and coincidence pulses and holds the analog values for digitization if a master trigger happens to be generated. If no master coincidence occurs within a specified time interval, the data present condition and the held analog values are released. The ommatidial board is then ready to process another signal.

Discriminator thresholds are determined by noise levels. Each phototube observes a relatively bright night sky resulting in an average phototube current of about 2000 photoelectrons per microsecond. Statistical fluctuations in this "dc" current result in noise. The bias on the discriminators is automatically adjusted by computer such that individual photomultiplier count rates are fixed at about 100 counts per second. Thus, as the night sky brightness level drops, the sensitivity of the Fly's Eye increases and vice-versa. The triggering sensitivity is always optimal and the dead time is fixed at a value of the order of 1%.

An event trigger consists basically of a given number of photomultiplier triggers in a single mirror followed by a given number of mirror triggers. The first logic pulse generated by an ommatidial board is a coincidence pulse which is transmitted to a mirror coincidence board (see Fig. 9), physically adjacent to each set of ommatidial boards servicing a given mirror cluster of photomultipliers. This board produces an output coincidence pulse in any one or all of four time channels upon receipt of two or more coincidence inputs from its associated ommatidial boards. We call this situation a "local" coincidence. A local coincidence thus consists of two or more photomultiplier triggers which occur within any one or all of four possible time spans ranging from several hundred nanoseconds (appropriate for nearby events) to several microseconds (for remote events). The number of photomultipliers required for a mirror coincidence can be set to either 1, 2, or 3. We feel that 3 is probably too "tight" a requirement, while "1" is certainly too loose. It is virtually impossible for a real event to pass through the field of view of only one tube. (If it passes through 2, 3, or 4 tubes, but only one is triggered, the event might well be marginal at this particular stage of shower development). Occurrence of a local coincidence initiates two actions: (1) a local "SAVE" command is generated which is passed to all ommatidial boards in the triggered mirror cluster and the "holding time" of all analog values is extended to 100 μsec. Thus, if a real event is in the process of development (subsequent mirrors may not yet be triggered), the data generated in this particular mirror will be safely held; (2) a local coincidence pulse is sent to the master coincidence unit. The master coincidence can be programmed to respond to anywhere from 1-16 local coincidences. For example, we could require that four mirrors produce local coincidences to generate a master coincidence. Remember, that all logic is duplicated for four different time channels; in this case, at the master coincidence level, time scales range from several microseconds to several hundred microseconds.

Upon occurrence of a master coincidence, two actions are initiated: (1) the data "SAVE" condition is asserted by a flip/flop for all photomultipliers in all mirror clusters; (2) an event flip/flop is sent which interrupts the computer...a PDP 11/34. The 11/34 waits for about 200 μsec to insure that the shower completely develops from this point on, allowing the developing shower to completely run its course, thus generating data in all possible mirror clusters. The 11/34 computer then asserts a master "HOLD-OFF" condition. Hold-off gates off all ommatidial boards from their respective photomultipliers. The Fly's Eye is now blind and the computer initiates data gathering operations.

As might be guessed, some sort of automatic calibration and system monitoring mechanism is obviously required in order to calibrate and check such a prodigious number of data channels. We have built a programmable pulse generator to carry out this task. Each phototube preamplifier has two inputs; one is obviously from the photomultiplier tube itself which is capacitively coupled into the input emitter stage of the transistor preamplifier. The other input is through an external BNC connector and a precision 34.8Ω resistor. (One can think of the phototube preamplifier as a current to voltage converter, i.e., a resistor of value 34.8KΩ). Thus, a voltage pulse can be sent out from the centralized programmable pulse generator to each photomultiplier in the system via coaxial cable. This voltage pulse is converted into a current pulse by the 34.8KΩ input resistor and amplified by the preamplifier in precisely the same fashion as a real photomultiplier current pulse. The voltage pulse that returns from each photomultiplier is then identical to the input calibration pulse. The programmable pulse generator can thus be programmed to generate a sequence of pulses that vary in amplitude from $100\text{mv} \leq V \leq 5\text{v}$ and in width from $100 \text{ nsec} \leq \tau \leq 15 \text{ } \mu\text{sec}$ which can be used for both calibration as well as system checking purposes. Indeed, each ommatidial board in the system is calibrated with a well-defined pulse sequence and the response is logged on a system disk. Hence, at any stage of data taking the response of any data channel may be quickly checked.

The programmable pulse generator thus provides a quick check on the entire electronic system response. Calibration and checking the optical response of the light collecting elements is equally important. Such a check necessitates exposure of each mirror and associated light funnel-photomultiplier cluster to a well-defined light source. The light source for each mirror consists of an argon flash tube located physically outside of the mirror and optically coupled to it via four optical fiber bundles (see Fig. 5). The fiber bundles are mounted in small holes drilled into the phototube cluster mounting plate such that a light pulse emitted from the argon flash tube is carried essentially to the focal plane of each mirror. This light flash illuminates the mirror; a plane wave of light is thus reflected into all 14 photomultiplier light collecting funnels which pipe the light into each photomultiplier. The optical pulse is reproducible from pulse to pulse[8] to a few 1%. Each mirror has its own flash tube and fiber bundle network which can be triggered on computer command. About four times each night during data taking, a series of 100 optical pulses can be used to monitor the combination of mirror-funnel light collection efficiency degradation and photomultiplier gain shift.

As a final check on photomultiplier performance, a small Cs^{137} source and associated NaI crystal are permanently bonded to one photomultiplier in each cluster. (The light obscuration is only a few % and the count rate is so low that this source does not interfere with "real" pulses). Pulse integral distributions for this particular tube can be generated at any time during data taking. Cross checking these distributions with distributions obtained from the triggered argon flash tube source permit the separation of light collection efficiency degradation from photomultiplier gain shift. Thus, the gains and efficiencies of all system components can be normalized to absolute values previously measured in the lab on the Utah campus.

d. Future Operational Plans

Electronic units are being assembled in groups capable of handling eight mirrors and associated photomultipliers. We are now on the air with first eight units. By February, 1979, we should have

24 units operational. At that point, we would basically halt all construction and production efforts (with the exception of phototube preamplifiers and ommatidial boards) and commence a period of data taking primarily to clean up all system bugs and quite possibly to do two significant experiments: (1) measure the primary cosmic ray spectrum in the energy range $10^{16} \leq E \leq 10^{18}$ eV. (All other experiments require geometrical resolutions, obtainable only with an almost fully complete Fly's Eye.) (2) intercalibrate shower measurements with Professor Ted Turver's Cherenkov array. Professor Turver, from the Durham, England, group, will have his Cherenkov detectors fully deployed and operational this winter at the base of Little Granite Mountain upon which the Fly's Eye sits. Fig. 11 shows a projection of the field of view of the Fly's Eye as seen from the zenith. Each projected mirror aperture is labelled with a number. The lightly shaded region shows those 25 mirror units which have been completely installed and which should be fully operational this winter. It is obviously no accident that their field of view is directed towards Turver's Cherenkov array about 1 km away. (The darkly shaded region represents those mirror units we are currently installing and plan to bring into operation next). A portion of those showers seen by Turver's array should also be visible to the Fly's Eye. Thus, our intercalibration can be performed á la the Volcano Ranch experiment, and our measurements of shower sizes can then be directly compared to those made by Haverah Park--an experiment of crucial importance, if future discrepancies are to be answered with confidence.

One final point should be mentioned. As previously pointed out, most of the experiments to be performed by the Flý's Eye require almost a fully implemented array in order to obtain good shower geometry. In particular, the measurement of the proton-air nucleus total cross section and the separation of primary particle composition depend rather crucially on the ability of the Fly's Eye to accurately locate shower trajectories and in particular, shower "starting points". It may well prove to be the case that deployment of a second peripheral Fly's Eye, about 7 km from the first, might provide far more valuable data than that obtained with a fully completed single Fly's Eye. The trade-off would be loss of overall data rate versus the increased geometrical redundancy of stereoscopic view. This trade-off would be most important for the remotest, most energetic and, therefore, the most intrinsically interesting events. In addition, for nearby events seen by both eyes (occurring at the rate of several per hour) the attenuation length of light through the atmosphere could be directly determined. A knowledge of this number is, again, most crucial for correct interpretation of remote events. Furthermore, a direct experimental separation of Cherenkov light and scintillation light could be obtained. For these reasons, we anticipate at the moment, that instead of completing the final 18 units during the summer of 1980, that we would instead begin set-up of those 18 units on another hilltop about 7 km away. By then, we should be able to ascertain with more confidence whether or not this possibility is more desirable than simply completing the main Fly's Eye complex.

RATES FOR MEASURING THE PRIMARY SPECTRUM

The geometry of a detector like the Fly's Eye is quite complicated. In order to simulate the response of such a detector to EAS, we developed a Monte Carlo program, which generates showers at random which are then "observed" by the detector. The program determines the minimum energy E which a shower of specified geometry must attain in order to "trigger" a sequence of N photomultiplier tubes, each of which is a certain minimum number of standard deviations above noise. Both N and the number of standard deviations can be chosen to optimize data rate limited only by chance rate and dead time considerations. The simplest trigger thus envisioned requires that at least four phototubes out of the entire Fly's Eye Array have signals at least five standard deviations above noise. The effect on data rate of any arbitrarily chosen triggering scheme can be tested by the program.

Showers were synthesized with random geometrical parameters and the energy of the shower is scaled up or down until the shower triggers the detector. The event rate is then

$$\text{Rate} = \int_{A\Omega} I(>E) \, d(A\Omega)$$

where A is the detector aperture and I(>E) is the integral cosmic ray spectrum. Clearly, obtaining I(>E) is one of the design goals of the Fly's Eye. Hence, a way of measuring this spectrum involves fitting the observed data rates to the above integral function. Here, in order to a priori estimate data rates, we carry out the inverse process. We use the spectrum I(>E) given by Greisen 1965[9]. However, we normalize our rate calculation to the experimental results obtained with the prototype Fly's Eye detector operating in coincidence with the Volcano Ranch Array (VRA) at Albuquerque, N.M., during November 1976. There we achieved an event rate of 0.5/hr. and a size threshold of $N_e \approx (0.5-1.0) \cdot 10^8$ electrons at a distance of $R_p \approx 1$ Km. We then plot in Fig. 12 the results of our calculation normalized to that result. The rates have been readjusted to correspond to the brightness of the night sky at our Dugway, Utah, experimental site which is not as severe as at Albuquerque, N.M., where the calibration experiment was performed.

We see from Fig. 12 that we can expect to see EAS at distances of $0.2 \lesssim R_p \lesssim 50$ km. The corresponding pulse widths range from $.07 \lesssim \Delta t \lesssim 17$ µsec. In order to trigger the detector, showers 50 km distant would have to have a size of $N_e \approx 4 \cdot 10^{11}$ electrons (at ground impact) or an energy near 10^{21} eV. Such showers would be observed at a rate of about 1/yr. Showers impacting within several hundred meters of the detector would require a size of about $N_e \approx 10^7$ electrons or an energy of about $2 \cdot 10^{16}$ eV. Such showers would occur at a frequency of about 10^6/yr. Hence, it should be possible to map out the cosmic ray spectrum in the energy range $10^{16} \leq E \leq 10^{21}$ eV with significant data rates.

RATES FOR CROSS SECTION AND PRIMARY SEPARATION MEASUREMENTS

In order to carry out the cross section measurement and the separation of protons from the heavy component, significant data cuts will have to be made. These two measurements each require that a reasonably precise location of the starting point of an EAS be determined. Obviously, the shower's starting point cannot be directly observed. However, the Fly's Eye can be expected to observe the ¼ maximum point which typically occurs at an atmospheric depth roughly 300g cm^{-2} beyond the depth of primary interaction. This occurs somewhere between 300-600g cm^{-2}. The distribution of the ¼ maximum points may be used as a measure of the distribution of interaction points. Unfortunately, many of the showers are seen early in their history traveling more or less towards the Fly's Eye. Hence, the optical emmission angles are small and a sizable portion of the shower's trajectory may be contained in a single photomultiplier's field of view. For the purposes of estimating the rate of events acceptable for making the above two measurements, we have accepted only Monte Carlo events whose trajectories had no PMT fields of view containing slant depth bins greater than 120g cm^{-2} at slant depths beyond 300g cm^{-2}. This requirement allows the ¼ maximum point for all selected events to be located with precision $\approx \pm 25$g cm^{-2}. The result of this selection process leaves us with about 10,000 events/yr. in the energy range $10^{16} \leq E \leq 10^{19}$ eV useful for measuring the p-air cross section and separating protons from heavy primaries.

The method of separation is fairly straightforward. Essentially, it involves a multiparameter fit to the composite "interaction point distribution" obtained after making data cuts as outlined above. We have approximated such an analysis based upon a Monte Carlo-generated sample of showers using standard composition at lower energies.[10,11] The primary nuclei were grouped into four regions of composition as shown in Table II. The Monte Carlo program sampled this composition, choosing nuclei at random with probabilities based on their abundance on an energy per particle basis. The nuclear groups [3-6] and [15-23] were ignored due to their low abundances. The relative weights of the four "included" nuclear groups, shown in Table II, were obtained after conversion to an energy per particle basis by the factor $A^{\gamma-1}$ with $\gamma = 1.7$. The cross section for nucleus-air collisions are based on the results of Heckman et al, 1975.[12] However, the proton-air interaction length was taken to be 75g cm^{-2} corresponding to a total inelastic cross section of 325 mb. The composite interaction lengths for the remaining nuclear groups are given in Table II. The resultant distribution of depths of first interactions for 5,500 events generated by the Monte Carlo program was collected into 50 bins of 10g cm^{-2} width. This distribution is pictured in Fig. 13. (For the sake of clarity only the final fit to the distribution is shown.)

TABLE II

Standard Composition of Low-Energy Cosmic Rays Weighted by $A^{\gamma-1}$

Z	Group	Interaction Length of cm^{-2}	Fraction of Nuclei at a Given Energy/Nucleon
1	H	75	0.395
2	He	49	0.175
6-14	CNO	26	0.179
>24	Fe	19	0.175

A very simple fitting procedure was employed in order to extract information from the Monte Carlo-generated composite distribution. The He and the CNO groups were combined assuming that half of the resultant group had an interaction length of 49g cm^{-2} and that the other half had an interaction length of 26g cm^{-2}. The Monte Carlo distribution was then assumed to be the sum of four exponentials of the form.

$$f(x) = w_1 \frac{e^{-x/\lambda_1}}{\lambda_1} + w_2 \frac{e^{-x/\lambda_2}}{\lambda_2} + w_3 \frac{e^{-x/\lambda_3}}{\lambda_3} + w_4 \frac{e^{-x/\lambda_4}}{\lambda_4} \quad (2)$$

where the λ_i are the interaction lengths and the w_i are the relative weights of the relevant nuclear groups.

The above function was fit to three "centering points" for the distribution (X = 50, 100 and 250g cm^{-2} indicated in Fig. 13) varying only the proton interaction length λ_1 while holding the others fixed. This particular procedure was carried out primarily for ease of calculation. However, these fits would not be expected to be nearly as sensitive to the interaction lengths of the heavier nuclei primarily for two reasons: (1) rather large centering depths were chosen to optimize sensitivity to the more penetrating particles and (2) the interaction lengths of the heavies do not scale linearly with the proton interaction lengths. A minimum χ^2 best fit (shown in Fig. 13) was obtained for λ_1 = 78g cm^{-2} compared to an input value of 75g cm^{-2}.

In order to evaluate the accuracy of determining the proton cross section with this method, ten independent Monte Carlo runs were analyzed giving an average value of 73.4g cm^{-2}. The r.m.s. error in a single test is 5.5g cm^{-2}. The average weight w_1 obtained for proton showers was 0.42 compared to 0.395 in the input. The r.m.s. error in this quantity for a single test is 0.10. These results imply that a proton component in the presence of a "mixed" composition is indeed detectable and that the interaction length can be determined to order ~10%. In the real measurements, a full-blown multiparameter fitting routine will be used over the whole range of data. Moreover, the measured p-air cross sections could then be used to recalculate the cross sections of the heavier nuclei. Thus, some iteration should be possible which should lead to better accuracies in extracting the relative weights of the various components. It appears then that protons can be detected and their cross section measured if they make up 10-20% of the primaries in our accessible energy range. The possible absence of low mass primaries is also detectable and would be an interesting discovery in itself.

ACKNOWLEDGEMENT

We wish to thank Professor J. C. Boone for his assistance in designing the optical system. We also gratefully acknowledge the technical assistance of Mr. A. B. Larsen and Mrs. J. W. Keuffel, and we thank Mrs. Isabel Stout for preparation of this manuscript.

This work has been supported by the National Science Foundation, Washington, D.C.

REFERENCES

1. K. Greisen, in "Proceedings of the Ninth International Conference on Cosmic Rays, London, 1965" (The Institute of Physics and The Physical Society, London, England, 1965), Vol. 2, p. 609.

2. L. G. Porter et al., Nucl. Instrum. Methods $\underline{87}$, 87 (1970).

3. T. Hara et al., Acta Phys. Acad. Sci. Hung. $\underline{29}$, Suppl. 3, 369 (1970).

4. G. Tanahashi et al., in "Proceedings of the Fourteenth International Conference on Cosmic Rays, Munich, West Germany, 1975 (Max-Planck-Institut fur Extraterrestrische Physik, Garching, West Germany, 1975), Vol. 12, p. 4385.

5. P. B. Landecker, Ph.D. thesis, Cornell University, 1968 (unpublished); A. N. Bunner, Ph.D. thesis, Cornell University, 1967 (unpublished).

6. J. Linsley, L. Scarsi, and B. Rossi, J. Phys. Soc. Jpn. $\underline{17}$, Suppl. A-I 91 (1962); J. Linsley, in "Proceedings of the Thirteenth International Conference on Cosmic Rays, Denver, Colorado, 1973 (University of Denve Denver, Colorado, 1973), Vol. 5, p. 3212.

7. E. C. Flowers, R. A. McCormick, and K. R. Kurfis, J. Appl. Meteorol, $\underline{8}$ 955 (1969).

8. Q. A. Kerns and R. F. Tusting, "Constant Amplitude Light-Flash Generator for Gain Stabilization of Photosensitive Systems", UCRL-10895, Lawrence Radiation Lab, Berkeley, California (1963).

9. Greisen, K. 1965, Proc. 9th Int. Conf. on Cosmic Rays (London) $\underline{2}$, p. 609.

10. Ryan et al. 1972, Phys. Rev. Let. $\underline{28}$, p. 985.

11. Elbert, J. W. et al. 1975, Phys. Rev. D. $\underline{12}$, p.660.

12. Heckman et al. 1975, Proc. 14th Int. Conf. on Cosmic Rays (Munich) $\underline{7}$, p. 2319.

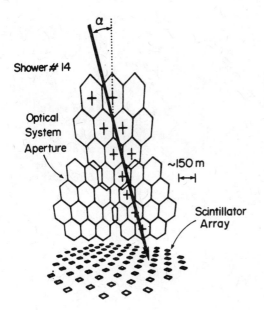

Fig. 1. Projection of the optical detector's aperture onto a vertical plane above the center of the Volcano Ranch Scintillator Array. A reconstructed shower trajectory is indicated by the heavy line. Crosses denote phototube apertures in which a signal was detected.

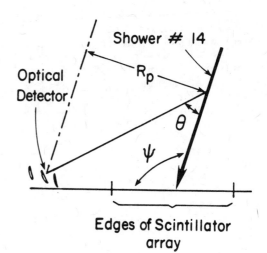

Fig. 2. View of the plane defined by a shower and the optical detector.

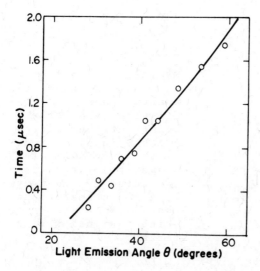

Fig. 3. Times at which phototubes triggered as Shower #14 passed through the field of view. The curve shows the best fit to these data.

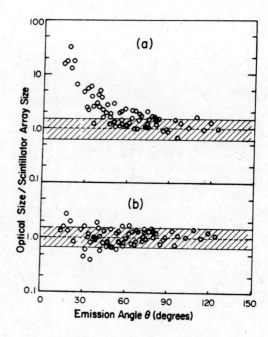

Fig. 4. Ratios of sizes obtained from the optical data to sizes measured by the Volcano Ranch Array using (a) computed scintillation light only and (b) estimated light from all sources. Data are plotted for each phototube in all 15 reconstructed showers. The shaded bands display the uncertainty due to systematic effects in both size measurements.

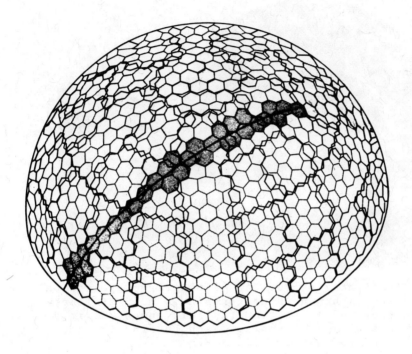

Fig. 5. Projection of the Fly's Eye field of view on the celestial sphere. Each small hexagon represents the field of view of a single "eye" while each darkly outlined region represents the field of view of a mirror unit and its associated 14 "eyes". Also shown is an extensive air shower whose path through the atmosphere projects as a great circle on the Fly's Eye. The darkened eyes represent those eyes which saw this particular shower.

Fig. 6. Mirror housing for one of the Fly's Eye optical units. Each housing contains a 1.59 m diameter mirror and 14 photomultipliers and associated Winston light funnels. Each housing is rock bolted to the mountain top.

Fig. 7. Field of view as seen by one of the horizontal-looking mirror units. The human eye can typically see mountains on the horizon about 120 km away.

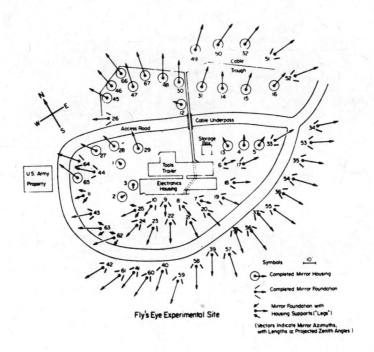

Fig. 8. A top view schematic of the Fly's Eye experimental site. The symbols representing the 67 mirror units are explained in the legend. Also indicated are the tools trailer, electronics housing, two completed cable spurs, and the cable underpass as required by Dugway to keep the access road free.

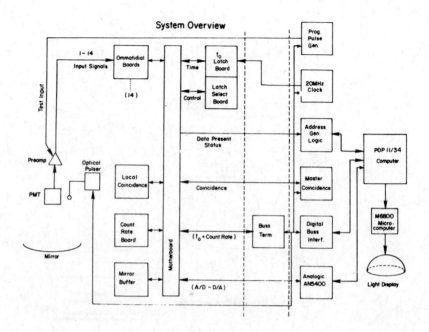

Fig. 9. Schematic of the Fly's Eye electronics processing system. The units to the left of the dashed line service an individual mirror and must be duplicated 67 times. Approximately 4 buss terminators (shown between the dashed lines) will be used and the units to the right of the dashed line are common to all 67 units. The ommatidial boards are the front end signal processing boards which service each photomultiplier "eye" and essentially measure optical light yields. The t_o latch boards measure light arrival times.

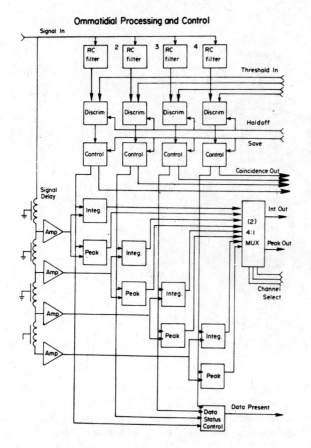

Fig. 10. Schematic of the main signal processing board (ommatidial board) for each photomultiplier. This board measures pulse integrals and peaks in four different time channels ranging from 200 nsec to 15 µsec gate widths. Each channel also has a different filter to optimize signal to noise for pulses of different widths.

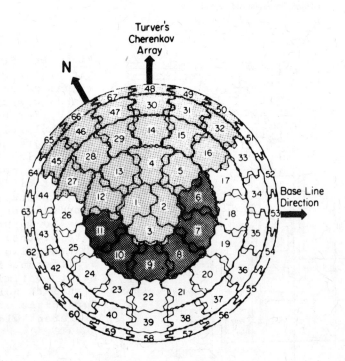

Fig. 11. Projected field of view of each Fly's Eye unit as seen from the vertical. The gray areas represent completely installed units which will soon be taking data. The darkly shaded units are those to be brought next into operation. An arrow denotes the direction of Turver's (the Durham, England, group) Cherenkov array located about 1 km away. We have centered initial deployment on that array to undertake intercalibration studies.

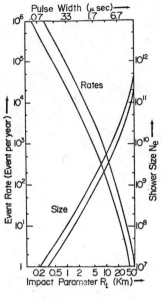

Fig. 12. Rates (left scale) and observed shower sizes N_e (right scale) vs shower impact parameter R_1 (lower scale) and corresponding pulse widths (upper scale).

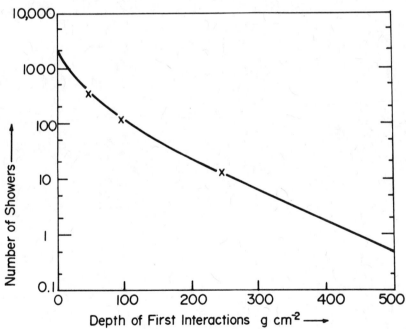

Fig. 13. Monte-Carlo generated distribution of shower starting points for 5,500 events. Primary constituents were selected from nuclear groups H, He, CNO, and Fe. The curve represents a four-component best fit to the Monte-Carlo generated distribution.

AKENO AIR SHOWER PROJECT

K. Kamata
Institute For Cosmic Ray Research, University of Tokyo
3-2-1, Midoricho, Tanashi, Tokyo, 188 Japan

Abstract

The new air shower array has been built at Akeno, Yamanashi Prefecture, Japan as an inter-university project. The array aims at the simultaneous observation of electrons, muons, core energy flow, and air Cherenkov light for air showers with energies 10^{14} ev-10^{18} ev. The outline of this array will be introduced in this paper.

1. INTRODUCTION

Our knowledge on extensive air showers(EAS) has been greatly improved in these two decades due to a number of works so far done both experimentally and theoretically. Nevertheless, we are still meeting difficulties in the interpretation of EAS data and understanding high energy interactions at EAS energies.
 The difficulties arise from the fact that:
(a) The nature of primary particles which initiate EAS is not known.
 Therefore, in many cases, it is difficult to separate the parameter related to particle interaction from that connected with primary particle.
(b) Observed parameters always suffer from large fluctuations due to complicated stochastic processes which develop showers in the atmosphere. Then, a serious question always arises whether the observed parameters is due to the fluctuation or not.
(c) The energy dependence of each shower array is different from one to the other. Each shower detector has its own detection bias. Accordingly, an extreme care should be paid for the comparison with data obtained with different shower array, especially for the different energy range.
 In order to reduce ambiguity and to minimize option in the interpretation of EAS data, an attempt should be made to observe the characteristic of each observed shower, not the average value over many showers, especially the longituginal development of each air shower observed.
 The best way to do this is to find the starting point of each shower in the atmosphere. The most promising way to this approach is at the present level of technique the observation of air flourescent light, which is planned by Fly's eye project of Utah group[1]. However this method can be applied only to very large showers due to signal to noise ratio.
 Taking into account the recent development of accelerators, we planned to make a different approach. Our target is to observe the different component simultaneously for each shower, and not only to study each component with an improved accuracy, but also to study the internal relationship between diffrent component for each

event.

The experimental parameters to be observed at Akeno are the lateral distribution and the arrival time of electrons, the lateral distribution of muons, the total energy flow at core region, and the lateral distribution and the pulse profile of air Cherenkov light.

The array is designed to observe showers with energies 10^{14}-10^{18} ev. The area covered by the array is nearly 1 km^2, which gives the upper limit to the shower energy observed. At the lower end of observable energy, we expect that accelerators will give us good calibration data in near future.

The construction of the array has been started in 1975 at Akeno highland(900 m a.s.l.), which is located at about 100 km to the west of Tokyo. The experimental site is in the midst of farmers land. The shape of the land is a gentle hill with a slope of $11°$ with respect to the horizon.

The outline of each detectors will be given in the following sections.

2. ELECTRON DETECTORS

Electron array is composed of scintillation counters of four different sizes, which are shown in the following table.

Table I. Characteristics of scintillation counters

Area of plastic scintillator(m^2)	2	1	0.25	0.02
Thickness of scintillator(cm)	5	5	5	0.3
PMT diameter(inch)	5	5	2	2
Dynamic range including amp.	10^4	10^5	10^4	10^4
Rise time of preamp.output(ns)	100	20	80	80

The arrangement of 1 m^2 and 2 m^2 detectors is shown in Fig.1. These detectors will be used to measure the lateral distribution of electrons from the core to about 5 Moliere Unit from the core. As is shown in Fig.1, there are four densely arranged areas, where 16 scintillation counters are put in lattice shape with mutual distance of 30 m. These densely arranged detectors are used to measure the lateral distribution near the core when a core hits in this area. They act together as large area density detectors when a core falls at a distant place.

Around these areas, counters are arranged with 60 m spacing, and in outmost area detectors are placed with 120 m intervals. All the scintillation counters are directly, or via sub-stations, connected with the central computer by coaxiable cables.

One hundred 0.25 m^2 detectors, which are not shown in Fig.1, are at the moment placed at the top of shielded detector(ME 1 in Fig.1) in a closely packed arrangement to observe the detailed

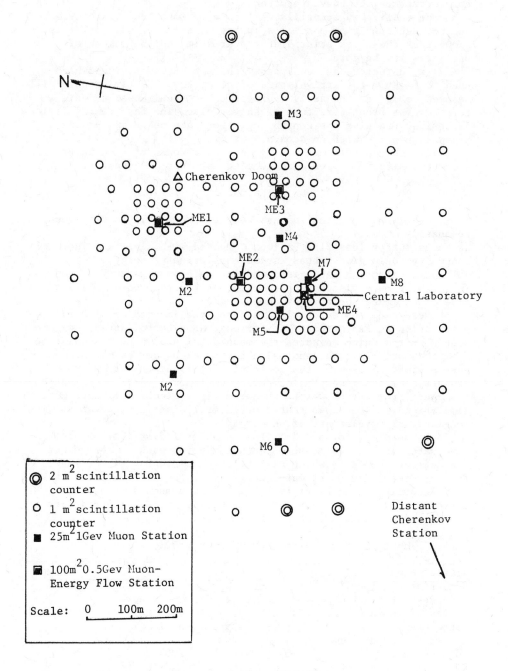

Fig.1 Arrangement of Akeno EAS Array

core structure, but they are movable.
The small, thin scintillator(1/50 m^2, 3 mm thick) is to measure the correct number of particles in the vicinity of core, as it is known that the thick scintillator overestimates the number of particles at the core.

1 m^2 detectors are equipped with fast timing devices, which is used to measure the arrival time of electrons. The timing data will be used to decide the arrival direction of each shower as well as to study the front structure of shower disc. The fast timing device is designed to give the zenith angle for each shower with size greater than 10^5 with an accuracy of $1.5°$.

All the counters are monitored with a constant time interval, and the monitor data are also sent to the central computer.

3. MUON DETECTORS

Increasing attention has recently been paid to muon component because;
(a) Muons suffer less fluctuations compared with electrons. Therefore muon size could give better estimate of primary energy.
(b) Muon to electron ratio is a model-sensitive parameter.
(c) Ratio of muon fluctuation to electron fluctuation may give a clue to estimate the primary composition.

These examples evidently indicate the importance of muon observation in EAS studies. So far, most of electron arrays has one muon detector, which measures the muon density at a certain distance from the core. Then, the muon size, the total number of muons, has been estimated assuming the constant lateral distribution of muons. However, some experiment[2] indicated that muon lateral distribution fluctuate from one shower to the other, like electrons. Monte Carlo simulation[3] also shows that muon lateral distribution varies with longitudinal development of showers.

It is needless to say that if muon lateral distribution does fluctuate, it must be measured accurately for each shower to get the correct muon size, which will be the basic quantity for the further discussion on primary energy, muon to electron ratio etc..

Accordingly, in Akeno array, eight 1 Gev muon stations and three 0.5 Gev muon stations are being built. The characteristics of these muon stations are shown in table II.

Table II. Characteristics of muon stations

station code	M	ME
number of stations	8	3
thickness of absober	2 m concrete	1 m concrete
energy of muons	1 Gev	0.5 Gev
detectors	proportional counters	
number of proportional counters per station	50	200
area of proportional counter array per station	25 m^2	100 m^2

The arrangement of muon stations is shown in Fig. 1, where M 1 - M 8 show 1 Gev muon stations, and ME 1 - ME 3 are the wide area muon stations. Out of M stations, M-5 is equipped with three layers of 25 m^2 proportioan counters array, and in M-7 eight layers of neon flash tubes with an area of 8 m^2 are installed. These devices are used for the calibration. The structure of M station is shown in Fig.2.

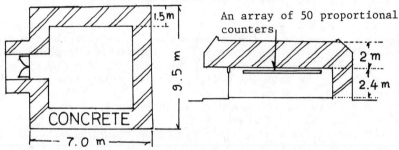

Fig.2. Structure of 1 Gev muon station (M st'n)

Each proportional counter used for muon detectors is made of a rectangular iron pipe with size 10 cm X 10 cm X 5 m. The thickness of iron wall is 2.3 mm. The central wire is a Tungsten wire coated with gold with 100 micron diameter. Each counter is filled with PR gas (90 % Argon and 10 % Methane) to a pressure of 690 mmHg at 15°C.

4. ENERGY FLOW DETECTORS

The energy flow of shower particles is a measure of shower development at the level of observation. Particularly, the energy carried by core particles is interesting to know how active the shower is at the observation level. If the relation of core energy flow to the other observed parameter is known, it could give us an important information on the longitudinal development of the shower observed.

At Akeno, energy flow detectors of two different types are being constructed. One is a large area, multilayer calorimeter located in the central laboratory building (ME4 in Fig.1). This is composed of alternative layers of concrete slab and arrays of proportional counters. Each of top three layers of concrete slab has a thickness of 50 cm, and the bottom layer 125 cm. Between each layer of concrete, an array of proportional counters of an area of 90 m^2 will be inserted. The total thickness of concrete is then 2.75 m which is equivalent to 8 nuclear m.f.p..

The other energy flow detector is the same as the large area muon detector (ME station), which has a concrete roof of 1 m thick corresponding to 3 nuclear m.f.p.. Therefore, ME stations function as a large area muon detector when the core hit far away, while they work as single layer calorimeter when hit by the core. The energy calibration will be made by the central calorimeter.

We shall start with the observation of total energy carried by core particles including high energy hadrons and electron-photon component. In future an attempt will made to put heavy material on the top of concrete layers to observe separately the hadrons and electron-photon component.

5. AIR CHERENKOV LIGHT DETECTORS

It is well known that the atmospheric Cherenkov light emitted by shower particles in the upper atmosphere reaches the observation level without suffering catastrophic interaction if observed at moonless clear night. Therefore, the observation of Cherenkov light could give us the data of quite different nature from that of charged particles observed at ground level. Then it is obviously quite useful to add Cherenkov light observation to Akeno data to improve the quality of shower data.

The detectors of two different types are being built at Akeno to observe optical Cherenkov radiation. One is to observe the total intensity of optical light at several distances from the core. One of this type is a dome shaped station, the location of which is shown in Fig.1. This dome is equipped with seven Fresnel lenses, behind each of which six photomultipliers are set at the focal point[4]. Beside this station, eight Cherenkov detectors with single photomultiplier will be arranged in Akeno shower array. These will be used to measure the lateral distribution of optical Cherenkov radiation for showers with size greater than a few times 10^5.

The other detector is to observe the pulse profile of optical light emitted by shower particles. This observation will be made at 1.7 km to the South-West from the center of array. It is planned to have seven spherical mirrors with 2 m diameter, each of which is viewed by 19 photomultipliers. With these photomultipliers, successive intensity of light will be recorded with 100ns time resolution for each shower. Due to the signal to noise ratio and available time resolution, the observation of this type shall be limited to showers with energies higher than several times 10^{16} eV. The plan of optical detectors at Akeno is shown in Table III.

Table III. Plan of Optical Detectors

Experiment	Total light intensity		Time sequence
Apparatus	single PMT	Doom	Distant Chrenkov St'n
Location	100-400m	500m	1.7 km
Light collection	Fresnel lense	Fresnel lenses	Mirror
Number of unit	8 st'n	7 lenses	7 mirrors
PMT/unit	1	6	19
Angle/PMT		$14.4°$	$4.5°$

6. RECORDING SYSTEM

When all detectors are prepared, data of 3600 channels have to be recorded for one shower. Therefore a special care has been taken to design the automatic recording system to achieve constant and reliable data collection.

The data signal from all the detectors except those from the distant Cherenkov station are collected by the Mitsubishi loop Ⅱ data transport system which is connected to the central computor (Mitsubishi Melcom 70/25 with 64 kw memories). The block diagram of this system is shown in Fig.3.

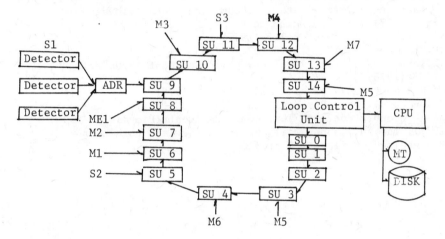

Fig.3   Block Diagram of Data Collection System

The signal from each detector is once stored in automatic digital register(ADR). These data are collected successively via station unit(SU), and fed into the central computor through loop line. The EAS data are then stored in magnetic tape and monitor data in magnetic disk. The time required to collect data of 3000 channels is about 10 s.

The data of distant Cherenkov station is stored independently by a FACOM U-200 computor when a shower signal is sent from the central station.

7. TIME SCHEDULE

The construction of Akeno EAS array was started in 1975, and the regular operation will be started in the middle of 1979, although the construction of a part of array will still be continued. The arrangement of all the electron detectors was finished by the end of 1978, and all the M stations will be ready in March.

The construction of the array has been made by the collaboration of the following universities.

Institute For Cosmic Ray Research of University of Tokyo,Hirosaki University,Ashikaga Institute of Technology,Saitama University, Meisei University, Tokyo Institute of Technology, Yamanashi University, Nagoya University, Osaka City University, Konan University, and Kobe University. Besides these groups,a number of physicists at Ehime University,Kochi University,and The Institute of Physical and Chemical Research joined actively in the discussion on the planning and preparation.

References
(1) G.W.Mason et al.Proc. 15th Int.Cosmic Ray Conf.(Plovdiv) $\underline{8}$ 252 (1977), and preceding talk.
(2) C.Aguirre et al. ibid. $\underline{8}$ 213 (1977)
J.N.Stamenov et al. ibid. $\underline{8}$ 1o2 (1977)
S.Miyake et al. Proc. 14th Int.Cosmic Ray Conf.(Munchen) $\underline{8}$ 2783 (1975)
S.M.Rozhdestvensky et al. ibid. $\underline{8}$ 2790 (1975)
(3) Dixon et al. Proc.Roy.Soc.Lond. $\underline{A339}$ 133 (1974)
(4) Akeno group. Proc.15th Int.Cosmic Ray Conf.(Plovdiv) $\underline{8}$ 503 (1977). The full paper will appear in the late volume of Proceedings of Plovdiv Conference.

Chap. 8 New Accelerator Facilities

QUARK - ANTIQUARK COLLIDING BEAM MACHINES

D. Cline
Fermilab, Batavia, Illinois
and
University of Wisconsin, Madison Wisconsin

ABSTRACT

A progress report on the proton-antiproton storage rings being constructed at CERN and Fermilab is given. Techniques of phase space cooling and of obtaining high luminosity in these machines using relativistic electron cooling are discussed. Some of the possible physics goals and experimental detectors for such machines are discussed. The possibility of high luminosity machines of very high energy is briefly discussed.

INTRODUCTION: $\bar{q}q$, e^+e^- AND $\bar{p}p$ MACHINES

An ideal type of high energy machine could be made if free quarks and antiquarks existed and could be made to collide. These hypothetical machines might use u, d, $\bar{u}$, and $\bar{d}$ quarks in separate rings or the same ring so that both weak and electromagnetic processes of the type

$$u + \bar{u} \to \gamma \to \mu^+ + \mu^-$$

$$u + \bar{u} \to Z^\circ \to \mu^+ + \mu^-$$

$$\to \nu + \bar{\nu}$$

$$\to \text{(heavy lepton pairs)}$$

$$\to \text{(heavy quark pairs)}$$

$$u + \bar{d} \to W^+ \to \mu^+ + \nu_\mu$$

$$\to \text{(heavy quark pairs)}$$

$$u + \bar{d} \to W + \text{(Higgs Boson)}$$

would be observable. One of the potentially serious problems in these machines might be synchrotron radiation if the u and d quarks are actually nearly massless. Another problem would be the different charges of the u and d quark which would require separate storage rings for collisions.

Of course quarks are probably not free and this hypothetical situation can probably never be realized. The closest direct realization of such machines is found in e^+e^- storage rings. In this case the reactions

$$e^+ + e^- \to \gamma \to \mu^+ + \mu^-$$

$$e^+ + e^- \to Z^\circ \to \mu^+ + \mu^-$$

$$\to \text{(heavy quark pairs)}$$
$$\to \text{(heavy lepton pairs)}$$

would be possible but not the direct single W^+ production because of the $e^\pm$ charge. Of course e^+e^- storage rings of great energy emit enormous amounts of synchrotron radiation and are therefore very expensive and there is a definite practical energy limit of a few hundred GeV for such machines.

Another possibility is that effective quark - antiquark machines can be obtained from proton - antiproton storage rings since there is compelling evidence that each hadron consists of basically a small number of either quarks or antiquarks. It was pointed out in 1976 by Cline, McIntyre and Rubbia that such machines could be constructed using existing synchrotrons with counter rotating beams and would have adequate luminosity for the study of weak interactions at extremely high energy.[1] They further pointed out that the phase space cooling techniques invented by Budker[2] and Van-Der Meer[3] would be necessary in order to accumulate adequate numbers of antiprotons for the beam of antiprotons. The suggestion was made that an adequate mechanism of accumulating $\bar{p}$'s was to construct a small storage ring near a high energy accelerator that incorporates one of the phase space cooling techniques.[1] The components of these machines are shown in Fig. 1.

$\bar{p}p$ machines will provide copious $\bar{q}q$ collisions without the very large synchrotron energy loss incurred in e^+e^- machines. Furthermore, there is no definite energy limit on such machines which could reach 10 - 20 TeV someday.

We can roughly estimate the $\bar{q}q$ center of mass energy using the formula:

$$S_{\bar{q}q} = <X>_{\bar{q}} <X>_q S_{\bar{p}p}$$

for $\bar{p}p$ machine $<X>_{\bar{q}} = <X>_q$

thus

$$\sqrt{S_{\bar{q}q}} \sim <X>_q \sqrt{S_{\bar{p}p}}$$

for $<X> \sim 0.2$ (as determined from neutrino data) we find

$\sqrt{S_{\bar{q}q}}$	$\sqrt{S_{\bar{p}p}}$	machine
108 GeV	0.54 TeV	SPS at CERN
400 GeV	2 TeV	ED/s at Fermilab
1200 GeV	6 TeV	UNK in the USSR
2000 GeV	10 TeV	Site Filler Storage Ring at Fermilab

In comparison with e^+e^- machines it is clear that these machines may some day reach a limit of 200 GeV in the center of mass (LEP for example). High energy $\bar{q}q$ collisions can be reached in either the ED/s machine at Fermilab or a future 3 TeV ring in the USSR or a 5 TeV machine perhaps to be constructed at Fermilab someday. Clearly the $\bar{q}q$ machines will take the energy lead and the central question will be the luminosity of these machines and the cleanliness of the experimental signals for new processes.

In the past two and one half years both CERN and Fermilab have developed designs for $\bar{p}p$ machines that are practical.[4,5] Furthermore there is now a firm commitment at CERN to construct such a machine and some progress in this direction is being made at Fermilab. The machine at CERN will collide 270 GeV p on 270 $\bar{p}$ GeV in the SPS whereas the machine at Fermilab will collide 1000 GeV p on 1000 GeV/c $\bar{p}$ in the super conducting energy doubler ring. The initial luminosity of these machines will be $\sim 10^{29} - 10^{30}$ cm^{-2}sec^{-1}. Future developments, to be discussed in sections 4 and 5 show promise that a luminosity of $\sim 10^{31}$ cm^{-2}sec^{-1} is possible with the addition of relativistic electron cooling of the high energy beams.

ANTIPROTON PRODUCTION AND COLLECTION TECHNIQUES

We can estimate the ideal yield of antiprotons into an element of phase space from a point target as

$$\frac{N_{\bar{p}}}{N_p} = \frac{1}{\sigma_a} \left[E \frac{d\sigma}{dp^3} \right] \left[\frac{dp_{11}}{E} \right] \left[\pi dp_\perp^2 \right]$$

where $Ed\sigma/dp^3$ is the invariant cross section for the production of $\bar{p}$, dp_{11} is the longitudinal phase space element and $dp_\perp^2$ is the transverse phase space element and σ_a is the proton absorption cross section. dp_{11}/E is the momentum bite of the collector. For ideal yields we assume a complete collection of $p_\perp$ out to 0.3 GeV/c where the cross section falls off by $1/E$. Presently designed $\bar{p}$ sources use different values of dp_{11}/E and $dp_\perp^2$ and collect antiprotons at different energies thus different factors enter into 1. It appears that the cross section for $\bar{p}$ production is a maximum for Feynman x = 0.[6]

It is clear that there is little gain in cross section above $\sqrt{s} \sim 15$ GeV. Thus proton energies of 30 - 100 GeV are adequate to produce the antiprotons provided $\delta p/p$ and $dp_\perp^2$ are adjusted for the decreasing $\bar{p}$ production cross section at the low energy. The <u>ideal</u> yields for the CERN and Fermilab $\bar{p}$ collectors are

$$\frac{N_{\bar{p}}}{N_p} \text{ Fermilab} = 2.5 \times 10^{-5} \qquad \delta p/p = \pm 0.15 \times 10^{-2}$$

$$\frac{N_{\bar{p}}}{N_p} \text{ CERN} = 2.8 \times 10^{-5} \qquad \frac{\delta p}{p} = \pm 0.7 \times 10^{-2}$$

We can compare these ideal yields with the expected "real" yields for the CERN or Fermilab collectors which includes real p acceptance, effects of finite length, targets, etc.[6]

$$\frac{N_{\bar{p}}}{N_p} \text{ Fermilab} = 2 \times 10^{-7} \qquad \frac{\delta p}{p} = \pm 0.15 \times 10^{-2}$$

$$\frac{N_{\bar{p}}}{N_p} \text{ CERN} = 8 \times 10^{-7} \qquad \frac{\delta p}{p} = \pm 0.7 \times 10^{-2}$$

A comparison of the ideal and realistic yields indicates that there are about one to two orders of magnitude of p̄ yields lost due to various factors, the largest of which is the poor depth of focus of the beam transport systems from a finite length production target. At the LBL - Fermilab workshop a scheme was worked out to improve the target efficiency using a "field immersion" lens.[6] It appears that various improvements are possible to increase the intensity of the p̄ source in the future.

PHASE SPACE COOLING OF LOW ENERGY BEAMS BY ELECTRON AND STOCHASTIC COOLING

We will consider electron cooling at low energy since this is an example of phase space cooling to be used in one of the actual schemes (Fermilab).[2,4] Whenever protons stop in matter they give up their energy to electrons in matter. The central idea of Budker was to apply this cooling to finite energy beams by accelerating the electrons to an energy such that the mean velocity of the electrons is the same as that of the protons or antiprotons, then the cooling takes place in the co-moving system. The scheme is illustrated in Fig. 2. For realistic electron currents the cooling time can be as fast as $\sim 10 - 100$ ms to cool a proton beam of 200 MeV with an energy spread of ±300 KeV. Note that for protons stopping in matter with 600 KeV initial energy the stopping time is $\sim 10^{-14}$ sec. The density of electrons in matter is $\sim 10^{13}$ greater than that possible for an electron beam and very crudely the cooling time scales as the density of electrons.

Thus it is not unreasonable that electron cooling times of order fractions of seconds are possible with energetic electron beams. The central problem of electron cooling is maintaining a cold electron beam for a finite distance. Space charge tends to cause the beam to diverge in the long straight section where cooling takes place. In order to overcome the beam blow up, a strong solenoid magnetic field was introduced in the early Novosibirsk experiments. A surprising result of this innovation was an enhanced cooling time due to coherent effects (see Fig. 3). As illustrated in Fig. 2 the electron beam and proton beam form a plasma and screening effects are very important. An example of a realistic electron cooling system is indicated in Fig. 4. The electron gun, adapted from a PEP klystron gun will deliver 26 amps of current. Cooling times of 50 - 100 ms are expected from this system. With such short cooling times a scheme of fast cycling collection of antiprotons becomes feasible. In essence it is possible to capture a small $\delta p/p$ bite of antiprotons at an optimal $\bar{p}$ production energy, decelerate to low energy where electron cooling times of 50 - 100 ms are possible and then collapse the beam, store the beam and stack the beam in a small aperture. The collector is then ready for for another injection of antiprotons. This is the scheme that has been worked out in detail at Fermilab.[4]

Stochastic cooling of beams operates on an entirely different principle. Liouville's theorem is only valid for an infinite number of particles; whereas real beams have a finite number of particles. Thus fluctuations in density in the beam phase space can occur and be detected by suitable pickups. A correction for this fluctuation can be applied by activating a suitable kicker. A picture of the ICE Machine at CERN is shown in Fig. 5.[7] Stochastic cooling of Betatron oscillations and the momentum spread in the beam has been conclusively demonstrated in this machine.[7]

RELATIVISTIC ELECTRON COOLING OF HIGH ENERGY BEAMS TO IMPROVE $\bar{p}p$ LUMINOSITY AND LIFETIME

There is another interesting application of electron cooling recently discussed at the LBL - Fermilab workshop.[8] At high energy proton and antiproton beams in a storage ring will naturally heat up (blow up) due to beam-beam interactions and due to multiple coulomb scattering with the residual gas in the machine. In addition the beam sizes may not be optimal to obtain the maximum luminosity for a given number of antiprotons or protons in the bunches. If the beams can be "cooled" by high energy electrons some of these problems might be overcome. The idea is to bring the electrons in the long straight section of an electron storage ring in conjunction with protons or antiprotons. The protons cool, the electrons heat up. The electrons are then "cooled" by synchrotron radiation. If this cooling source were infinite the proton (antiproton) beam area (σ^2_p) would cool to

$$\frac{\sigma^2_p}{\sigma^2_e} \sim \frac{M_e}{M_p}$$

where σ^2_e is the electron beam area. Thus the proton (antiproton) beam would damp down to extremely small size. The limit in luminosity would now come from the beam - beam interactions that blow up the beam which is expressed by a tune shift.

In order to see the effects of relativistic electron cooling we must first turn to the calculations of luminosity for $\bar{p}p$ machines.

The luminosity for the head-on collision of N_B bunches with n_p, $n_{\bar{p}}$ particles per bunch is given by

$$L = \frac{n_p\, n_{\bar{p}}}{2\pi [\sigma^2_p + \sigma^2_{\bar{p}}]^{1/2}_H [\sigma^2_p + \sigma^2_{\bar{p}}]^{1/2}_V} fN_B$$

where $[\sigma_p, \sigma_{\bar{p}}]_{V,H}$ are the beam sizes and f is the revolution frequency. The tuneshift caused by beam-beam interactions is given by

$$\Delta\nu = 3\frac{r_o}{\gamma}\left(\frac{n_{p,\bar{p}}}{\varepsilon}\right) \propto \frac{n_p n_{\bar{p}}}{\sigma^2}$$

where ε is the beam emittance, and r_o the proton radius. In order to see clearly the limitations and possibility for increase in luminosity we can write the luminosity in two forms (keeping only the important factors)

$$L \propto \frac{n_p n_{\bar{p}}}{\sigma^2} fN_b \qquad (I)$$

where we have set $\sigma_{p_{V,H}} = \sigma_{\bar{p}_{V,H}} = \sigma$. The second form is

$$L \propto (\Delta\nu)^2 f\gamma\varepsilon_o \left(\frac{N_B}{\beta^*}\right) \qquad (II)$$

where ε_o is the invariant emittance of the beam and β^* is the Beta of the beam at the collision point in the machine. In order to increase luminosity the following possibilities exist:

1. Increase n_p or $n_{\bar{p}}$
2. Decrease the beam size (σ^2)
3. Decrease the β^* at the interaction point
4. Increase γ (higher energy collisions)
5. Increase the maximum ($\Delta\nu$) at which the beams blow up

Note that as the beam size is decreased the luminosity increases but so does the tuneshift. Therefore, the gain in luminosity is offset by a more unstable beam. Nevertheless this is what happens in an e^+e^- machine where the beam is damped by synchrotron radiation and the acceptable tuneshift $\Delta\nu]_{max} \simeq 0.02\text{-}0.05$ is much larger than for corresponding proton-proton machines (ISR $\Delta\nu]_{max} \simeq 0.005$). The possibility of increasing luminosity using the above 5 improvements appear to be feasible as discussed at the LBL - Fermilab workshop,[6] in particular the improvements include:

1a. Increase of n_p by bunch compression ($n_p = 2\times10^{10}$ to $np \sim 2\times10^{11}$/bunch)

1b. Increase $n_{\bar{p}}$ by improvements to the antiproton source - i.e. better target efficiency or increased collector aperture.

2. The beam size can be decreased somewhat by high energy electron cooling, in principle the luminosity could increase by the ratio
$[\sigma_p^2/\sigma_e^2]^{-1} \sim \frac{m_p}{m_e} \sim 2000$ but this limit will
not be reached because of the effects of heating of the electrons by the protons and the very large tuneshift caused by such small beams. Nevertheless, a gain of a factor of 10 in luminosity may be possible. Fig. 7 shows the possible luminosity gain for the 1000 GeV $\bar{p}$ storage ring at Fermilab.

3. The β^* of the beams may be as small as 1 meter - the present plans call for $\sim (2.5 - 5)$m and thus the increase could be 2.5 - 5.[6]

4. In the energy doubler $\gamma \sim 1000$ compared to $\gamma \sim 150 - 250$ for the SPS and Fermilab main ring. Thus there is a "natural" increase of 4 - 6 in the luminosity, all other things being equal.

5. It appears that if electron beam cooling times of ~ 1000 sec can be achieved it will be possible to operate at $\Delta\nu] = 0.02$, instead of 0.005. Thus the luminosity increase can be as much as
$\left(\frac{0.02}{0.005}\right)^2 = 16.$[6]

We now return to the prospects of high energy electron cooling.[6] The formula for the damping rate of the transverse velocity is

$$1/\tau_p = \frac{3\pi e^4 L}{m_p m_e} \frac{\eta_p}{\beta^4 \gamma^5} \frac{I_e/e}{a_e^2 \theta_{\perp e}^3}$$

where m is the root mass of the particle, L is the coulomb logarithm, η_p is the ratio $1/c_p$, where c_p is the proton ring circumference, I_e is the electron beam current within the bunch, and a_e is the electron beam radius. The heating time τ^-_e for the electron beam is given by a similar formula with m_e replacing m_p, thus the heating time is (m_p/m_e) faster and this is a potential problem in the operation of such machines. Because of the rapid electron heating which blows up the beam it is necessary to supply a great deal of synchrotron radiation to keep the electrons cool. In fact the time evolution of the emittance of the two beams follow the equations[8]

$$\frac{dE_p}{dt} = \frac{-2}{\tau_p} [E_p - \frac{m_e}{m_p} E_e]$$

$$\frac{dE_e}{dt} = \frac{-2}{\tau_e} [E_e - \frac{m_p}{m_e} E_p]$$

If we include the effects of gas scattering on the proton (or antiproton) beam the resulting diffusion equations are

$$\frac{dE_p}{dt} = D_p - K_p \frac{E_p - \frac{m_e}{m_p} E_e}{E_e [\frac{E_e}{\beta_e^*} + \frac{E_p}{\beta_p^*}]^{3/2}}$$

$$\frac{dE_e}{dt} = D_e - \frac{2}{\tau} E_e - K_e \frac{E_e - \frac{m_p}{m_e} E_p}{E_p [\frac{E_e}{\beta_e^*} + \frac{E_p}{\beta_p^*}]^{3/2}}$$

where τ is the synchrotron radiation damping time

$$K_p = \frac{6\pi e^3 \, L\eta_p \, I_e}{m_p m_e c^4 \beta^4 \gamma^5 \beta_e^*}$$

$$K_e = \frac{6\pi e^3 \, L\eta_e \, I_p}{m_e^2 c^4 \beta^4 \gamma^5 \beta_p^*}$$

where β_e^*, β_p^* are the wavelength of the electron and proton beams at the cooling point and η_e is the fraction of the electron ring in the cooling region. The cooling time in order to give a constant proton (antiproton) emittance is given by

$$\tau_o = \tau \left(\frac{m_p}{m_e}\right)^2 \left(\frac{\eta_e}{\eta_p}\right) \left(\frac{\beta_e^*}{\beta_p^*}\right) \left(\frac{I_p}{I_e}\right)$$

the factor of $(m_p/m_e)^2$ is of great importance in designing the appropriate electron storage ring. Since η_e/η_p will be larger than or equal to 1 and $(m_p/m_e)^2 \sim 4 \times 10^6$ it is clear that (I_p/I_e) must be extremely small ($\sim 10^{-3}$) if we want $\lesssim 10^3$ sec. Let us take an extreme example - assume electron machines of 100 amps are possible and that a damping time τ of 10^{-3} sec is possible. Both of these requirements are well beyond the present state of the art but are not inconceivable. Then

$$\tau_o \sim 40 \text{ sec.}$$

A schematic of an electron cooling device is shown in Fig. 8. In case of such rapid damping of the beam it is clear that the beams would shrink to a size determined entirely by the blow up of the beams from beam-beam interactions. We could anticipate a large increase in the luminosity of $\bar{p}p$ machines in this case.

HIGH CURRENT ELECTRON STORAGE RINGS FOR RELATIVISTIC ELECTRON COOLING

The first question of interest is whether the electron beam in a small storage ring (a_e) is as small as the proton beam in a high energy synchrotron. Recently a measurement of the beam size at 100 MeV of the Tantalus e^- storage ring at the University of Wisconsin was carried out. The beam size that was measured is very well matched to the proton beam in the Fermilab machine at 200 GeV.

Is it possible to construct a very high current electron storage ring for high energy cooling? There have been several small electron storage rings constructed in the past 10 years where reasonably large electron currents have been achieved. However the emittance of the electron beams at high currents is another question. It seems that the best procedure would be to construct a small ~ 100 MeV electron storage ring with the amount of radiative cooling needed for high energy cooling. Using very high current injections to fill the ring could lead to high currents and/or the observation of a new series of instabilities that limit the current.

SPECIFIC $\bar{p}p$ MACHINES AT CERN AND FERMILAB

(a) $\underline{\bar{p}p \text{ Collisions in the SPS}}$

The general scheme for $\bar{p}p$ collisions in the SPS is as follows (see Fig. 9 and 10):[5]

1. 3.5 GeV/c $\bar{p}$ are produced by 27 GeV/c protons from the CPS ($\delta p/p = \pm 0.7 \times 10^{-2}$ is collected).

2. The $\bar{p}$ are transferred to a small storage ring (APR) and cooled rapidly in momentum space (few seconds) and slowly in transverse phase space by stochastic cooling.

3. After ~ 10^{11} - 5×10^{11} $\bar{p}$ are collected they are transferred into the CPS and accelerated to 26 GeV/c, bunched and injected into the SPS.

4. Protons are injected into the SPS at 26 GeV/c.

5. p and $\bar{p}$ are accelerated to 270 GeV/c and collide at section F where a large experimental detector will be located. The center of mass energy will be 540 GeV.

The expected parameters for the system are given in table 1. The proposed $\bar{p}$ collector ring is shown in Fig. 10. This storage ring has an extremely large aperture to diameter ratio, which is needed in order to collect the large $\bar{p}$ phase space.

larger p̄p cross section allows for a search
out to much higher masses. Furthermore the
larger cross section in the p̄p case may provide a larger signal to background ratio.

Thus in this extreme case the larger cross section for
p̄p production of the W, Z would be of considerable advantage in the search for the IVB.

We now turn to considerations of the observation
of the $W^{\pm}$ in the presence of large strong interaction
backgrounds. Table 2 lists the expected decay modes
of the $W^{\pm}$ and Z (assuming the existence of six left handed
quarks and six leptons) and the relative decay weights
for the various channels. In order to be specific we
focus on two channels of the $W^{\pm}$ decay and the expected
backgrounds, i.e.

$$W^{\pm} \to e + \nu$$
$$\to \mu + \nu$$

The experimental signature for this decay is the observation of an electron or μ and the missing neutrino energy.
The process is schematically represented in Fig. 17.
Suppose the p̄p collision has 2 TeV of center of mass
energy and the neutrino carries 40 GeV energy away -
thus 5% of the initial energy is missing in the final
state. The direct detection of this energy imbalance
seems nearly impossible since so much energy is carried
down the beam pipe. In the absence of any model for
the background processes it would be necessary to attempt
to measure this 5% energy imbalance. Consideration of
transverse momentum imbalances are more promising.[1,10]
We can consider the alternative; what is the most likely
background for this process in our current theoretical
understanding of hadronic processes. This background
is illustrated in Fig. 17 where a high $p_\perp$ jet consists
of one 40 GeV π and the other jet is missed either due
to detection inefficiency or because it appeared at a
very small angle and went down the beam pipe. Fortunately
kinematics restrict the angle of the second jet and in
the parton model a small angle high $p_\perp$ jet from a qq̄
collision is extremely unlikely since it implies that
the initial q̄-q state has a very large effective x as
shown in Fig. 18. The most recent QCD calculation expected for the cross section for jet-jet production compared to $W^{\pm}$, Z production has been given by Feynman and
is shown in Fig. 19.[11]

We can estimate the signal to background for this process in the following way,

$$\frac{\text{Signal } (\mu\nu, e\nu)}{\text{Background}} = \frac{R_W(p = 40 \text{ GeV})B(\ell+\nu)}{R(\text{Jet}\to\pi)p_\ell \, p_{\text{Jet}_2}}$$

where R_W is the relative cross section for W production to the Jet x Jet production with $p_\perp$ = 40 GeV/c (for the jets), $B_{(\ell+\nu)}$ is the W → $\ell + \nu$ branching fraction, $R(\text{Jet} \to \pi)$ is the probability that one jet is realized as a single π with $p_\perp$ = 40 GeV/c, p_ℓ is the probability that a π is misidentified as an electron or muon and p_{Jet_2} is the probability that the second jet goes to an angle less than θ.
"Reasonable" numbers for these processes are:

$R_W \sim 10^{-3}$, $B_{(\ell+\nu)} \sim 10^{-1}$, $R(\text{Jet} \to \pi) \sim 10^{-2}$,

$p_\ell \simeq 10^{-3}$, $p_{\text{Jet}_2} < \theta \sim 10^{-2}$ for $\theta \sim 17°$; therefore

$$S/B = \frac{10^{-3} \times 10^{-1}}{10^{-2} \times 10^{-3} \times 10^{-2}} \sim 10^3 - 10^4$$

Thus in spite of the very large jet production cross section the $(\ell + \nu)$ channel seems especially free of background.[12] We therefore expect that the new machines will lead to the observation of the $W^\pm$ provided an experimental detector with large solid angle, lepton identification and missing transverse energy balance are available.

EXPERIMENTAL DETECTORS: UAI DETECTOR AT CERN

The designs for large detectors to operate at the high energy pp and p̄p colliding beam machines are still in a state of evolution. A large dipole detector for the SPS has been proposed and the detailed study of the detector is quite advanced. The central problem that the detector design must face is illustrated in Fig. 16 where the angular distributions of the various decay products of the $W^\pm$ are shown. It is clear that the study of weak interactions at high energy requires a very large solid angle detector. In addition background suppression very likely requires a large solid angle so that the details of the background events can be studied. Furthermore, good lepton identification is essential and this requires some form of shower counters and likely a magnetic

TABLE 1
COMPARISON OF CERN-FERMILAB $\bar{p}p$ SCHEMES

	CERN	FERMI
Proton Energy on Target (GeV)	26	80
Protons/cycle	10^{13}	2.5×10^{13}
Cycle Time (Secs.)	2.6	3
Protons/sec.	3.85×10^{12}	8.3×10^{12}
$\bar{p}$ Momentum (GeV/c)	3.5	5.2
Momentum Bite (Sp/p)	15×10^{-3}	3×10^{-3}
Limiting Acceptance (mm-mrad)^2	$100\pi \times 100\pi$	$40\pi \times 20\pi$
Corresponding Momentum (GeV/c)	3.5	.644
$\bar{p}$/sec.	1×10^7	3.3×10^6
$\bar{p}$/hr.	3.6×10^{10}	$1.2 \times 10^{10} \to 3 \times 10^{10}$
$(\beta_H^* \beta_V^*)^{1/2}$ @ Low Beta (m)	62.2	$2.5 \to 1$
No. of Proton Bunches	6	84
No. of Protons/Bunch	6×10^{10}	$2 \times 10^{10} \to 2 \times 10^{11}$
Normalized Emittances (mm-mrad)	$\sim .02\pi$	$.02\pi$
Rev. Frequency	43	48
Luminosity (cm^{-2} sec^{-1})	2.0×10^{29}	$2.2 \times 10^{28} \to 1.4 \times 10^{30}$

TABLE 2

W/Z Decay Modes

Mode	Weight	Technique to Identify
$W^+ \to \mu^+ \nu$	1	muon & missing neutrino
$\to e^+ \nu_e$	1	electron & missing neutrino
$\to \tau^+ \nu$	1	
$\to u\bar{d}$	3	Hadronic Jets
$\to c\bar{s}$	3	Jets with Flavor
$\to t\bar{b}$	3	Jets with Flavor cascade
$Z \to \mu^+ \mu^-$	1	Dimuons
$e^+ e^-$	1	Dielectrons
$\tau^+ \tau^-$	1	
$\Sigma \nu \bar{\nu}$	$\Sigma' \eta_i$=Neutrinos in nature	missing energy
$\bar{u}u$	3	⎫
$\bar{d}d$	3	⎬ Jets
$\bar{s}s$	3	⎫
$\bar{c}c$	3	⎪
$\bar{b}b$	3	⎬ Jets with Flavor
$\bar{t}t$	3	⎭

REFERENCES

1. C. Rubbia, P. McIntyre and D. Cline, "Producing Massive Neutral Intermediate Vector Bosons with Existing Accelerators", Proceedings of the International Neutrino Conference, Aachen, 1976, H. Faissner, H. Reithler, and P. Zerwas, editors, p. 683. See also proposal P-492, Fermilab (1976).

2. G. J. Budker, Atomic Energy 22, 346 (1967).

3. S. Van Der Meer, CERN-ISR-PS/72-31, August, 1972.

4. D. Cline, P. McIntyre, F. Mills and C. Rubbia, "Collecting Antiprotons in the Fermilab Booster and Very High $\bar{p}p$ Collisions", Fermilab TM-689, (1976). See also D. Cline, "Possibility for Antiproton-Proton Colliding Beams at Fermilab", CERN $\bar{p}p$ Note 08, May, 1977. E. Gray et al., "Phase Space Cooling & $\bar{p}p$ Colliding Beams at Fermilab", 1977 Accelerator Conference Proceedings.

5. Design Report for Proton-Antiproton Storage Ring at CERN.

6. For a review see Proceedings of the Joint Fermilab - LBL Workshop on High Luminosity $\bar{p}p$ Storage Rings, F. Cole, editor, Fermilab (1978); the report of D. Cline et al.

7. G. Carron et al., Phys. Lett. 77B, 353 (1978).

8. Proceedings of the Joint Fermilab - LBL Workshop on High Luminosity $\bar{p}p$ Storage Rings, F. Cole, editor, Fermilab (1978). See the reports by C. Rubbia, M. Month, F. Mills, and A. Ruggiero.

9. C. Quigg, Rev. Mod. Phys. 49, 297 (1977); F. Halzen, "Can One Really Observe Signatures?", Phys. Rev. D15, 1929 (1977); M. Chase and W. J. Stirling, unpublished; P. V. Landshoff, "Parton Physics and Hunting the W with Hadron Beams", in Proceedings Varenna Summer School, July, 1977 (North Holland Publisher); R. Palmer et al., Phys. Rev. D14, 119 (1976); J. Finjord, Nordita Preprint 76/22 (1976); Yu. A. Golukhov et al., Soviet J. Nucl. Phys. 18, 203 (1977); Okun-Voloshin, ITEP Preprint 111 (1976); see also refs. (1) and (18); R. F. Peierls et al, Phys. Rev. 16, 1397 (1977).

10. D. Cline, "Observation of the $W^{\pm}$ In Kinematically Forbidden Background Regions," Wisconsin Preprint (unpublished).

11. See the talk of R. Feynman, ref. (6).

12. See also F. Halzen, D. Scott, "Signature for the Intermediate Vector Boson", University of Wisconsin preprint (1978).

13. CERN-SPS Experiment UAI at CERN, A. Astbury, B. Aubert, A. Benvenuti, D. Bugg, A. Busierre, Ph. Catz, S. Cittolin, D. Cline, M. Corden, J. Colas, M. Della Negra, L. Dobrzynski, J. Dowell, K. Eggert, E. Eisenhandler, B. Equer, H. Faissner, G. Fontaine, S. Y. Fung, J. Garvey, C. Ghesquiere, W. R. Gibson, A. Grant, T. Hansl, H. Hoffman, R. J. Homer, J. Jobes, P. Kalmus, I. Kenyon, A. Kernan, F. Lacava, J. Ph. Laugier, A. Leveque, D. Linglin, J. Mallet, T. McMahan, F. Muller, A. Norton, R. T. Poe, E. Radermacher, H. Reithler, A. Robertson, C. Rubbia, B. Sadoulet, G. Salvini, T. Shah, C. Sutton, M. Spiro, K. Sumorok, P. Watkins, J. Wilson.

473

Elements in the p̄p Storage Ring Scheme

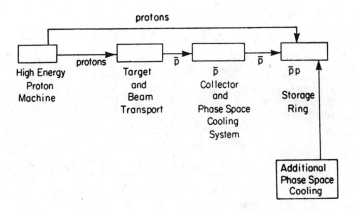

Fig. 1. Schematic of p̄p Storage Ring System.

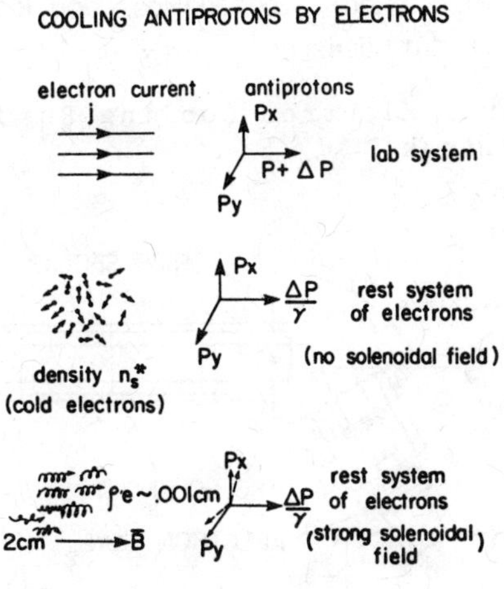

Fig. 2. Schematic of Electron Cooling Process.

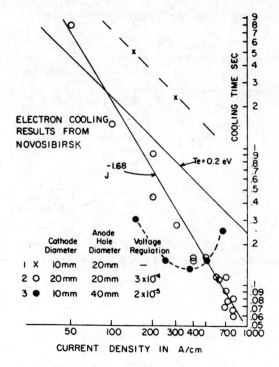

Fig. 3. Experimental Electron Cooling Results from Novosibirsk.

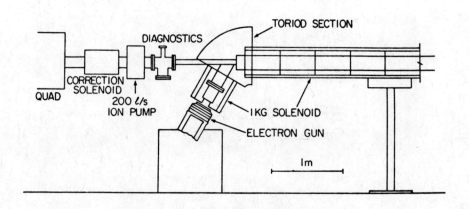

Fig. 4. Schematic of the Electron Cooling Gun and Straight Section at Fermilab.

Fig. 5. The ICE Storage Ring at CERN.

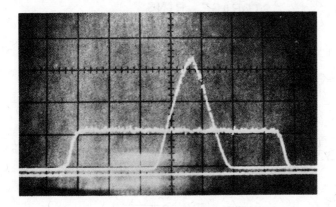

Fig. 6. Beam Momentum Profile Before and After Stochastic Cooling.

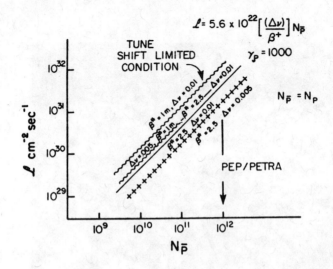

Fig. 7. Tune Shift Limited Luminosity vs $N_{\bar{p}}$ for 1 TeV at FNAL.

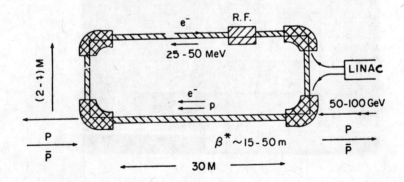

Fig. 8. High Energy Electron Cooling Storage Ring.

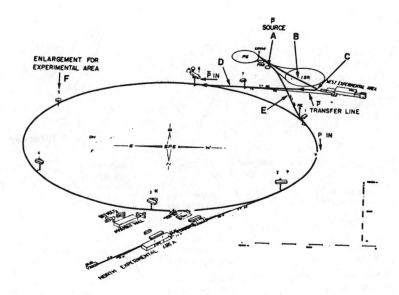

Fig. 9. General Layout for the CERN $\bar{p}p$ Collider.

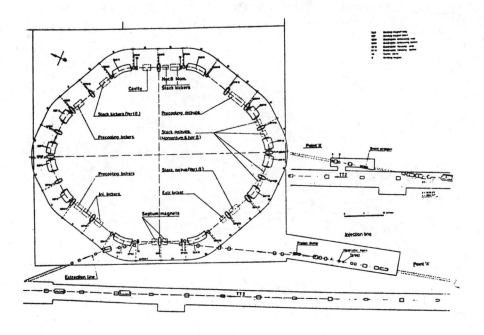

Fig. 10. General Layout of the Antiproton Accumulator at CERN.

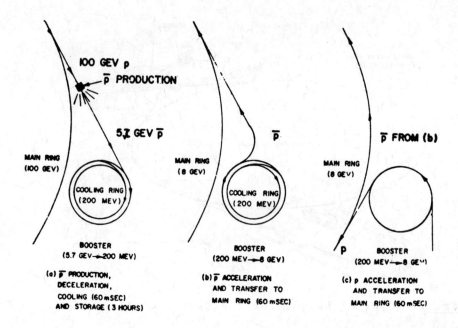

Fig. 11. Schematic Plan of the p̄p Colliding Beams at Fermilab.

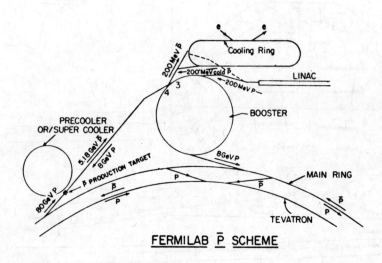

Fig. 12. Overview of Fermilab p̄p Scheme.

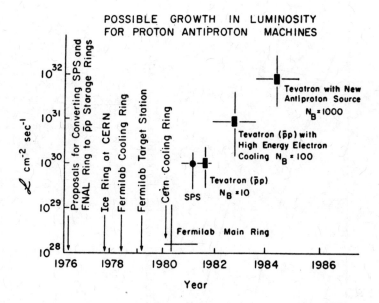

Fig. 13. Possible Increase in Luminosity of $\bar{p}p$ Machines with Year.

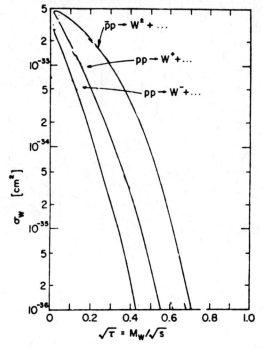

Fig. 14. Cross Section for Production of $W^{\pm}$ by $\bar{p}p$ and pp Collision vs. τ.

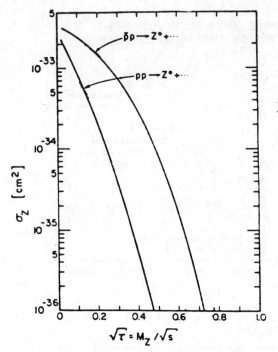

Fig. 15. Cross Section for Production of Z° by $\bar{p}p$ and pp Collision vs. τ.

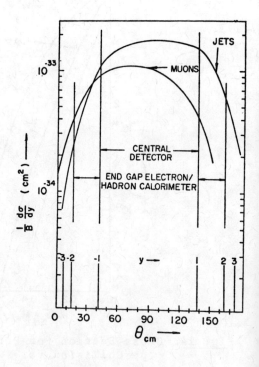

Fig. 16. Angular Distribution of $W^\pm$ Decay Products.

**DETECTION of W$^\pm$ THROUGH
ELECTRON + MISSING NEUTRINO**

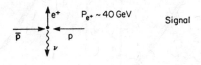

Fig. 17. Signature for W→ℓ+ν Detection and Background.

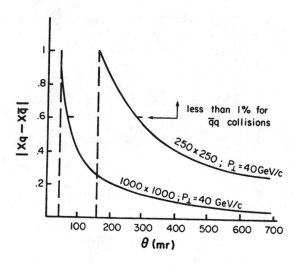

Fig. 18. Kinematic Limits on $\bar{q}q$ Scattering vs. q Scattering Angle.

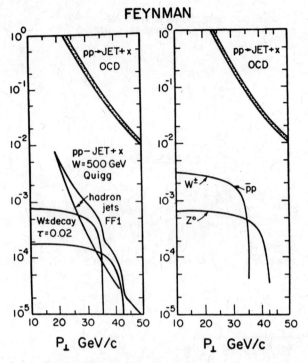

Fig. 19. Estimated Jet — Jet Cross Section for QCD and W,Z Decays.

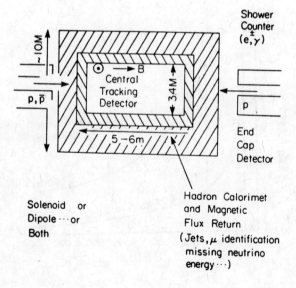

Fig. 20. Schematic of Large $\bar{p}p$ Detectors.

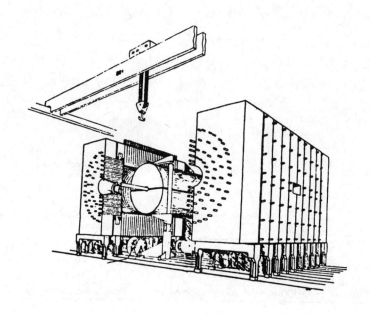

Fig. 21. Artist View of the P92 Detector at CERN.

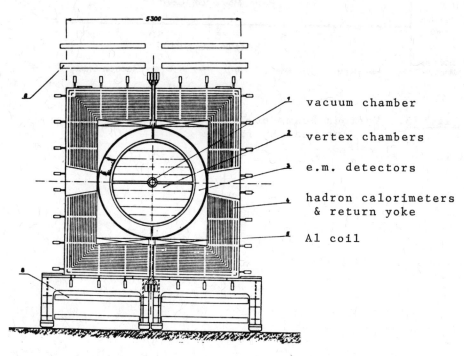

1. vacuum chamber
2. vertex chambers
3. e.m. detectors
4. hadron calorimeters & return yoke
5. Al coil

Fig. 22. Cut View of the P92 Detector at CERN.

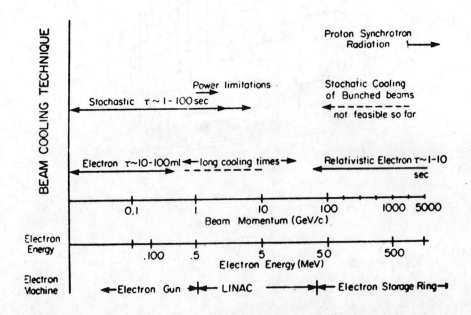

Fig. 23. Various Beam Cooling Techniques for Different Beam Energies.

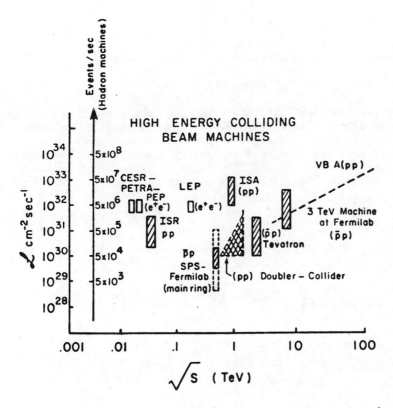

Fig. 24. Luminosity of High Energy Machines in the Planning and Construction Stage.

ISABELLE*

D. Hywel White
Brookhaven National Laboratory, Upton, N.Y. 11973

INTRODUCTION

Brookhaven National Laboratory on Long Island has been the site of two major particle accelerators, the Cosmotron and the Alternating Gradient Synchrotron. A further development in high energy physics is now being made with the construction by the Department of Energy of a new facility called ISABELLE. It is located in the northwest corner of the Brookhaven site which is shown in Fig. 1. Two rings

Fig. 1. ISABELLE Placement

---
*Work performed under the auspices of the U.S. Department of Energy.

of superconducting magnets will guide protons circulating both clockwise and anticlockwise crisscrossing at six points on the periphery. These rings are located in a common tunnel with special areas around the crossing points designed to facilitate the performance of experiments on the colliding particles. Each proton beam will be accelerated at energies up to 400 GeV, and the head-on collisions at the intersecting regions give a center-of-mass energy of 800 GeV. At the present time construction has begun and we summarize here the general features of the design and the status of the project.

Figure 2 shows a schematic of the ISABELLE-AGS layout. The AGS is a conventional proton accelerator that has been in operation since 1961. Injection of the beam to the AGS is accomplished with a 200 MeV linac and then protons are accelerated to an energy of 30 GeV. These protons are used to sustain a vigorous high energy experimental program, and will also be utilized as a source of injected beam for ISABELLE. The AGS is a very well understood accelerator, and it lends itself very well to the delicate task of injecting protons into a pair of superconducting rings. The extracted beam branch to the North Area is presently used for neutrino physics and this beam has operated in a very clean manner over a five year period in a configuration close to that required to inject into ISABELLE.

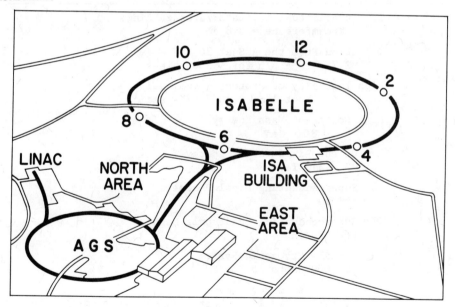

Fig. 2. ISABELLE Schematic

Figure 2 also shows the layout of the experimental areas, six in number. The design of each of these areas is being varied so as to exploit the experimental opportunities of ISABELLE. We will discuss this in more detail below. Each of the areas is designated by the appropriate clock face number 2,4,6,8,10,12. Injection of the 30 GeV beams and ejection at all energies takes place near area 6; this puts some constraints on the design of area number 6, the Wide Angle Hall.

The major features of ISABELLE are summarized in Table I. As we have indicated, there are two interlaced accelerator storage rings with a circumference of 3.8 km. The bending radius of 400 GeV protons by a 5 T magnetic field is ~ 270 m, with the extra perimeter being taken up with straight sections for experimental areas and the space for quadrupoles and instrumentation. The AGS is capable of accelerating more than 10^{13} particles per pulse; we expect to restrict the injection pulses to about 2.7×10^{12} particles thereby providing a clean beam of low emittance. As we will see below, some of the physics goals of ISABELLE can only be met with a high luminosity $10^{32} - 10^{33}$ cm^{-2} sec^{-1}. This will require approximately 7×10^{14} protons in each ring, a current of 8 A. The center-of-mass energy is just the sum of the energy of the two beams in a symmetric collision of the type envisaged at ISABELLE; this c.m. energy can be attained with protons of 3.4×10^{14} eV (340 TeV) on a stationary proton target.

Table I. Major Features of ISABELLE

Two Interlaced Accelerator/Storage Rings
  Circumference = 3.8 km

Injection From the AGS at 30 GeV
  2.7×10^{12} p/p for ~ 300 pulses

High Intensity of Protons - Luminosity
  8 A $\to$ L of $10^{32} - 10^{33}$ cm^{-2} sec^{-1}

High Center-of-Mass Energy
  60 - 800 GeV

Equivalent Fixed Target Energy
  2 - 340 TeV

732 Superconducting Dipole Magnets
  50 kG

348 Superconducting Quadrupole Magnets
  6 kG/cm

Large Scale Cryogenic System
  $\sim$20,000 watts at 3.8 K

Demanding System Performance
  Vacuum $\sim 3 \times 10^{-11}$ Torr,   400 Power Supplies, Current
  Regulation $\sim 10^{-5}$
Six Experimental Areas
  1 Wide Angle Area
  2 Major Facility Areas
  1 Small Angle Area
  2 Open Areas

The magnets needed to provide the 5 T guide field are superconducting dipoles, 732 in all each 4.75 m long. In addition, there are 348 superconducting quadrupoles to provide a satisfactory strong focusing lattice with a peak gradient capability of 6 kG/cm. These magnets are cooled by 3.8 K helium at 5 atm in the gaseous phase. This cooling requires a refrigerator power of 20 kW at the low temperature. The compressors needed to provide this power are rated at 14 MW. Two levels of vacuum are necessary for the machine, the insulating vacuum at 10^{-5} Torr and the main vacuum for particle transport at 3×10^{-11} Torr.

A cross section of the tunnel with magnets in place is shown in Fig. 3. The superconducting magnet enclosures are shown with the beam pipe center 1.27 m above the floor. Each dipole vessel is 5 m long and about 60 cm in diameter. The beam lines are nearly 1 m apart in the tunnel region away from the crossovers and the whole assembly is enclosed by the 2.5 m radius tunnel. The utilities both cryogenic and electrical are mounted at the top of the enclosure which is 3.35 m from the floor.

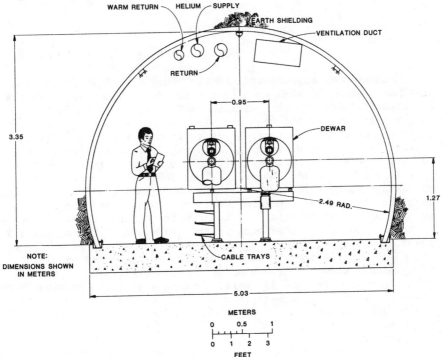

Fig. 3. Section of Tunnel

The space between the legs of the magnet stand has been set aside for magnets for a possible electron option in the future. Figure 4 is an isometric drawing of a dipole pair, but with an earlier design of the support frame.

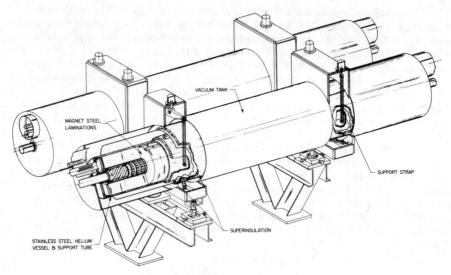

Fig. 4. ISABELLE Magnets

SUPERCONDUCTING MAGNETS

Conventional magnets utilize the boundary conditions at the steel as a dominant influence on the field configuration. At the elevated fields (5 T) that will be used at ISABELLE the saturation of the steel makes this technique inapplicable. The desired dipole field can be achieved by careful attention to the distribution of current density in the windings. In particular, if the current density depends on the azimuthal angle θ as cos θ then a uniform B field results inside the coil. The ISABELLE coils represent an approximation to this ideal. In Fig. 5 we show a schematic of the coil structure. The conductor that is used to fabricate these coils is made of a multifilament strand of NbTi in copper, braided so as to reduce eddy current effects in the conductor. The conductor is wound with spacers between the layers so that an approximation to the cos θ distribution is achieved. The wedge-shaped insulating pieces then hold the component coils rigidly under the magnetic forces. The coil is cooled by He gas at 5 atm passing through the helical cooling passages that are shown. Inside the main coil are correction coils which carry relatively modest current (∼ 200 A) which make the higher multipole terms appropriate for a storage accelerator.

The beam will circulate in a tube at room temperature for reasons we discuss below and so this tube of 8.8 cm i.d. is insulated from the inside of the magnet tube by superinsulation in insulating vacuum. A cold bore would act as an effective cryopump and it is expected that a circulating beam of the magnitude to be used in ISABELLE would cause emission of the adsorbed molecules raising the pressure to an intolerable level. This effect is sufficiently serious to warrant the addition of a warm bore with the necessary insulation. The coils that we are using in production are fabricated by industry and are inserted into the laminations shown in Fig. 5. The laminations contain the return flux and provide mechanical support for the coil. The laminations are held in place by a thick-wall stainless steel tube. The major difficulty in fabricating high field superconducting magnets rests with the need to prevent any motion of the coil under the magnetic forces. Motion of the coil induces eddy currents which causes heating and makes the magnet quench (go normal locally). ISABELLE magnets are designed to tolerate quenching without difficulty although the B field goes to zero and the heat must be extracted from the magnet before the B field can be reestablished.

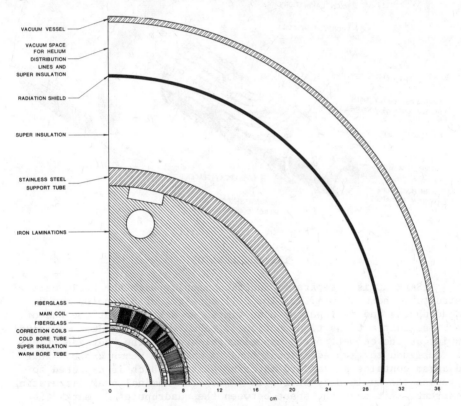

Fig. 5. Section of Coil

The assembly process is designed so that the coil is positively held in place after the magnet is cooled to operating temperature. This process is sufficiently well understood that magnets have now been produced that have an acceptable peak field after a few quenches to firmly set the coil in the laminations. Figure 6 shows the end of the dipole magnet where the coils form a saddle to return current on the opposite side of the bore tube. The cooling passages are also shown which run continuously throughout an entire sextant of the machine. The coil and laminations in the cast tube are mounted in the dewar with fiberglass straps as shown in Fig. 4. The total heat load from conduction and radiation at 3.8 K is close to 4 W, which is close to the design figure, giving confidence that the refrigeration is adequate.

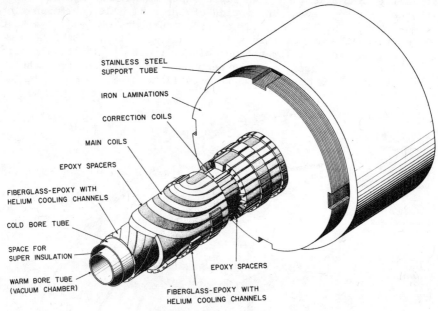

Fig. 6. End of Coil

THE ACCELERATOR

ISABELLE is a separated function machine with the basic cell structure shown in Fig. 7. The cell consists of 6 dipoles and 2 quadrupoles and is repeated nine times per sextant. At the end of each sextant a transition section is required to provide a crossing point at the center of the experimental areas. The "outside" ring is designed to give additional bend to make the beams cross. This diagram contains a standard matching section which is expected to produce the luminosity of 2×10^{32} with the nominal peak circulating current. Although the space between the quadrupoles Q_1 marks the easily accessible area available to experimenters, clearly for small angle scattering experiments detectors will be placed outboard of Q_1 towards the bending sextant itself.

Table II shows the operational sequence that is expected for ISABELLE. At injection 30 GeV pulses of $\sim 2.7 \times 10^{12}$ protons from the AGS will be injected into ISABELLE. The AGS normally accelerates 12 bunches of protons and one of them will be removed to allow for clean single turn ejection. These 11 bunches will be used to fill one fifth of the ISABELLE and this operation repeated five times to fill the ISABELLE circumference. This will be repeated for a total of 300 AGS pulses to fill ISABELLE to design current, taking about 10 minutes/ring. The beams will then be accelerated from 30 to 400 GeV in four minutes to keep the heat load on the magnets to an acceptable level.

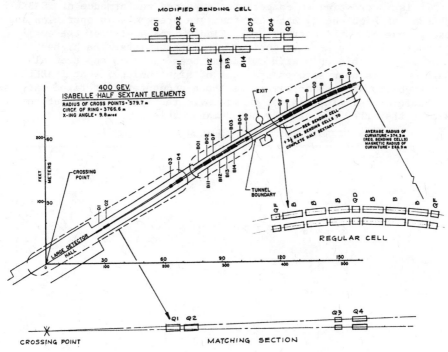

Fig. 7. 400 GeV ISABELLE Half Sextant Elements

Table II. ISABELLE Operational Sequence

Injection
    30 GeV protons from the AGS
    2.7×10^{12} p/pulse
    11 bunches

Stacking
    Each pulse fills 1/5 of ISABELLE
    Move into holding pattern
    Repeat 300 times/ring during 20 minutes
    $\rightarrow 6.3 \times 10^{14}$ protons/ring or 8 A

Acceleration
    Accelerate at 235 KHz
    $<dB/dt> \sim 250$ G/sec
    30 to 400 GeV in four minutes

The lifetime of the beams is expected to be many days, the limitation on the use of the beam coming from gradual deterioration of the beam quality rather than outright loss of particles. The loss of energy of the beam from synchrotron radiation is noticeable; it is about 1% per day and will probably be compensated for by lowering the magnetic field.

It is interesting to compare the expected performance of ISABELLE with that of high energy machines that are presently in operation and those planned for the near future. Figure 8 is a plot of the available energy ($\sqrt{s}$) vs the luminosity. It is clear that the highest energy is available only with colliding beam machines and that although much has been said on problems of high luminosity at ISABELLE, these problems are not as severe as those encountered at fixed target accelerators. This is not to minimize the difficulties but to indicate that they are clearly not unmanageable.

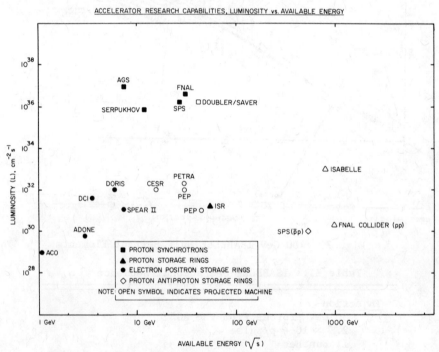

Fig. 8. Accelerator Research Capabilities

RESEARCH POSSIBILITIES AT ISABELLE

Table III shows a sample of the research possibilities to be expected at ISABELLE. The most spectacular goal for the new generation of high energy machines is the search for the intermediate vector bosons $W^{\pm}$ and Z^0. The assurance with which the gauge theories are now regarded leaves the experiments with unusually precise predictions for the masses of these objects whose importance is hard to overemphasize. The cross section for their production is also predicted and it is small $\sim 10^{-33}$ cm^2. A model in which the W is produced by the annihilation of quarks and antiquarks from the sea yields the production cross section in p+p collisions shown in Fig. 9. The arrow is the point where ISABELLE sits for a W mass of about 70 GeV/c^2. It is clear both that high energy is needed to investigate this mass region and high luminosity is needed to get many events.

Table III. Research Possibilities at ISABELLE

Search for W and Z	M $\sim$ 75 GeV/c	$\sigma \sim 10^{-33}$ cm^2
Search for massive hadrons and leptons		
Measure high p_T phenomena (see cosmic rays)		
Look for particle correlations (jets)		
Elastic scattering, total cross sections and measure the real part		

The success of the Υ discovery in pp collisions through the observation of μ pairs leads to the belief that the study of lepton pairs and even single leptons will be a major industry at ISABELLE. Again, the cross sections are low but there is little in the way of prediction in terms of what masses to expect, except that the quark masses so far go up by factors of 3 for each flavor.

This is a cosmic ray conference and many tantalizing hints have been seen of dramatic phenomena at ISABELLE energies. The existence of a change in the p_T behavior from the Tien Shan data leads us to put this item high on the list of initial experiments. We list also here the work-horse experiments that have yielded surprises and understanding in the past, namely elastic scattering, total cross sections and the forward scattering amplitude. Finally, this conference has made it clear that we do not understand high energy phenomena and there is a store of the unexpected at high energies.

EXPERIMENTAL FACILITIES

We show the layout of ISABELLE again in Fig. 10 with the disposition of the experimental halls. These halls have been designed so that a diversity of facilities is available to the experimenter. We shall discuss in general terms the expected function of each area and return to two of the areas whose detailed design is quite advanced.

The wide angle area has been designed so that there is sufficient lateral distance that particle identification at high p_T is possible. We have taken it as self-evident that this kind of study is likely to be interesting and that also the correlations between particles produced near 90° should also be accessible at ISABELLE. The constraints put upon this area by the nearby injection and ejection have made it convenient to put a wide angle/high p_T area in this location.

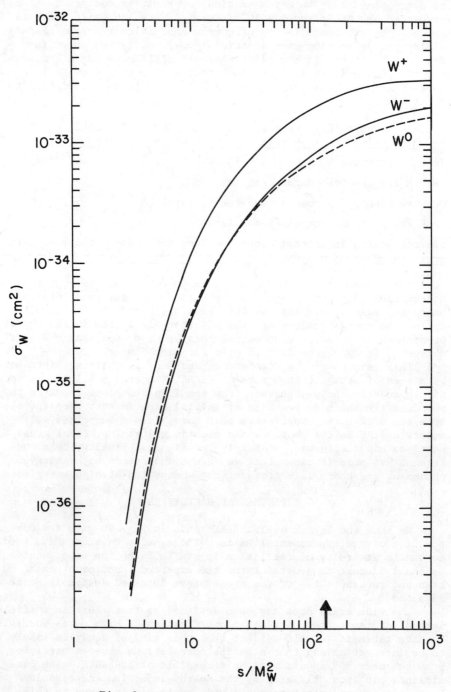

Fig. 9. W Production Cross Sections

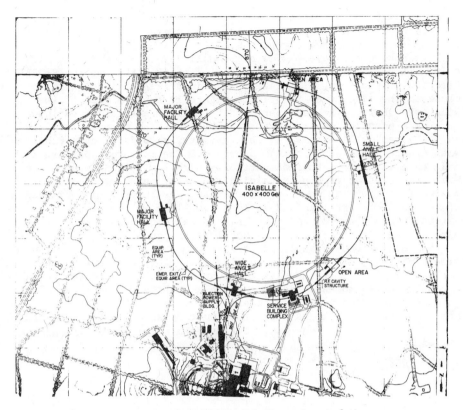

Fig. 10. ISABELLE Outline with Experimental Areas

The 4 o'clock open area is designed so that maximum flexibility for small experiments is preserved. In the initial operation of ISABELLE it is expected that there will be a need for a series of relatively modest experiments, modest in size and also in the requirements for beam time. This area is then planned to be convenient for short duration small size experiments to be installed and run.

The small angle hall at 2 o'clock reflects the demands of low p_T physics. The total cross section and forward scattering amplitude measurements as well as particle production at all x (Feynman) and low p_T can be serviced in this area. The spur that is visible in the diagram is a feature which the single arm spectrometer physics finds imperative.

The remaining three areas are presently in a fluid conceptual state. The history of recent particle physics has led to an emphasis on the power of the large and complicated electronic detector as a source of major physics discoveries. Detector development and conceptual design has occupied the attention of many people and these areas must reflect the demands of the large devices that are appropriate for ISABELLE. The need, however, for flexibility as well as restricting the cost of these areas themselves causes the design of these areas to be a delicate question. Delicate enough so that we have left the area at 12 o'clock as an open area for the time being, leaving a wide spectrum of options.

We return now to the Wide Angle Hall #6. We show in Fig. 11 this hall together with the injection lines from the AGS to ISABELLE. Ejection of the beams also takes place in this area into two dumps close to the outside of the ring. Figure 12 shows a plan of the Wide Angle Hall. The main part of the hall is a shielded enclosure; we show a piece of apparatus that we have used as a model in place in this area. This apparatus is a symmetrical pair of spectrometers with a wide angle and momentum aperture. The momentum aperture is made especially wide so that particles with low p_T can be correlated with particles at high p_T retaining a precise measurement of the momentum of the high momentum particles. The intersection diamond of the particles at ISABELLE is less than 1 mm in the vertical direction and about 25 cm in the horizontal direction. The large size of the diamond in the horizontal direction is influenced mainly by the small crossing angle of 11.2 mrad. The apparent horizontal size of the diamond is then about 100 times the beam size in the horizontal plane. If the particle source position is to be used as a constraint in momentum determination then bending must take place in a vertical plane. The combination of vertical bending and a wide momentum acceptance means that there must be an adequate clearance between the beams and the floor height. In this area the clearance is 4.3 m.

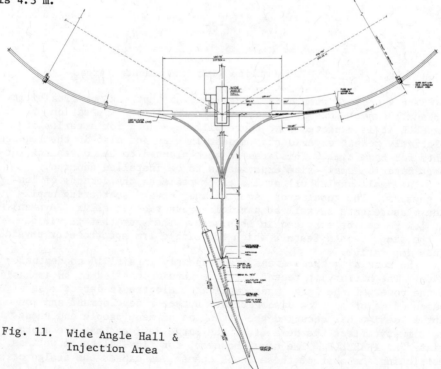

Fig. 11. Wide Angle Hall & Injection Area

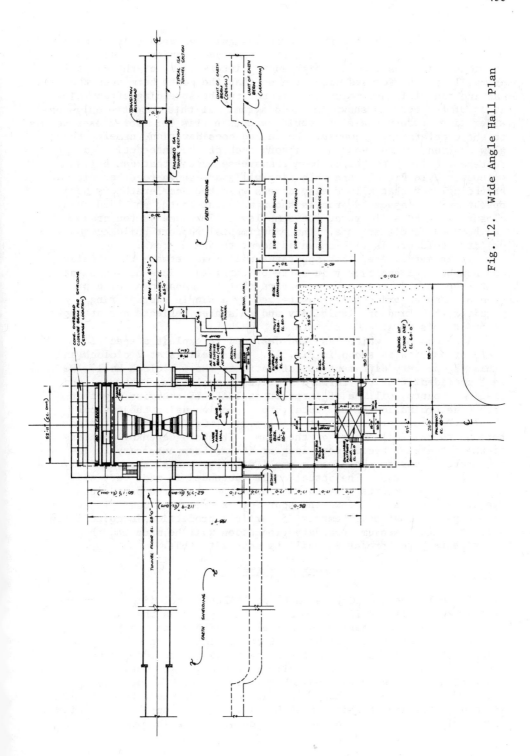

Fig. 12. Wide Angle Hall Plan

The shielding considerations in designing an area such as this divide into two parts. The muon flux that accompanies the decays of pions and kaons from the high energy interactions of the beams is primarily in the forward direction so that the earth berm over the machine tunnel is the major contributor to the shielding effect. The limit of the berm is shown in the diagram. If this were the only consideration, transverse slots could be cut in this berm for closer access to the accelerator. However, the second consideration, namely, the slow neutrons which eventually absorb much of the interaction energy represent a limit to the minimum transverse extent that can be tolerated. In Fig. 12 the shielding walls are shown together with a shielding door that allows easy access when the accelerator is not operating. An assembly area is shown adjacent to the shielded enclosure with both floors at the same level. The same 40 ton crane can be used in either area. The experimental support building and utility enclosure is also shown adjacent to these areas.

It is anticipated that apparatus will move between the shielded and nonshielded enclosures in spite of the large sizes of equipment that have been discussed. It appears that the management of a program of many experimental areas served by a single storage ring requires this kind of flexibility and this area reflects our attempts to build this in at an early stage.

A second hall which has been designed to exploit a specific field of physics is shown in Fig. 13. This hall is at 2 o'clock and exhibits one very different aspect ratio to the 6 o'clock hall. We feel confident that the forward scattering amplitude, both real and imaginary parts, will be of interest at ISABELLE energies together with a determination of the total cross section using the optical theorem. Particle production at small p_T ($\sim$ 1 GeV/c) can be studied here up to the full energy of the beam. Theshold Cerenkov counters at these energies become very long so it has been necessary to incorporate a spur to the area, a small diameter tunnel approximately tangent to the beam. The actual angle subtended is made very small by the use of Lambertson septa that bend the scattered particles away from the beam and down the spur for experimental use. It is also expected that measurements of particle production in coincidence with the high momentum secondary production will be made and the experimental area proper is built to cope with this.

COST ESTIMATE

Table IV shows a cost estimate of ISABELLE including the experimental areas but not including any of the apparatus that we expect to install for the experiments themselves. The costs are in 1978 dollars with the escalation added at the bottom of the table. The major part of the cost of the installation is in the accelerator components themselves and especially in the magnets. The time limitation on the construction schedule is determined entirely by the rate of funding. Although many scenarios for the rate of funding are made, the data at which ISABELLE will begin doing physics varies from early 1985 to mid 1986 depending on the fiscal assumptions.

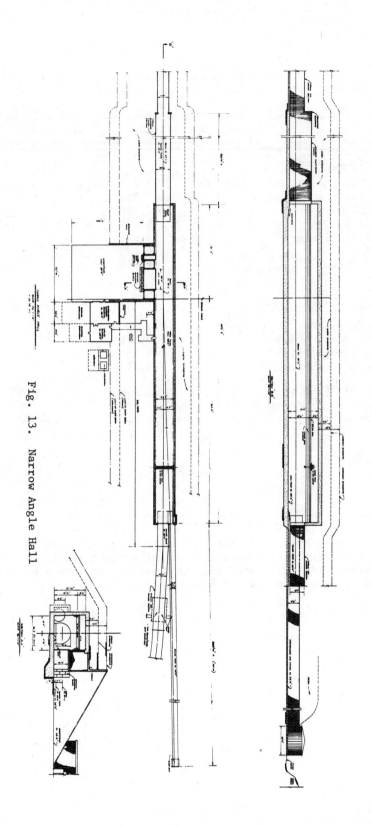

Fig. 13. Narrow Angle Hall

Although even earlier dates can be mentioned they are almost certainly unrealistic at this time.

Table IV. ISABELLE Cost Estimate (April 1978, 400 × 400 GeV, in Thousands of Dollars)

Accelerator Components		128,570
Ring Magnets and Power Supplies	72,740	
Refrigeration System	19,860	
Control System	7,810	
Vacuum System	9,570	
Injection System	6,840	
rf System	6,020	
Other Items	5,730	
Conventional Construction		31,450
Beam Enclosures	12,950	
Experimental Areas	8,010	
Service Buildings	4,180	
Site and Utilities	6,310	
Architect/Engineer		4,330
EDIA		27,170
	Subtotal	191,520
Contingency		36,890
Escalation		46,590
	TOTAL	275,000

We have here described some of the components of ISABELLE, a high energy physics resource that we expect to make available to enjoy understanding the tantalizing hints of new phenomena we have heard from the cosmic ray physics of the present.

Chap. 9   Conference Comments on Centauro

QUARK HOTSPOTS AS A CENTAURO SCENARIO

David G. Sutherland[+]
Dept. of Physics, Yale University, New Haven, CN. 06520

ABSTRACT

At this meeting the Centauro event(s)[1] emerged as probably the hardest feature of cosmic ray data to understand with conventional physics, should future experiments indeed substantiate their existence. Here we propose a simple scenario which seems compatible with the qualitative features of the event(s), while it maintains contact with present physics ideas.

We are motivated to present this despite the ugly features it involves from the viewpoint of most QCD theorists partly by the paucity of other explanations, and more by the fact that other cosmic ray and laboratory data are on the verge of disproving it, so that a modest improvement in technique could rule out this explanation. First, I sketch the scenario and show how it seems in qualitative agreement with Centauro event(s). Next I discuss the implications for other experiments and compare with present limits. At the end I discuss the probably very complicated theoretical future that would emerge if this explanation turned out to be true.

The scenario is that in collisions of nucleons at high energy $\gtrsim 1$ TeV, a small region of the colliding material gets heated to a high temperature, presumably $kT \gg 1$ GeV. We suppose, in accord with the spirit of many recent treatments of QCD and confinement,[2] that another phase dominates the dynamics at these temperatures and is such that quarks are able to escape (we discuss this further below). This hotspot could then materialize in the final state as a collection of Quarks coming from a massive fireball decaying. This would look quite like the Centauro event(s) and would solve the missing π° problems. The remaining feature of Centauro events is the very high multiplicity. Here we remark that as the quarks come from a different phase we might indeed expect a difference in multiplicity. At the end we consider a crude potential model and argue that a high multiplicity is to be expected.

Since we are anticipating a large number of quarks being produced at high energies, we must check that this is not in palpable conflict with known searches. We follow the useful review of Jones[3] in checking this. From the flux $\phi_p(E > 10^{15} \text{eV}) \sim 10^{-10} (\text{sr cm}^2 \text{ sec})^{-1}$ of nucleons

[+]On Study Leave from the University of Glasgow, Glasgow, G128QQ   Scotland.

and the indications, from ref. 1, that 5% of events are of Centauro type, and of a (quark) multiplicity of around 100 we find 5.10^{-10} (sr cm^2 sec)$^{-1}$ as the quark flux ϕ_q arising from Centauro events.

Looking first at laboratory searches for quarks we obtain a ratio quark/proton $\sim 10^{-20}$ to 10^{-23} depending upon assumptions on the depth of the mixing layer in the crust of the earth. This is to be compared with the limit set by Gallinaro et al[4] of 3.10^{-21}. Thus we see there is no conflict, but the limit is quite close.

Single quark searches give a limit on the flux of quarks, $\phi_q \lesssim 10^{-11}$ (sr cm^2 sec)$^{-1}$, apparently much lower than our expectation. However, most of these experiments are sensitive mainly to single quarks accompanied by no, or at most a few, hadrons, and furthermore much of the data is at lower energies where Centauro events have not been reported, presumably as the collisions are not yet sufficiently violent to induce the hotspots. Thus we do not feel these searches are in conflict with this picture.

A more serious worry is the searches in air showers which also give $\phi_q \lesssim 10^{-11}$ (sr cm2 sec)$^{-1}$. These are not subject to the above problems. To estimate whether the discrepancy here is serious, one has to consider more the geometry of the showers and the setup of a typical detector. Based on 'typical' kinematics of Centauro event(s) it seems the quarks might be spread over an area of $O(1$ m$^2)$ at ground level. A typical cloud chamber setup is of area around 3 m^2. The chance of detecting a quark is then essentially given by the inverse of the area in m^2 within which an incoming particle will trigger the cloud chamber. I have not been able to find a sure value for this area, but my impression is that it is $O(100$m$^2)$, based on the typical size of showers at this energy. In this case one would be on the verge of seeing quarks in these searches. It should be clear from the above that these are very crude estimates since the 'typicality' of Centauro event(s) is unclear, and the energy estimates in the events are unsure as they are based on experience with other more standard events and could be quite misleading. Still, we feel that with a modest increase in sensitivity or a change in triggering criteria, quarks should be detected easily in air showers, if indeed a more careful reanalysis does not already render this explanation untenable.

A further question which should be answered is the effect of interactions of quarks after production. Assuming quark proton cross sections are $O(10$ mb) as obtained for bound quarks,[5] there is a high probability that quarks will scatter at least once in coming through the atmosphere. If they undergo a large transverse momentum collision ($p_T \gtrsim 1$ geV/c), the scattered quark could be detected in a single quark search, or the quarks spread over an area $\gg 1$ m^2 in air shower searches. The probability of this is very small if qp scattering at high p_T were as rare as in pp scattering but there is reason to suspect that, starting with a particle q, high p_T scattering will be more common and perhaps a few percent of events will have $p_T > 1$ GeV/c. Thus again one gets an estimate $O(10^{-11})$ for ϕ_q, just on the verge of detection.

We turn now to the theoretical aspects of such a scenario. The first point is that confinement of quarks is required to be only temporary even in the laboratory (with $T \sim 0$), contrary to the beliefs and widely held aspirations of most QCD theorists. To try to illustrate to what extent this may be possible, and to show the price in simplicity to be paid, we use a simple potential model, though we are aware of the limitations of such an approach and certainly feel it does little justice to the complicated nonequilibrium situation at high T in our hotspot model.

We suppose V(r) at $T \sim 0$ to have the following form:

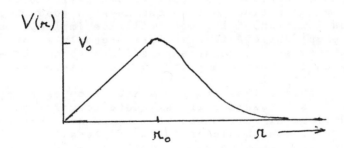

We suppose that above T_c the motion of quarks can be described by an effective potential which may be essentially coulombic, but at any rate lacks the linearly rising piece of V(r). It seems natural to expect that V_o, kT_c and the mass of the hotspot of quarks should be comparable and thus, from ref. 1, of order 100 GeV. Extrapolating from low energy phenomena such as charmonism then gives us $r_o \sim 10^{-11}$ cm.

With such a picture one can see some possibilities[+] for understanding the high multiplicity of quarks. In

---

[+]The most picturesque, if very nondynamical, possibility is to suppose each line in a duality diagram to give a quark. The ensuing doubling (roughly) of multiplcity seems not quite large enough to explain Centauro event(s).

the hotspot the dynamics of quarks is no longer determined by a stringlike picture, as in ordinary hadron physics at low T and modest energies, with most of the energy going into kinetic energy. Instead the hotspot could fall apart into many quarks with much of the energy going into the production of mass.+

A crude estimate suggests that, with the above parameters for V(r), the rate of tunnelling out of quarks, either from unexcited nucleons or as a result of absorbing a virtual boson, should be acceptably low. More work is certainly required to make a realistic estimate though.

From a theoretical point of view we have effectively introduced a new length scale, and a very large one, $\sim 10^{-11}$ cm, at that. Since the theory is now not confining, it should be, by the general analysis of t'Hooft[2], in a phase at T = 0 described by the Higgs model. The gluons would have a mass of order 1 MeV. Clearly only the presence of temporary, if imperfect, confinement could prevent this being in conflict with observation. We know of no theoretical indications for such behavior and it may be that, were it confirmed experimentally, it would be an indication that we should go beyond QCD. However exploration of the possibilities implicit in strongly coupled field theories are probably at an early stage, so that even such strange effects cannot be excluded, even if they seem unlikely from present experience.

Another theoretical problem arises from astrophysical considerations. Assuming the standard cosmological model, Zeldovic et al.[6] estimated many years ago that the ratio quark/proton should be $\sim 10^{-10}$ as a result of relic quarks from the early universe, provided that the universe was once very hot. While their calculations should be reconsidered for a V(r) such as ours (we believe $\sigma_2 \sim \pi r_o^2 \sim 10^5 \, m_p^{-2}$ would be more adequate than $\sigma_2 \sim m_p^{-2}$ as used by Zeldovic et al. and this could depress the quark/proton ratio by several orders of magnitude) it still seems much larger than the limit set in ref. 4. Thus it appears necessary that the universe should not have been very hot, i.e., $kT \lesssim 100$ GeV. In view of recent interesting work[7] on baryon number generation with $kT \sim 10^{15}$ GeV, this is not an encouraging feature of this picture.

+In order to avoid the reappearance of π^o's in the final state, it is necessary that the hotspot persist till it has a dimension $\geq r_o$. This appears analogous to a supercooling phenomenon in ordinary phase transitions. How likely such behavior would be in the turbulent environment of a hadronic collision, is not clear to us.

It is clear that, if Centauro events had not been seen, there would be little theoretical reason to invent them. If our scenario is correct there will be little difficulty in detecting quarks in new colliding beam, facilities, except that the energy may not be quite high enough for some years. In the meantime it should be possible to eliminate this scenario by refinement or reanalysis of existing quark searches, even if theoretical analyses do not give such a refutation.

ACKNOWLEDGMENTS

My interest in Centauro events was rekindled by a lecture by B. Schrempp. Subsequently my thoughts on this have been clarified (if not enough) by conversations with H. Kasha, G. Ringland, R. Shankar and especially T. Gaisser, who also encouraged me to write this up for the Proceedings. Receipt of a Fulbright Travel Grant is also acknowledged.

REFERENCES

1) "A New Type of Nuclear Interaction in the $\Sigma E_\gamma > 10^{14}$ eV Region," Brasil-Japan Emulsion Chamber Collaboration, these Proceedings, and references quoted therein.
2) G. t'Hooft, Nucl. Phys. B138, 1 (1978), and references therein to related work.
3) L. W. Jones, Rev. Mod. Phys. $\underline{49}$, 717 (1977).
4) G. Gallinaro, M. Marigelli and G. Morpurgo, Phys. Rev. Lett. $\underline{38}$, 1255 (1977).
5) J. J. J. Kokkedee, The Quark Model, W. A. Benjamin, (1969).
6) Ya. B. Zeldovic, L. B. Okun and S. B. Pikelner, Phys. Letts. $\underline{17}$, 164 (1965).
7) J. Ellis, M. K. Gaillard and D. V. Nanopoulos, CERN Preprint TH2596 (1978) and references therein to other recent papers on this idea.

WHAT IS CENTAURO?[*]

J. D. Bjorken and L. D. McLerran
Stanford Linear Accelerator Center
Stanford University, Stanford, California 94305

IS THE PRIMARY OF HIGH Z?

The celebrated Centauro event[1] found in the Mt. Chacaltaya emulsion exposures by the Brazil-Japan collaboration continues to resist rational interpretation. The event is described in terms of production of a leading "fireball", with the parameters listed in Table I. The main properties of this event have been inferred from the data, and are not here called into question.

TABLE I

Primary Energy	~	1000 TeV
Production Height	~	50 m above apparatus
	~	500 gm/cm^2 from the top of the atmosphere
"Fireball" Mass	~	200 GeV
"Fireball" Multiplicity	~	100
Multiplicity of π^o and e in the "Fireball"	~	0
$\langle p_\perp \rangle$ of Secondaries from the "Fireball"	~	1.7 ± .7 GeV

The absence of π^o's in the event would suggest a high-Z primary nucleus as the source. This interpretation, however, appears to be untenable since there is a negligible probability that a nucleus would penetrate so deeply into the atmosphere. Furthermore, the mean transverse momentum of the secondaries is much larger than the value typical of a nuclear fragmentation.

These objections, however, offer no obstacle to another interpretation: that the primary object initiating the collision is a glob of nuclear matter of very high density.[2,3] Since the $\langle p_\perp \rangle$ ~ 3-5 times the normal value typical of a nucleus, we might expect a density ~ 30-100 times that of ordinary nuclear matter. This highly compressed glob would have a radius 3-5 times smaller than that of an ordinary nucleus. If the larger transverse momentum is associated with the binding energies possessed by the constituents of the dense nuclear matter, these binding energies are much greater than those of conventional nuclear matter. Thus, were such globs of primordial

---

[*] Work supported by the Department of Energy under contract number EY-76-C-03-0515.

superdense nuclear matter to exist, the reduced geometrical cross
section and increased binding energy might allow these globs to
penetrate 500 gm/cm^2 of atmosphere and initiate events of the
Centauro type.

Nevertheless, there are still many major difficulties con-
fronting such a hypothesis. One difficulty is the nature of the
primary spectrum. If high energy, high-Z primaries exist with flux
sufficient to account for Centauro events ($\sim 10^{-2}$ m^{-2} yr^{-1} for E $\sim 10^6$
GeV at 500 gm/cm^{-2}), one would expect a much larger number at lower
energies. However, the flux of relativistic high-Z (Z $\gtrsim$ 30) primaries at the top of the atmosphere is known[4] to be $\sim 10^3$ m^{-2} yr^{-1}.
Assuming a power law integral spectrum, N(E) α E$^{-1.5}$, we need to cut-
off the flux of superdense globs at an energy E $\gtrsim$ 1 TeV. A possible
explanation for such a cut-off is that the superdense globs are
metastable and have some finite lifetime. The less energetic globs
would have a smaller time dilation factor and hence would have been
removed from the primary spectrum. Another possible explanation
would be that the process by which these globs are formed is very
energetic, and that the spectrum of produced globs is not close to
the 1/E$^{1.5}$ behavior typical of cosmic rays.

A second problem concerns the mechanism of fragmentation of the
primary in the actual Centauro event. It is a priori unlikely that
superdense nuclear matter exists in a free state. We might expect,
however, that globs of superdense matter could be metastable. If
the density of the glob is 30-100 times ordinary nuclear matter
density, we would almost certainly expect that the matter would be
in the quark phase. In a central collision of an air nucleus with
the metastable glob of quark matter, the glob might become sufficiently excited for it to "boil" or explode.

WHAT HOLDS THE GLOB TOGETHER?

The primary obstacle to a straightforward description of the
glob in terms of quark matter is the determination of a mechanism
which would hold the glob together. Such a mechanism might be
provided if the glob were in an exotic configuration involving
strange or charmed quarks. Although the energy of the exotic quark
glob could be higher than that of ordinary nuclear matter, the glob
might have to decay through an intermediate state of yet higher
energy. For example, if there were a mismatch in the exotic quantum
numbers of the quark glob and nuclear matter, then the intermediate
configuration might be a glob of quark matter with different quantum
numbers than those of the metastable quark glob. Such a configuration could be reached by the weak interaction Hamiltonian and the
glob would have a long lifetime.

Another mechanism is suggested by deRujula, Giles and Jaffe
(DGJ),[5] who have discussed the properties of quantum chromodynamics
when the color symmetry is slightly broken, so that the color field
is of large but finite range. Under these circumstances, the conventional conjecture of perfect confinement of quarks and other
color-bearing particles (e.g., diquarks and gluons) no longer holds.

The quark in such a model is expected to be a complex object of relatively large mass (indeed, as the range of the color force tends to infinity, so also does the quark mass) and large size (of order of the range of the color field). The long-range color field of such a highly massive quark should attract nucleons via an induced color dipole-moment. If such a quark is produced in a dense environment of nuclear matter, it might pick up an amount of nucleons comparable in mass to that of the original quark. Indeed, DGJ estimate that in the most stable configuration the nuclear matter has a mass ~ 80% of the quark mass. The presence of a quark excess in the conjectured superdense glob of nuclear matter may therefore provide an attractive force, and aid in compressing the glob to such high density, and in maintaining it in a stable state.

In the DGJ picture of a glob of quark matter bound by an extra quark, the nucleus would have the quantum numbers of a quark. In collisions with air nuclei, the glob will be stripped of its loosely bound matter, and in a central collision a Centauro might be produced. The additional quark necessarily penetrates to sea level. In the more conventional quark matter description, the glob need not penetrate to sea level, and might explode in a central collision with an air nucleus. In the DGJ model, the glob would be stable, and a low energy cut-off in the primary spectrum might be due to the production process for the globs.

EXPERIMENTAL IMPLICATIONS

The hypothesis we have made is so qualitative and so speculative that a detailed theoretical treatment is not likely to be fruitful. There are, however, fairly definite implications for observations. The flux of extremely relativistic high-Z ($Z \gtrsim 30$, $A \gtrsim 50$-100) primaries must be large enough to generate Centauro events. If these high-Z primaries have large penetrating power, as we have conjectured, then there could be a rate for Centauro events at sea level which is within an order of magnitude of the rate at Mount Chacaltaya. These events might be observable in terms of an unexpected amount of energy deposition. A detector with dimensions $\gtrsim 30m \times 30m$, exposed for at least a year, would be necessary to see such events.

In the absence of a catastrophic "Centauro" collision, we would expect that the globs lose per collision at most only a few percent of their energy. The reason for supposing this is best seen in the rest frame of the glob. The slowest particles produced in that frame (the isotropic component) become the fastest in the experimental frame of reference, and should dominate the energy loss. If the mass of this isotropic component or produced "fireball" is m and the mass of the glob is M, then the energy ΔE of the produced fireball in the laboratory frame is

$$\Delta E/E \simeq m/M \equiv k \quad . \qquad (1)$$

A reasonable range of values for m is ~ 3-30 GeV leading to $\Delta E/E \lesssim 1\%$ for M ~ 100-300 GeV. The differential energy loss of the glob is

$$\frac{d\langle E\rangle}{dx} \sim \frac{k}{\lambda} \langle E\rangle \qquad (2)$$

giving a range

$$R \sim \frac{\lambda}{k} \ln \frac{E}{M} \qquad (3)$$

Using $E \sim 10^6$ GeV and $\lambda \sim 500$ gm/cm^{-2}, we see that $R \sim 10^5$ gm/cm^{-2}. That is, in the absence of the catastrophic "Centauro" collisions, these globs could penetrate 100–1000 m of rock before being stopped. Thus underground experiments might also be relevant. However, again because of the low rate, it is not clear to us that existing underground data offer evidence refuting our hypothesis.

Data from measurements of horizontal air showers, however, appear to provide a stronger constraint.[6] The rate for showers of size $> 10^4$ and with zenith angle $> 70°$ (corresponding to a depth $\gtrsim 3000$ gm/cm^{-2}) is comparable to the rate for Centauro events at Mount Chacaltaya. The penetration length for catastrophic collisions, therefore, should be $\lambda \lesssim 500$ gm/cm^2. We conclude that most globs do not survive catastrophic collisions to arrive at sea level.

Assuming the glob contains a quark, we can estimate the concentration of quarks in the earth from this source. Taking 10^8 years as a characteristic time for quarks to be "stirred" by geological processes within a layer of depth ~ 100 m, we find a mean density of quarks

$$\mathcal{N} \lesssim 10^{-2} \text{ cm}^{-3}, \qquad (4)$$

safely within experimental limits.[7]

We conclude that the hypothesis we have made, while rather contrived and qualitative, is not obviously in contradiction with what is known. If the practical difficulties are not unreasonable, it might be worthwhile to test this hypothesis experimentally. This testing might be done by improving searches for penetrating relativistic high-Z primaries at high altitudes, and by searching at sea level or underground for unusual energy deposition which might be initiated by such a particle.

ACKNOWLEDGEMENTS

We thank Y. Fujimoto, T. Gaisser, R. Giles, P. B. Price, D. Ritson, and L. Susskind for very helpful discussions.

REFERENCES

1. Brazil-Japan Collaboration, "A New Type of Nuclear Interaction in the $\Sigma E_\gamma > 10^{14}$ eV Region," these proceedings.

2. T. D. Lee and G. C. Wick, Phys. Rev. $\underline{D9}$, 2291 (1974).

3. For a recent review of superdense quark matter and nuclear matter, see G. Baym, Les Houches Summer School (1977), unpublished.

4. M. Israel, P. B. Price and C. J. Waddington, Phys. Today $\underline{28}$, (1975), and references therein.

5. A. DeRujula, R. Giles and R. Jaffe, Phys. Rev. $\underline{D17}$, 285 (1978).

6. For example, cf., P. Catz et al., Proceedings of the 14th International Cosmic Ray Conference, Munich, 1975, Vol. 6, p. 2097.

7. For a reviews, cf., L. Jones, Phys. Rev. $\underline{D17}$, 1462 (1978).

AIP Conference Proceedings

		L.C. Number	ISBN
No.1	Feedback and Dynamic Control of Plasmas (Princeton) 1970	70-141596	0-88318-100-2
No.2	Particles and Fields - 1971 (Rochester)	71-184662	0-88318-101-0
No.3	Thermal Expansion - 1971 (Corning)	72-76970	0-88318-102-9
No.4	Superconductivity in d- and f-Band Metals (Rochester, 1971)	74-18879	0-88318-103-7
No.5	Magnetism and Magnetic Materials - 1971 (2 parts) (Chicago)	59-2468	0-88318-104-5
No.6	Particle Physics (Irvine, 1971)	72-81239	0-88318-105-3
No.7	Exploring the History of Nuclear Physics (Brookline, 1967, 1969)	72-81883	0-88318-106-1
No.8	Experimental Meson Spectroscopy - 1972 (Philadelphia)	72-88226	0-88318-107-X
No.9	Cyclotrons - 1972 (Vancouver)	72-92798	0-88318-108-8
No.10	Magnetism and Magnetic Materials - 1972 (2 parts) (Denver)	72-623469	0-88318-109-6
No.11	Transport Phenomena - 1973 (Brown University Conference)	73-80682	0-88318-110-X
No.12	Experiments on High Energy Particle Collisions - 1973 (Vanderbilt Conference)	73-81705	0-88318-111-8
No.13	$\pi-\pi$ Scattering - 1973 (Tallahassee Conference)	73-81704	0-88318-112-6
No.14	Particles and Fields - 1973 (APS/DPF Berkeley)	73-91923	0-88318-113-4
No.15	High Energy Collisions - 1973 (Stony Brook)	73-92324	0-88318-114-2
No.16	Causality and Physical Theories (Wayne State University, 1973)	73-93420	0-88318-115-0
No.17	Thermal Expansion - 1973 (Lake of the Ozarks)	73-94415	0-88318-116-9
No.18	Magnetism and Magnetic Materials - 1973 (2 parts) (Boston)	59-2468	0-88318-117-7
No.19	Physics and the Energy Problem - 1974 (APS Chicago)	73-94416	0-88318-118-5
No.20	Tetrahedrally Bonded Amorphous Semiconductors (Yorktown Heights, 1974)	74-80145	0-88318-119-3
No.21	Experimental Meson Spectroscopy - 1974 (Boston)	74-82628	0-88318-120-7
No.22	Neutrinos - 1974 (Philadelphia)	74-82413	0-88318-121-5
No.23	Particles and Fields - 1974 (APS/DPF Williamsburg)	74-27575	0-88318-122-3
No.24	Magnetism and Magnetic Materials - 1974 (20th Annual Conference, San Francisco)	75-2647	0-88318-123-1
No.25	Efficient Use of Energy (The APS Studies on the Technical Aspects of the More Efficient Use of Energy)	75-18227	0-88318-124-X

No.	Title		
No. 26	High-Energy Physics and Nuclear Structure - 1975 (Santa Fe and Los Alamos)	75-26411	0-88318-125-8
No. 27	Topics in Statistical Mechanics and Biophysics: A Memorial to Julius L. Jackson (Wayne State University, 1975)	75-36309	0-88318-126-6
No. 28	Physics and Our World: A Symposium in Honor of Victor F. Weisskopf (M.I.T., 1974)	76-7207	0-88318-127-4
No. 29	Magnetism and Magnetic Materials - 1975 (21st Annual Conference, Philadelphia)	76-10931	0-88318-128-2
No. 30	Particle Searches and Discoveries - 1976 (Vanderbilt Conference)	76-19949	0-88318-129-0
No. 31	Structure and Excitations of Amorphous Solids (Williamsburg, Va., 1976)	76-22279	0-88318-130-4
No. 32	Materials Technology - 1975 (APS New York Meeting)	76-27967	0-88318-131-2
No. 33	Meson-Nuclear Physics - 1976 (Carnegie-Mellon Conference)	76-26811	0-88318-132-0
No. 34	Magnetism and Magnetic Materials - 1976 (Joint MMM-Intermag Conference, Pittsburgh)	76-47106	0-88318-133-9
No. 35	High Energy Physics with Polarized Beams and Targets (Argonne, 1976)	76-50181	0-88318-134-7
No. 36	Momentum Wave Functions - 1976 (Indiana University)	77-82145	0-88318-135-5
No. 37	Weak Interaction Physics - 1977 (Indiana University)	77-83344	0-88318-136-3
No. 38	Workshop on New Directions in Mössbauer Spectroscopy (Argonne, 1977)	77-90635	0-88318-137-1
No. 39	Physics Careers, Employment and Education (Penn State, 1977)	77-94053	0-88318-138-X
No. 40	Electrical Transport and Optical Properties of Inhomogeneous Media (Ohio State University, 1977)	78-54319	0-88318-139-8
No. 41	Nucleon-Nucleon Interactions - 1977 (Vancouver)	78-54249	0-88318-140-1
No. 42	Higher Energy Polarized Proton Beams (Ann Arbor, 1977)	78-55682	0-88318-141-X
No. 43	Particles and Fields - 1977 (APS/DPF, Argonne)	78-55683	0-88318-142-8
No. 44	Future Trends in Superconductive Electronics (Charlottesville, 1978)	77-9240	0-88318-143-6
No. 45	New Results in High Energy Physics - 1978 (Vanderbilt Conference)	78-67196	0-88318-144-4
No. 46	Topics in Nonlinear Dynamics (La Jolla Institute)	78-057870	0-88318-145-2
No. 47	Clustering Aspects of Nuclear Structure and Nuclear Reactions (Winnepeg, 1978)	78-64942	0-88318-146-0
No. 48	Current Trends in the Theory of Fields (Tallahassee, 1978)	78-72948	0-88318-147-9
No. 49	Cosmic Rays and Particle Physics - 1978 (Bartol Conference)	79-50489	0-88318-148-7